Case-based Approach in Exfoliative Cytology

Case-based Approach in Exfoliative Cytology

Pranab Dey
MBBS MD MIAC FRCPath
Professor
Department of Cytology and Gynecologic Pathology
Postgraduate Institute of Medical Education and Research
Chandigarh, India

The Health Sciences Publisher
New Delhi | London | Panama

Jaypee Brothers Medical Publishers (P) Ltd

Headquarters
Jaypee Brothers Medical Publishers (P) Ltd
4838/24, Ansari Road, Daryaganj
New Delhi 110 002, India
Phone: +91-11-43574357
Fax: +91-11-43574314
Email: jaypee@jaypeebrothers.com

Overseas Offices

J.P. Medical Ltd
83 Victoria Street, London
SW1H 0HW (UK)
Phone: +44 20 3170 8910
Fax: +44 (0)20 3008 6180
Email: info@jpmedpub.com

Jaypee-Highlights Medical Publishers Inc
City of Knowledge, Bld. 235, 2nd Floor, Clayton
Panama City, Panama
Phone: +1 507-301-0496
Fax: +1 507-301-0499
Email: cservice@jphmedical.com

Jaypee Brothers Medical Publishers (P) Ltd
17/1-B Babar Road, Block-B, Shaymali
Mohammadpur, Dhaka-1207
Bangladesh
Mobile: +08801912003485
Email: jaypeedhaka@gmail.com

Jaypee Brothers Medical Publishers (P) Ltd
Bhotahity, Kathmandu
Nepal
Phone: +977-9741283608
Email: kathmandu@jaypeebrothers.com

Website: www.jaypeebrothers.com
Website: www.jaypeedigital.com

Inquiries for bulk sales may be solicited at: jaypee@jaypeebrothers.com

Case-based Approach in Exfoliative Cytology

First Edition: **2017**

ISBN: 978-93-86261-89-2

Printed at: Samrat Offset Pvt. Ltd.

Dedicated to

Shree Satyananda Giri, Rini
and
Madhumanti

PREFACE

In this book I have discussed many classical and interesting cases of exfoliative cytology. The book is written in the form of multiple choice questions (MCQs); and, therefore, it will help the students and the cytology practitioners to learn and self-evaluate. In each case I have discussed the diagnostic points and differential diagnosis along with salient immunocytochemistry. I hope that this case-based approach will be very effective in learning exfoliative cytology.

Pranab Dey

ACKNOWLEDGMENTS

This book should be considered the second part of the previous book *Case-based Approach in Fine Needle Aspiration Cytology*. I again honestly acknowledge that the main concept of the book belongs to Shri Jitendar P Vij (Group Chairman) and Mr Ankit Vij (Group President) of M/s Jaypee Brothers Medical Publishers (P) Ltd, New Delhi, India.

I wish to express my gratitude to *Late* Professor Subhash Kumari Gupta, and Professor Arvind Rajwanshi, Head of Cytology and Gynecologic Pathology, who taught me cytology.

I am grateful to Professor Uma Nahar Saikia and Dr Suvradeep Mitra for their assistance. They provided me the histopathology microphotographs. Dr Suvradeep Mitra deserves special thanks as he helped me throughout in the preparation of this manuscript.

My wife Rini and Madhumanti always gave me encouragement and support, and without their support it would not be possible to write this book.

I am grateful to God Almighty as the entire work is done by His blessings only.

CONTENTS

ABBREVIATIONS

AIDS	Acquired immunodeficiency disease
AGC	Atypical glandular cell
AGCUS	Atypical glandular cell of undetermined significance
ASC	Atypical squamous cell
ASCUS	Atypical squamous cell of undetermined significance
ASC-H	Atypical squamous cell high-grade SIL cannot be excluded
BCG	Bacillus Calmette-Guerin
BAL	Bronchoalveolar lavage
CSF	Cerebrospinal fluid
CIN	Cervical intraepithelial neoplasia
CD	Cluster differentiation
CK	Cytokeratin
CMV	Cytomegalovirus
DPAM	Disseminated peritoneal adenomucinosis
ERCP	Endoscopic retrograde cholangiopancreatectography
EUS TBNA	Endoscopic ultrasound guided transbronchial needle aspiration
EGFR	Epithelial growth factor receptor
EMA	Epithelial membrane antigen
FNAC	Fine needle aspiration cytology
FCM	Flow cytometry
GI	Gastrointestinal
GIST	Gastrointestinal stromal tumor
GLUT 1	Glucose transporter member 1
GM CSF	Granulocyte-macrophage colony stimulating factor
HSV	Herpes simplex virus
HSIL	High-grade squamous intraepithelial lesion
HPF	High power filed
HL	Hodgkin lymphoma
HIV	Human immunodeficiency virus
HPV	Human papilloma virus
HCG	Hyperchromatic crowded group
LC	Langerhans cell
LCC	Large cell carcinoma
LD body	Leishmania Donovani body
LCA	Leucocyte common antigen

LBC	Liquid based cytology
LSIL	Low-grade squamous intraepithelial lesion
MGG	May Grunwald Giemsa
NSE	Neuron specific enolase
NHL	Non-Hodgkin lymphoma
N/C	Nucleocytoplasmic ratio
PAS	Periodic Acid-Schiff
PMCA	Peritoneal mucinous carcinomatosis
PCP	Pneumocystis pneumonia
PMP	Pseudomyxoma peritonei
PAP	Pulmonary alveolar proteinosis
PLCH	Pulmonary Langerhans cell histiocytosis
R-S cell	Reed Sternberg cell
SCC	Small cell carcinoma
SQC	Squamous cell carcinoma
TdT	Terminal deoxynucleotidase
TBS	The Bethesda system
TTF	Thyroid transcription factor
TV	*Trichomonas vaginalis*
USG	Ultrasonograph
UCC	Urothelial cell carcinoma
WHO	World Health Organization
ZN	Ziehl Neelsen stain

CHAPTER 1

General Cytology

Introduction

A thorough cytological evaluation is needed for the proper diagnosis of any lesion. The following data should be checked:

Cytological Data

A. Cellularity
B. Cell arrangement or distribution
C. Cell morphology
D. Background non cellular materials

Cellularity

Overall assessment of cellularity is an essential requirement of the initial step of screening of the slide. Malignant lesions usually show good amount of cells in the smear.

Cell Arrangement

The following arrangements may be seen:

Single cells: Dissociated single cells are commonly seen in lymphoma and poorly differentiated carcinoma.

Clusters: Three dimensional clusters of cells indicate loss of polarity. Cell clustering is seen in most of the carcinomas and sarcomas.

Sheet: Flat monolayer arrangements of the cell is known as sheets and are usually seen in benign conditions such as fibroadenoma, lymphocytic thyroiditis, mesothelial cells, etc.

Follicular pattern: In this pattern, the monomorphic cells are arranged regularly forming small follicles. Occasionally the central lumen of the follicles contains colloid in the aspirate from thyroid.

Glandular arrangement: Here the cuboidal to columnar cells are arranged in gland like pattern. Complete gland like structures and fragments of glands are often seen in adenocarcinomas.

Rosettes: The cells in pseudorosette are arranged around central fibrillary structures. Cells are not attached with any basement membrane. This arrangement is commonly seen in neuroblastoma, peripheral neuroectodermal tumor, and retinoblastoma.

Papillary clusters: The cells are arranged as papillae like structure with central fibrovascular core. The peripheral cells in papillae are arranged perpendicularly to the axis of the papillae. Papillary clusters of cells are often present in papillary carcinoma of thyroid, breast, or renal cell carcinoma.

Fascicles: The cells are arranged in interwoven bundles. This pattern is commonly seen in leiomyosarcoma, fibrosarcoma, nerve sheath tumor and fibrous histiocytoma.

Individual Cell Morphology

Shape: Cell shape, cytoplasm and nucleus provide the very important information. The shape of the cell may be round, polygonal, spindle shaped or bizarre depending on the type of the lesion.

Cytoplasm: The following information of cytoplasm is needed for the diagnosis: volume of the cytoplasm, tinctorial character, vacuolation, cytoplasmic granularity, intracytoplasmic lumina, and cytoplasmic inclusions.

Nucleus: The following information about nuclei is helpful: nuclear size, shape, pleomorphism, margin, nucleoli, chromasia and chromatin pattern.

- *Nuclear pleomorphism*: Variation of nuclear shape and size is known as pleomorphism. Nuclear pleomorphism is one of the most important features of malignancy. However the following malignant tumors show minimal pleomorphism:
  - Follicular carcinoma of thyroid
  - Adenoid cystic carcinoma of salivary gland
  - Acinic cell carcinoma
  - Lobular carcinoma of breast
  - Mucinous carcinoma of breast
  - Carcinoid tumors.

  Similarly the following benign tumors may show marked pleomorphism:
  - Pleomorphic lipoma
  - Atypical lipoma
  - Ancient schwannoma
  - Paraganglioma
  - Angiomyolipoma
  - Adrenocortical adenoma.
- *Nucleoli*: Multiple, enlarged and pleomorphic nuclei indicate malignancy. Macronucleoli are usually seen in malignant melanoma, hepatocellular carcinoma, and germ cell tumors. However, large nucleoli are also seen in chronic inflammation and radiation exposure or post chemotherapy.
- *Nuclear margin*: Nuclear margin irregularity is seen in majority of cancer cells. Deep longitudinal nuclear groove is the characteristic feature of Langerhans cell histiocytosis, papillary carcinoma of thyroid, and granulosa cell tumor of ovary.

- *Chromatin pattern*: Abnormal chromatin pattern is possibly one of the most important criteria of malignancy.
  - Irregular clumped chromatin: Most of the malignancy.
  - Salt and pepper like chromatin: Peripheral neuroectodermal tumor, Ewing's tumor and carcinoid.
  - Vesicular chromatin: Adenocarcinomas.
  - Fine chromatin pattern: Leukemia, adenocarcinoma and many lymphoma cases.
  - Cartwheel appearance of chromatin: Plasma cells and the neoplasm arising from the plasma cells.
  - Clearing of nuclear chromatin: Papillary carcinoma of thyroid.
- *Intranuclear inclusion*: Intranuclear pseudoinclusions are the protrusion of the cytoplasm within the cell nucleus. Intranuclear inclusion is often seen in papillary carcinoma of thyroid, melanoma, and bronchoalveolar carcinoma.
- *Mitosis*: Frequent mitosis is seen in majority of malignant tumors. However, increased number of mitotic activity may also be seen in reactive lymphoid hyperplasia and regenerative condition. Qualitative abnormality of mitosis such as tripolar or multipolar mitosis is the more important characteristic feature of cancer cells.

Background of the Smear

The following background information is helpful in diagnosis:

- Mucinous: Mucinous adenocarcinoma
- Tigroid: Seminoma/dysgerminoma, rhabdomyosarcoma
- Fat: Lipomatous tumor, fat necrosis
- Myxoid: Myxoma, pleomorphic adenoma, chordoma
- Chondroid material: Cartilaginous tumor
- Colloid: Colloid goitre
- Necrosis: Tumor necrosis, inflammation
- Lymphoglandular bodies: Lymphoid cells
- Psammoma body: Papillary carcinoma of thyroid, meningioma, serous adenocarcinoma of ovary
- Amorphous globules: Collagenous spherulosis
- Reinke's crystal: Sertoli cell tumor.

QUESTIONS

Problem 1.1

Match the following (Fig. 1.1)

A.	1. Squamous cell
B.	2. Respiratory epithelial cells
C.	3. Columnar cell
D.	4. Urothelial cell

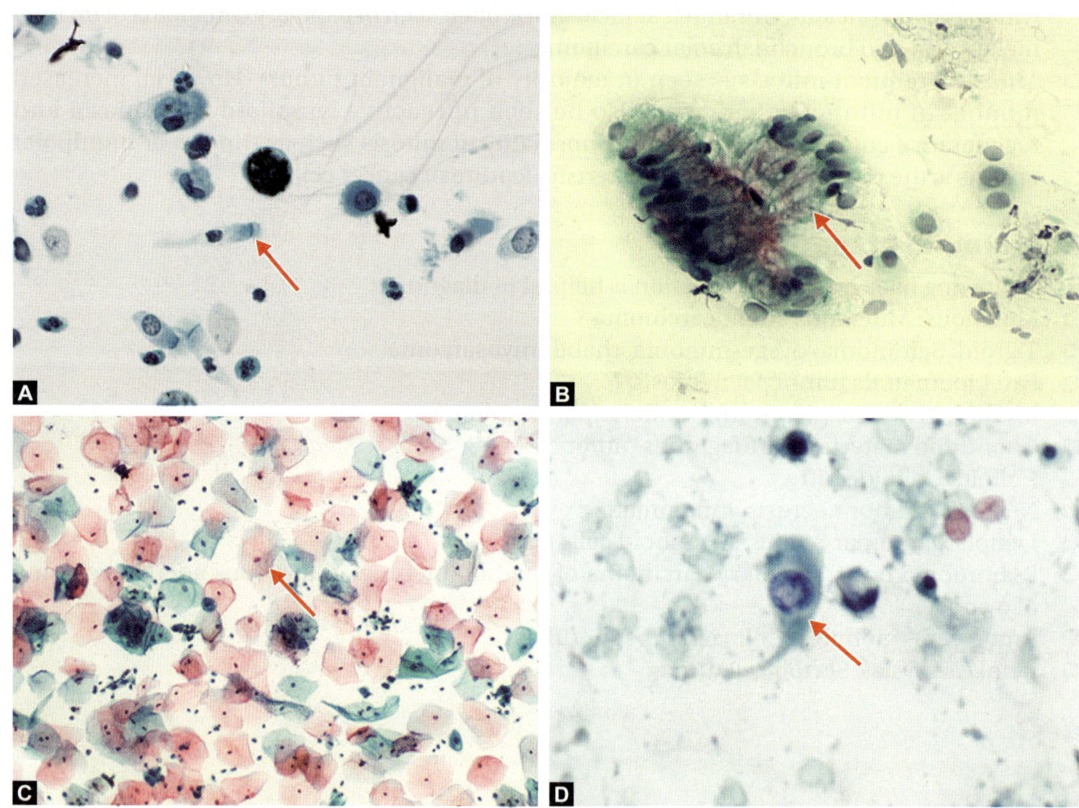

Figures 1.1A to D

Problem 1.2

Match the following (Fig. 1.2)

A.	1. Apocrine cell
B.	2. Foamy macrophages
C.	3. Hepatocytes
D.	4. Mesothelial cell

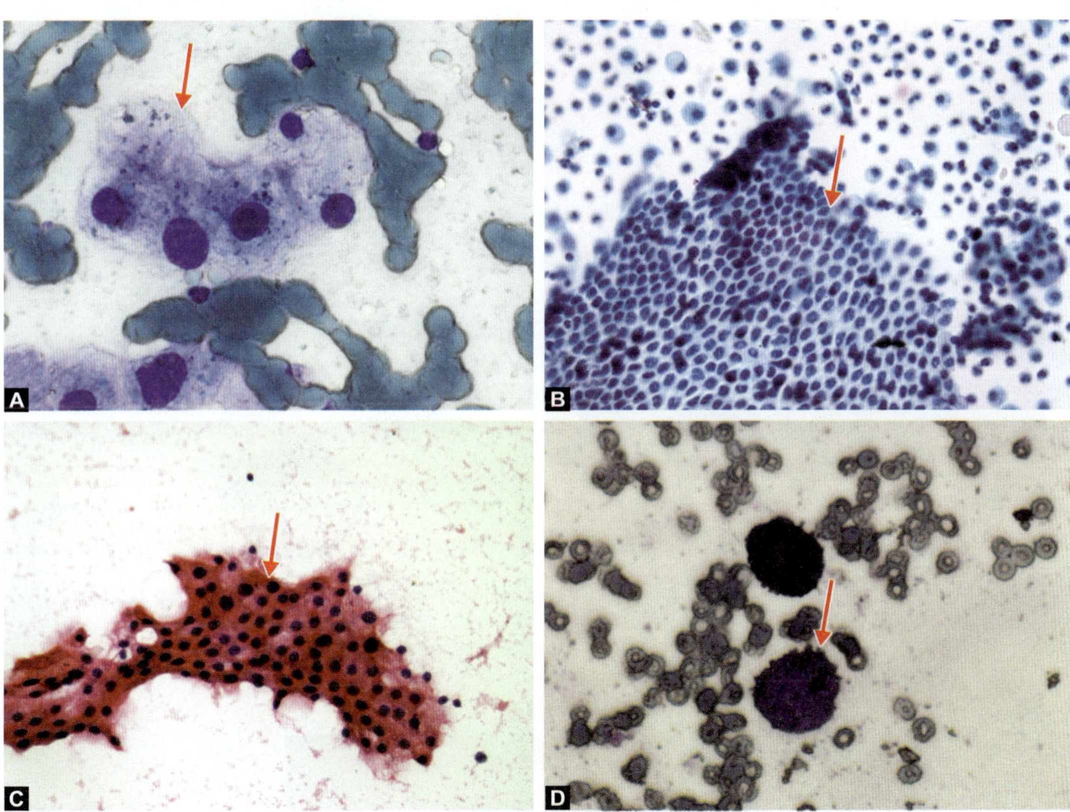

Figures 1.2A to D

Problem 1.3

Match the following (Fig. 1.3)

A.	1. Adenocarcinoma of stomach
B.	2. Chordoma
C.	3. Liposarcoma
D.	4. Chondrosarcoma
E.	5. HPV infection

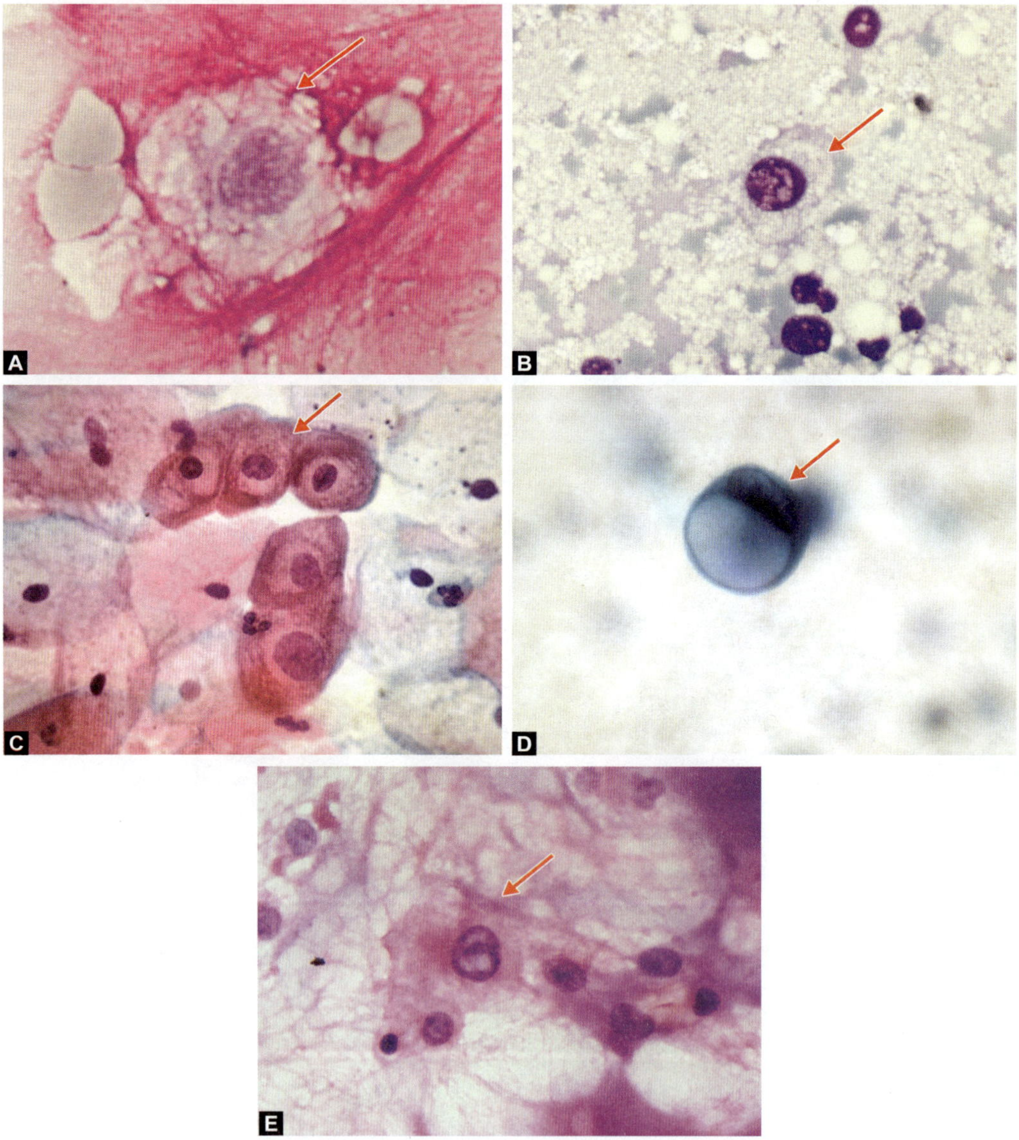

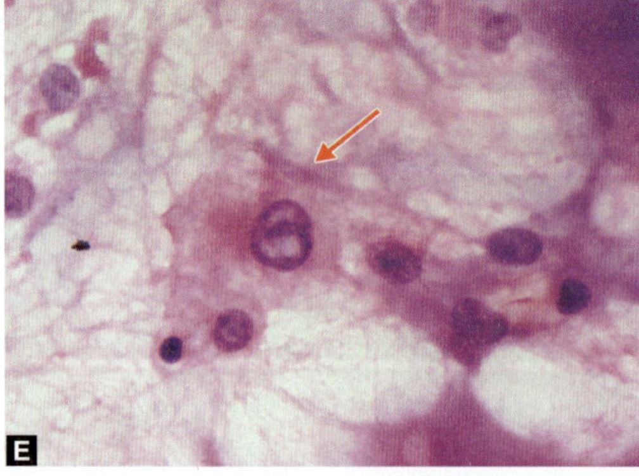

Figures 1.3A to E

Problem 1.4

Match the following (Fig. 1.4)

A.	1. Hodgkin lymphoma
B.	2. Squamous cell carcinoma
C.	3. Urothelial cell carcinoma
D.	4. Rhabdomyosarcoma

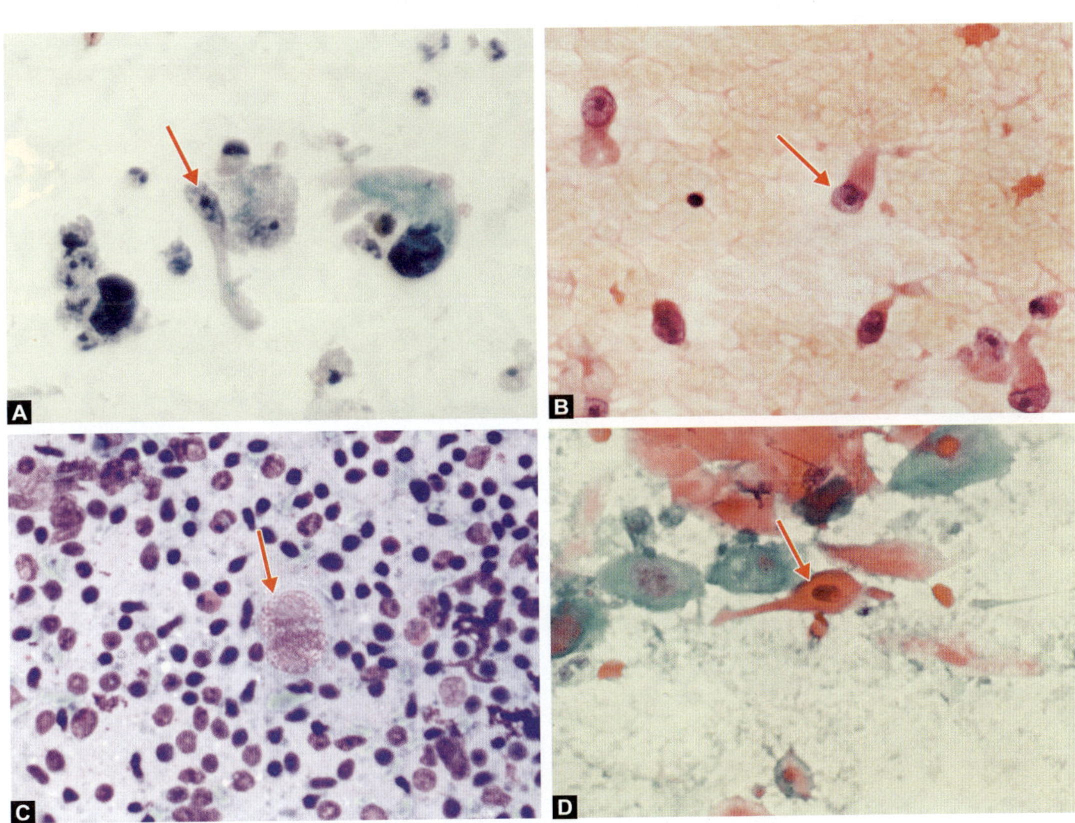

Figures 1.4A to D

Problem 1.5

Match the following (Fig. 1.5)

A.	1. Fat necrosis of breast
B.	2. Megakaryocyte in normal bone marrow
C.	3. Giant cell tumor of bone
D.	4. Pleomorphic sarcoma with tumor giant cell

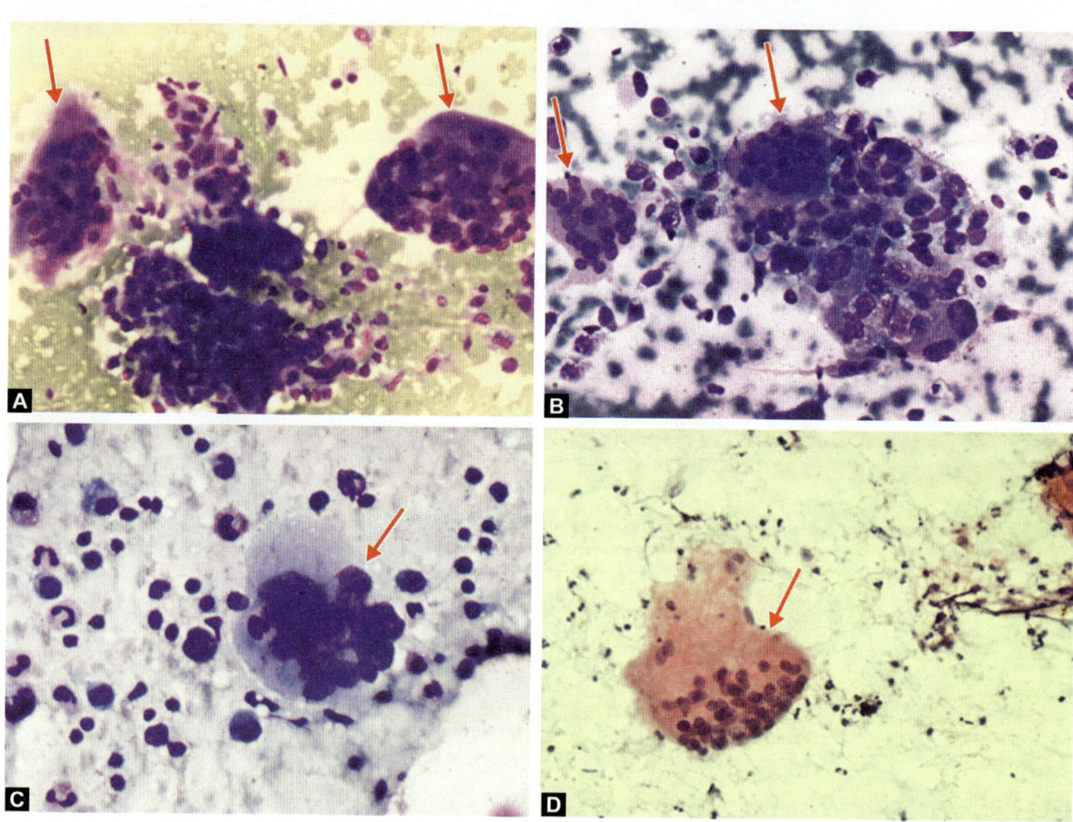

Figures 1.5A to D

Problem 1.6

Match the following (Fig. 1.6)

A.	1. Rosettes
B.	2. Glomeruloid body
C.	3. Microfollicles
D.	4. Papillae
E.	5. Vertebroid appearance

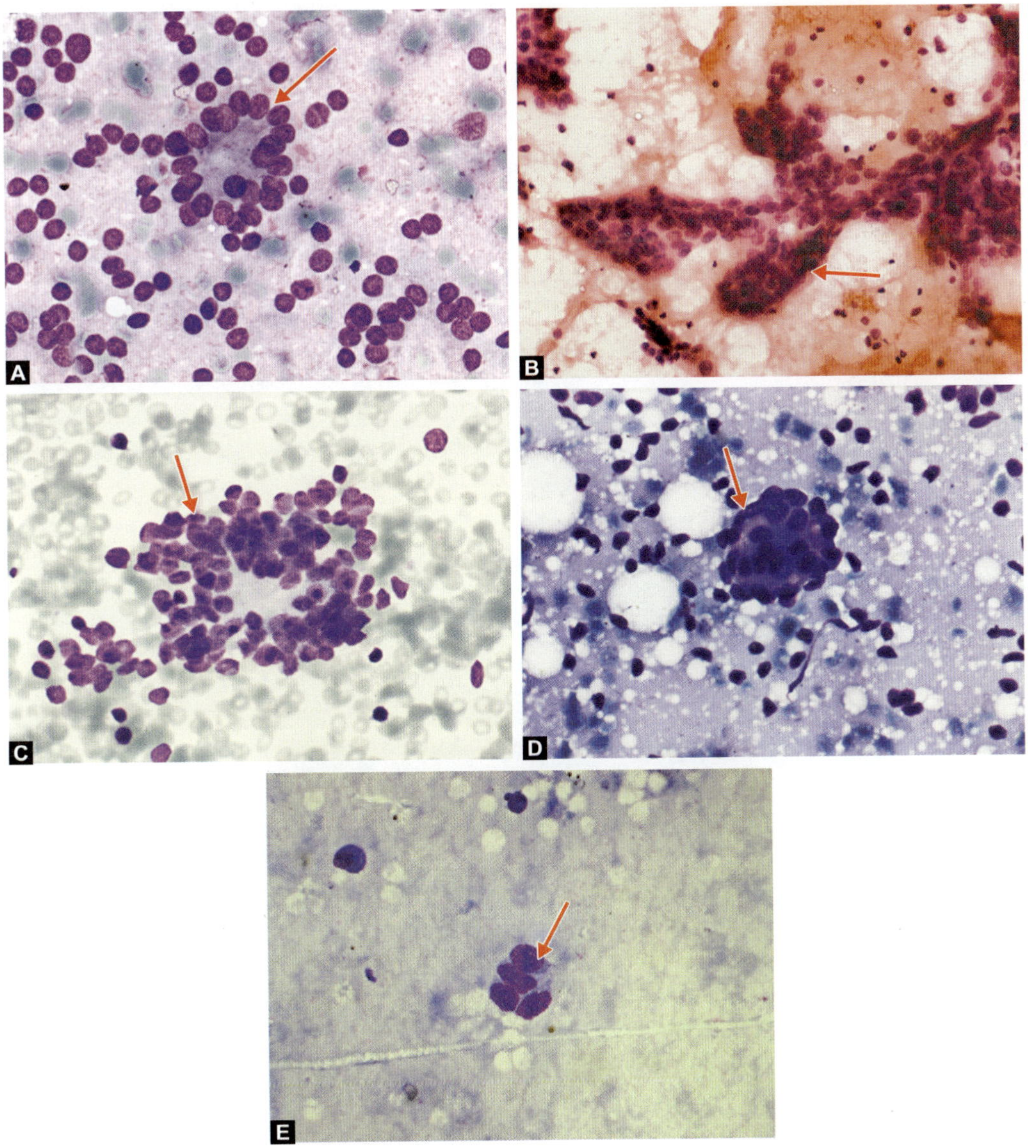

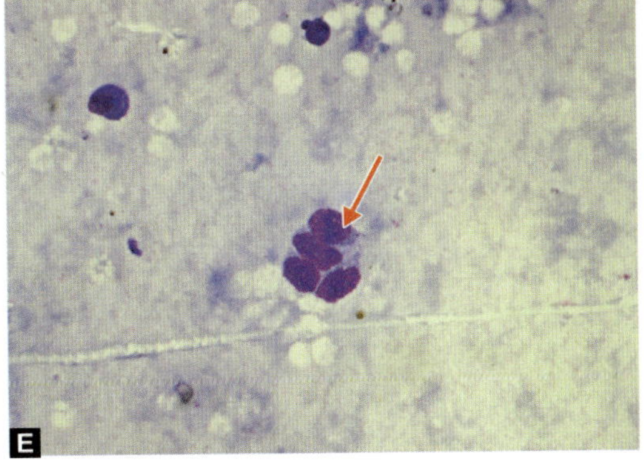

Figures 1.6A to E

Problem 1.7

Match the following (Fig. 1.7)

A.	1. Calcium crystal
B.	2. Necrosis
C.	3. Chondromyxoid material
D.	4. Mucinous material
E.	5. Synovial fluid

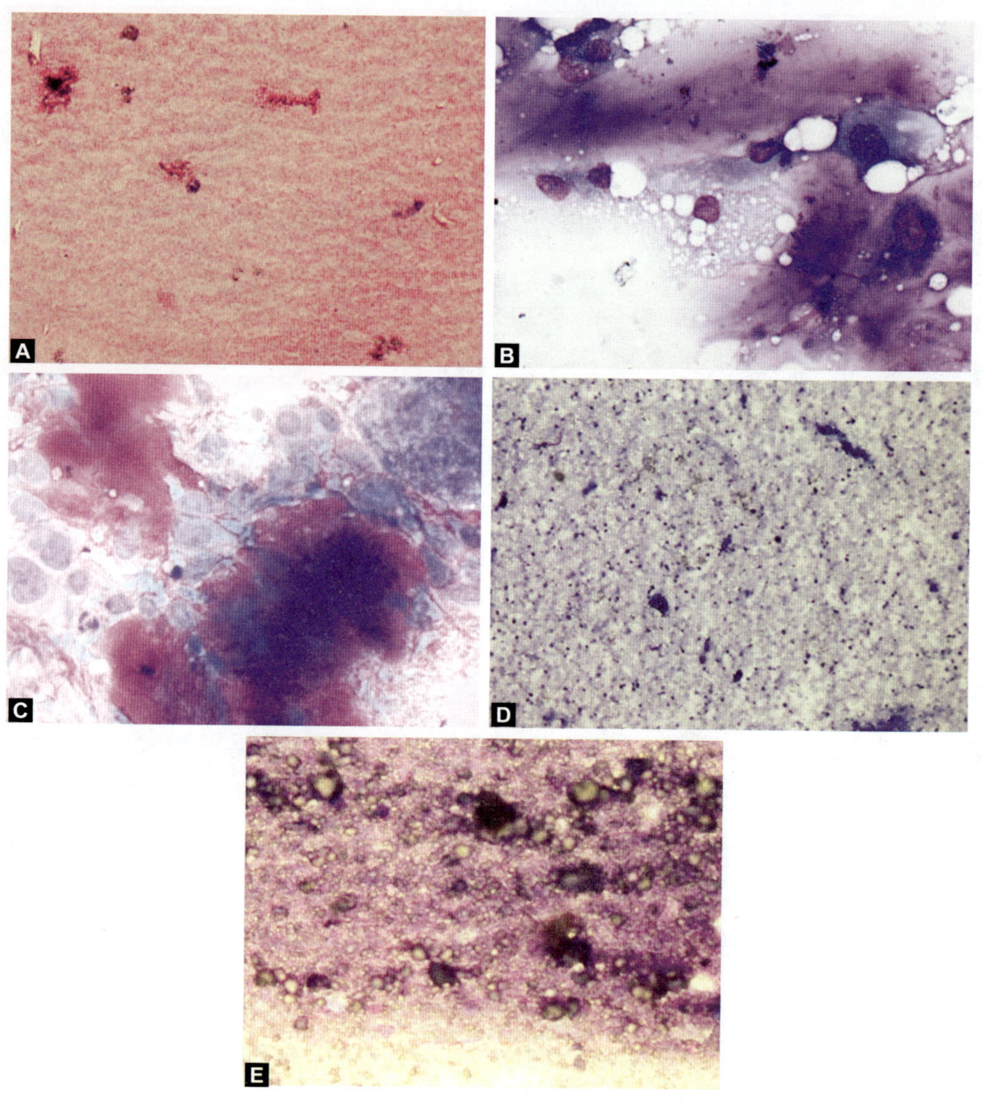

Figures 1.7A to E

Problem 1.8

Match the following (Fig. 1.8)

A.	1. Psammoma body
B.	2. Crystals in gout
C.	3. Melanin containing cells in dermatopathic lymphadenitis
D.	4. Melanin containing cells in melanoma

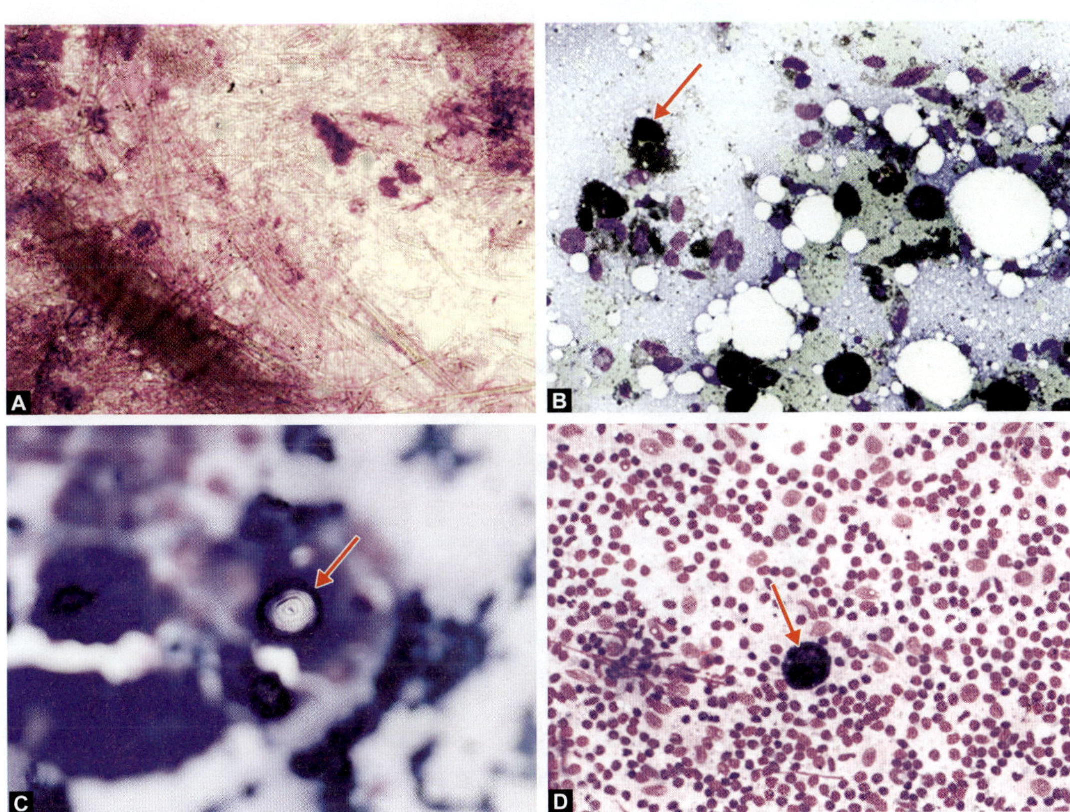

Figures 1.8A to D

Problem 1.9

Match the following (Fig. 1.9)

A.	1. Micronucleus
B.	2. Emperipolesis
C.	3. Cell cannibalism

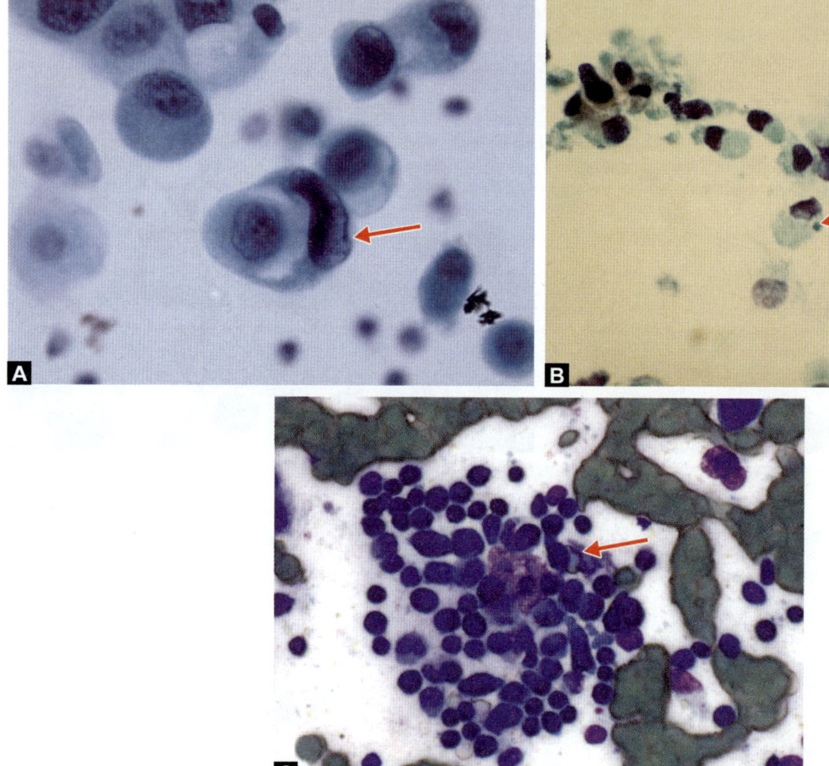

Figures 1.9A to C

Problem 1.10

Match the following (Fig. 1.10)

A.	1. Artifactual nuclear vacuolation
B.	2. Irregular nuclear margin
C.	3. Intranuclear inclusion

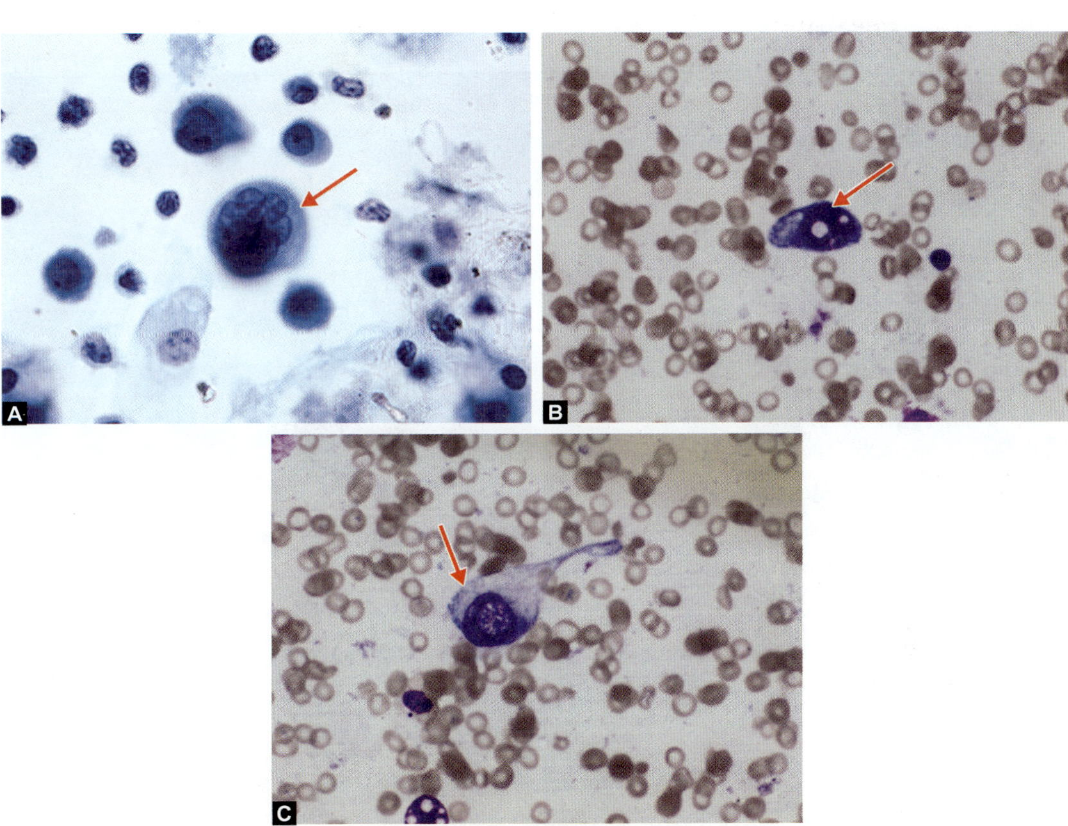

Figures 1.10A to C

Problem 1.11

Match the following (Fig. 1.11)

A.	1. Clumped chromatin
B.	2. Cartwheel
C.	3. Fine chromatin
D.	4. Salt and pepper

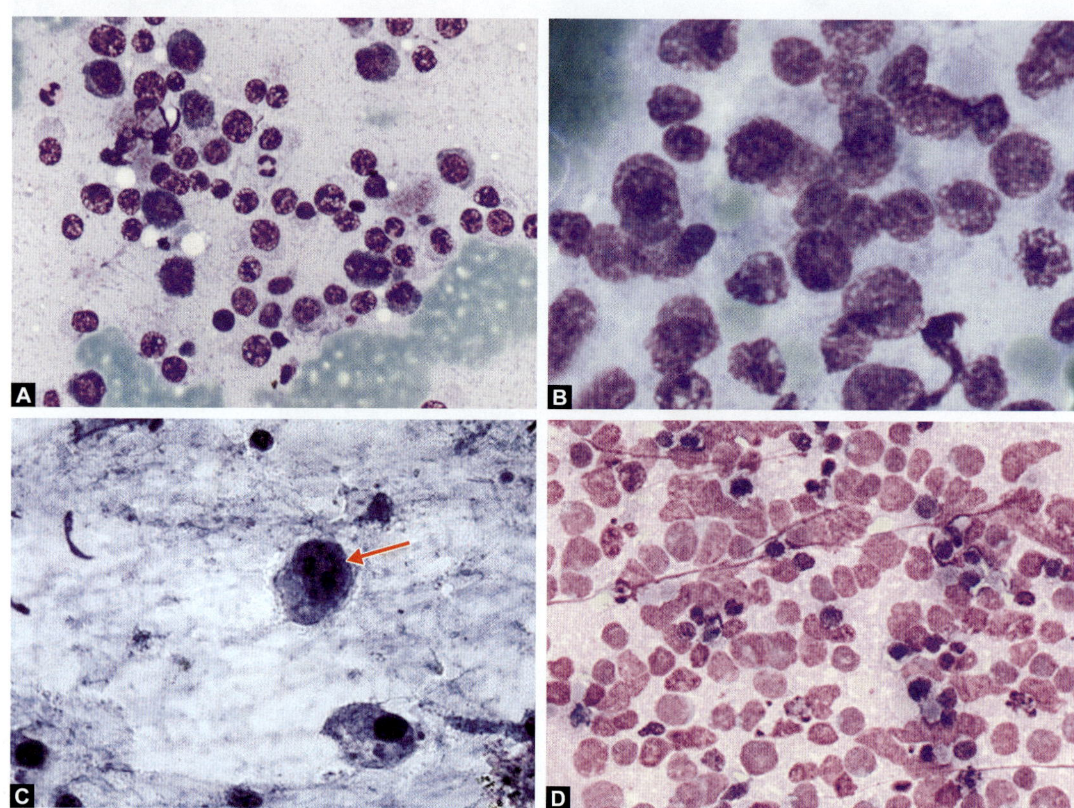

Figures 1.11A to D

Problem 1.12

Match the following (Fig. 1.12)

A. Inhibin	1. Adenocarcinoma of lung
B. Mic 2	2. Carcinoma of colon
C. Napsin	3. Peripheral neuroectodermal tumor
D. CK 20	4. Sex cord tumor of ovary

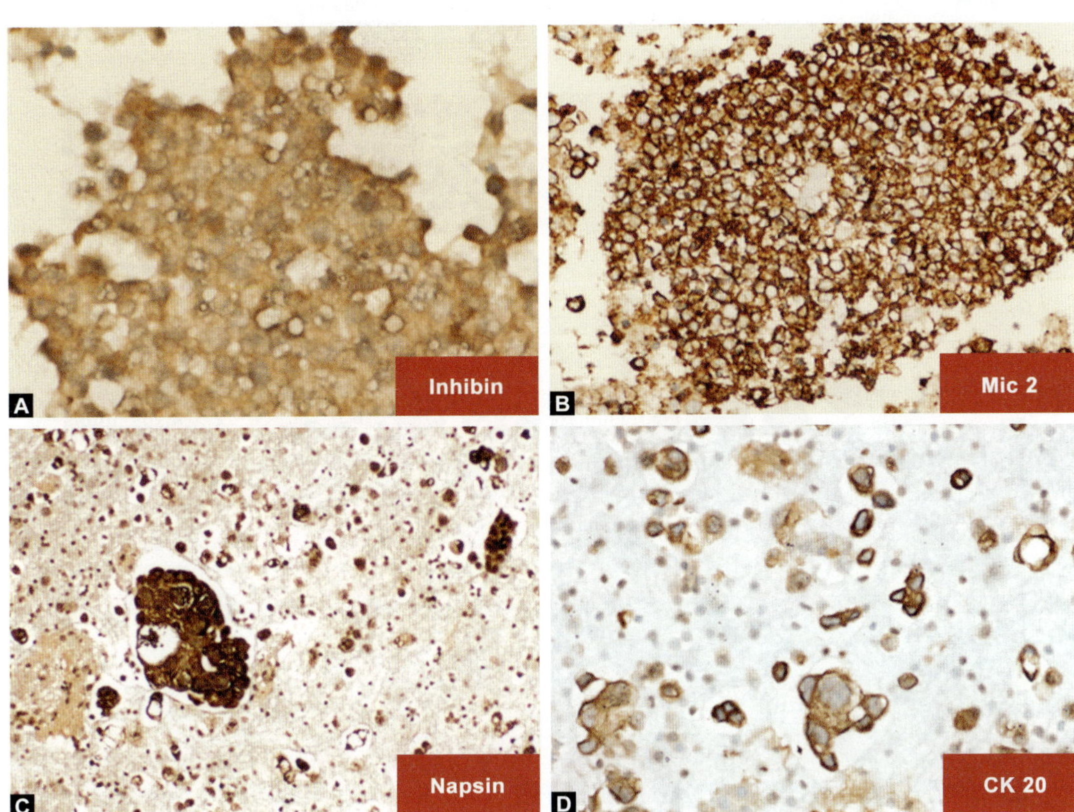

Figures 1.12A to D

Problem 1.13

Match the following (Fig. 1.13)

A. Desmin	1. Anaplastic large cell lymphoma
B. Vimentin	2. Hodgkin's lymphoma
C. CD 15	3. Histiocyte
D. ALK	4. Any soft tissue tumor
E. CD 68	5. Adenocarcinoma lung
F. CK 7	6. Skeletal muscle

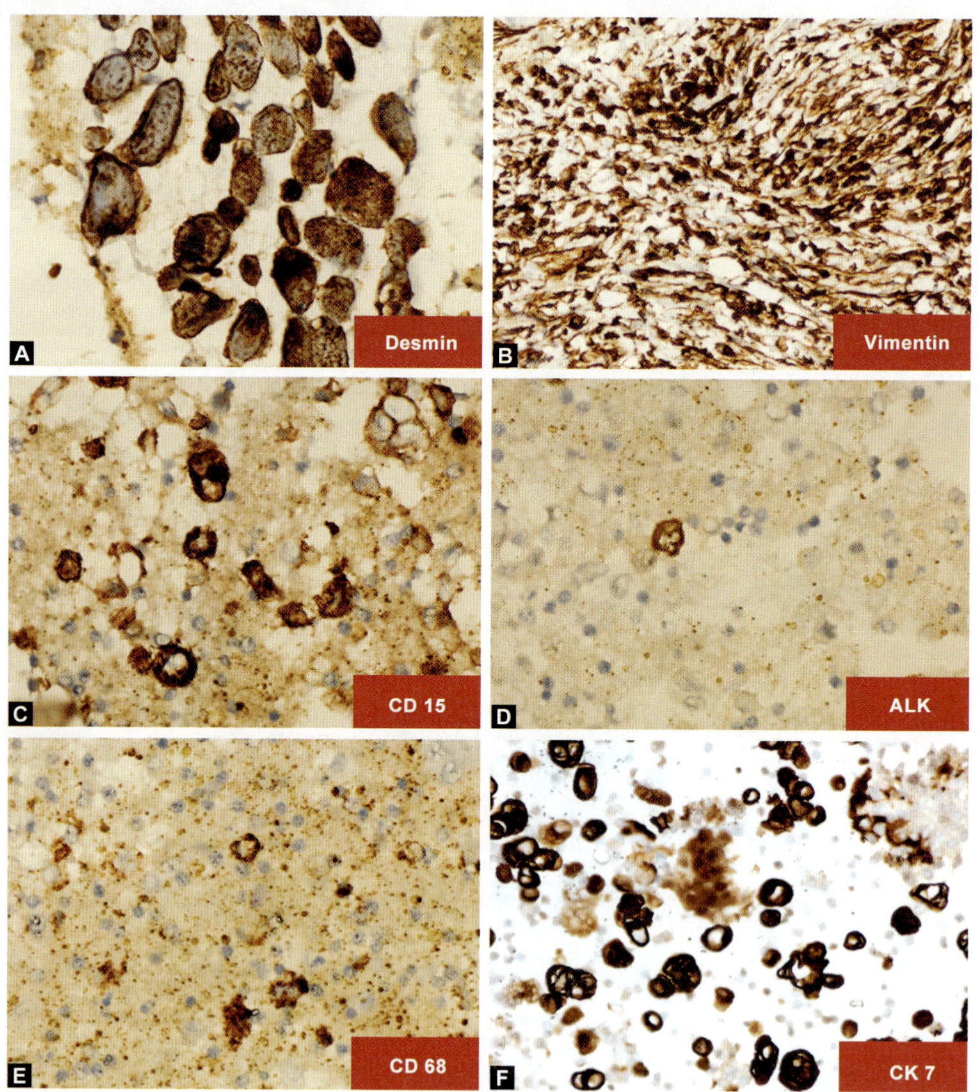

Figures 1.13A to F

Problem 1.14

Match the following (Fig. 1.14)

A.	1. B cell lymphoma
B.	2. All lymphoid cells express
C.	3. Lymphoblastic lymphoma
D.	4. Light chain restriction

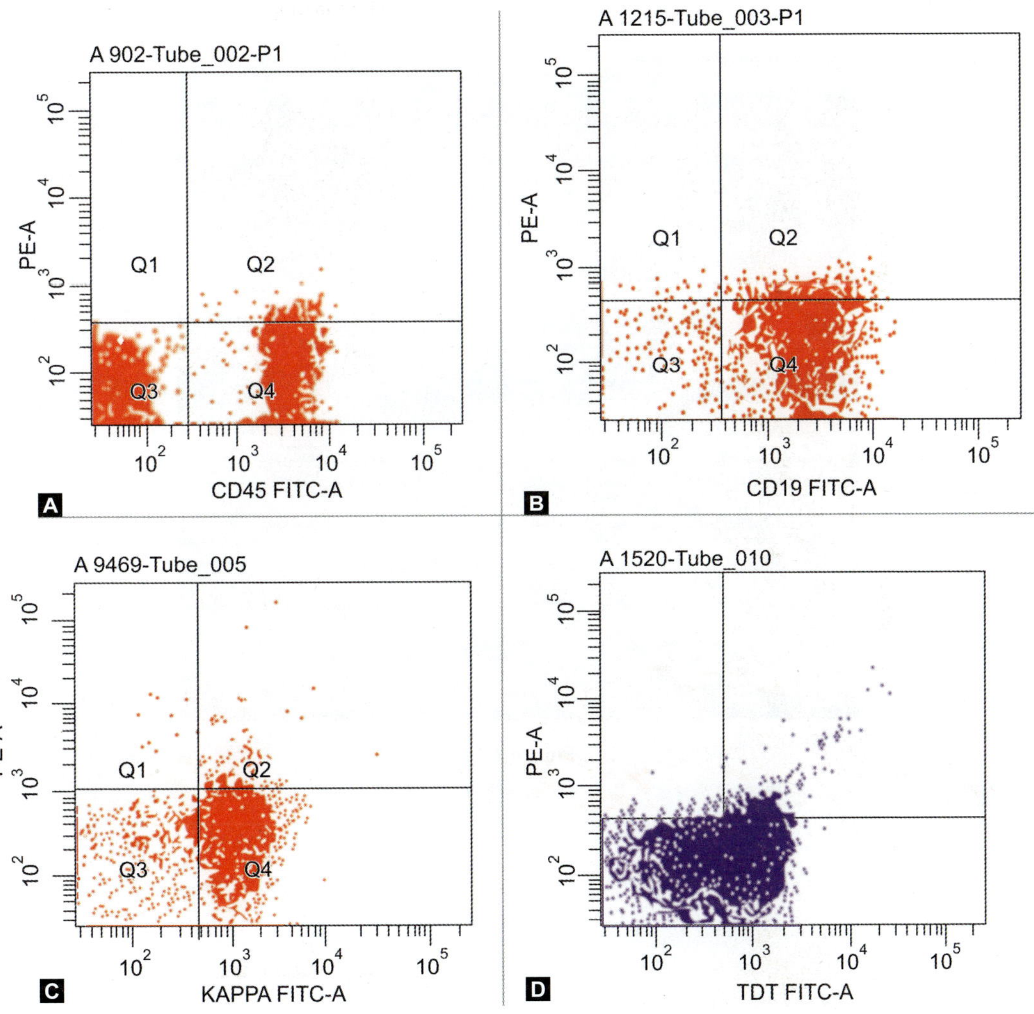

Figures 1.14A to D

Problem 1.15

Match the following (Fig. 1.15)

1.	A. Heterochromatin
2.	B. Mitochondria
3.	C. Euchromatin
4.	D. Rough endoplasmic reticulum
5.	E. Nuclear pore

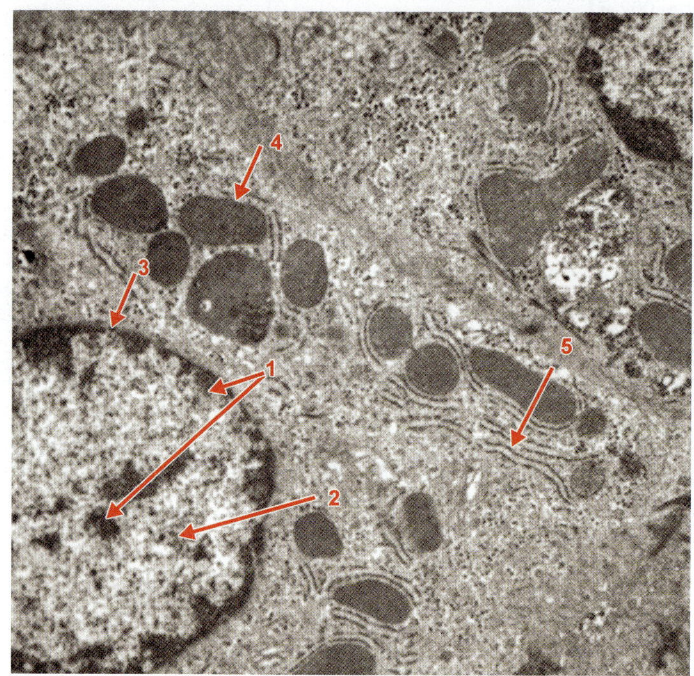

Figure 1.15

Problem 1.16

Match the following (Fig. 1.16)

A.	1. Nocardia
B.	2. Candia
C.	3. Aspergillosis
D.	4. Mucormycosis

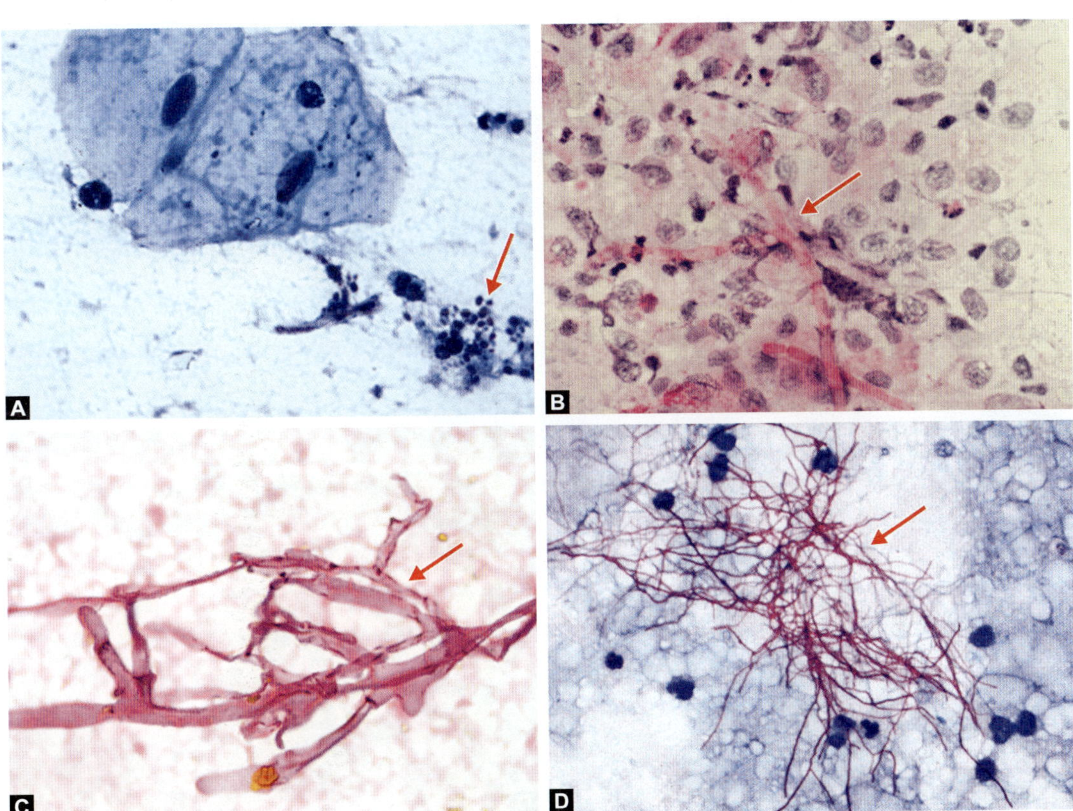

Figures 1.16A to D

Problem 1.17

Match the following (Fig. 1.17)

A.	1. *Trichomonas vaginalis*
B.	2. Leishmania donovani body
C.	3. Cryptococci
D.	4. Histoplasma

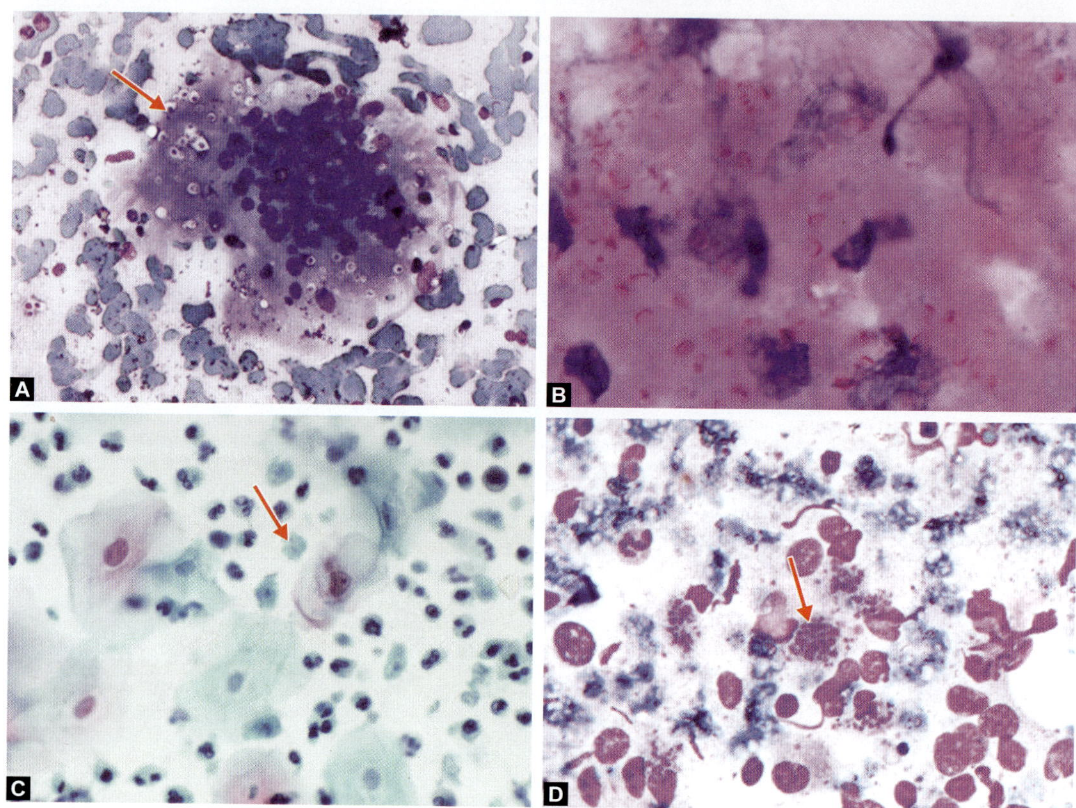

Figures 1.17A to D

Problem 1.18

Match the following (Fig. 1.18)

A.	1. Microfilaria
B.	2. Hydatid
C.	3. Actinomycosis
D.	4. Cysticercosis

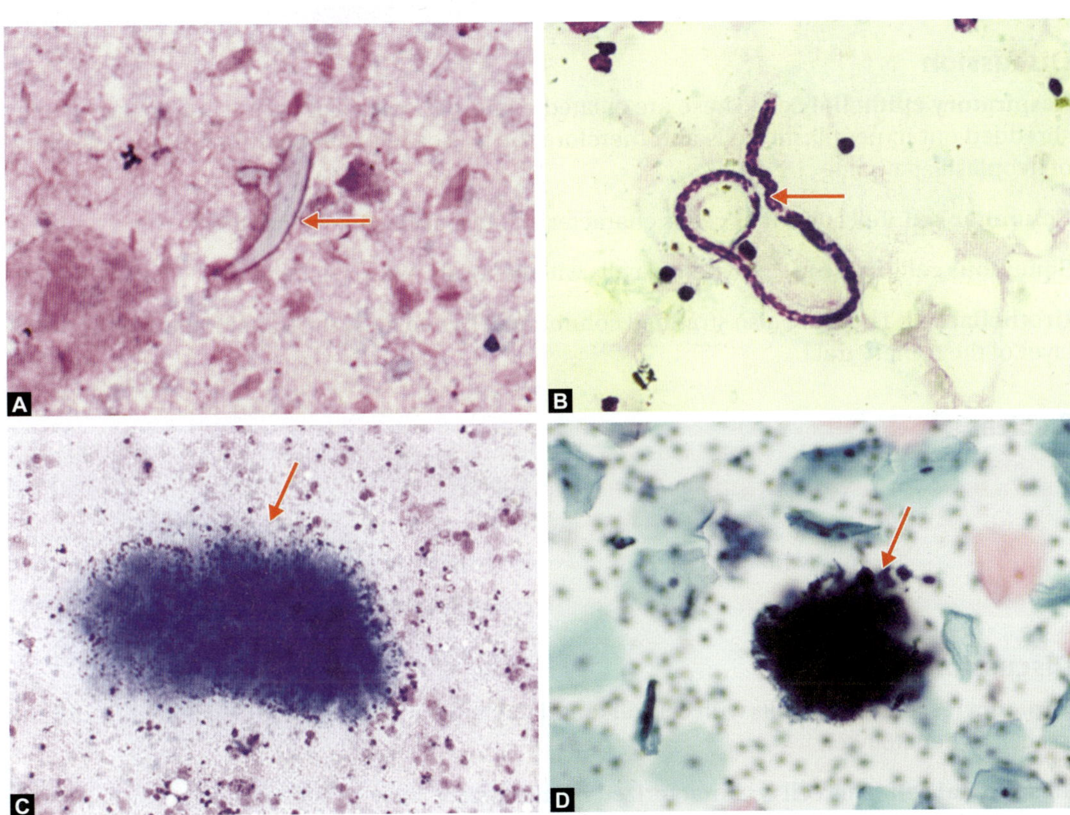

Figures 1.18A to D

ANSWERS

Problem 1.1

A.	2. Respiratory epithelial cells
B.	3. Columnar cell
C.	1. Squamous cell
D.	4. Urothelial cell

Discussion

Respiratory epithelial cell: These are ciliated columnar epithelial cells. The cilia are usually shredded out in neoplastic cells and therefore the presence of cilia eliminates the possibility of neoplastic pathology.

Columnar cell: Tall columnar cell is characterized by basally placed nuclei.

Squamous cell: These are polyhedral cells with centrally placed nuclei.

Urothelial cell: These are also stratified columnar cells and form the transitional epithelial cell layer of the urinary tract.

Problem 1.2

A.	3. Hepatocytes
B.	4. Mesothelial cell
C.	1. Apocrine cell
D.	2. Foamy macrophages

Discussion

Hepatocytes: Polygonal cells with central round nuclei. The presence of bile pigments indicates the cell as hepatocyte.

Mesothelial cell: This cell is characterized by monolayer sheet of cells with window pattern in between the cells.

Apocrine cells: The cells show abundant eosinophilic cytoplasm with central monomorphic nuclei.

Foamy macrophages: The cell contains abundant cytoplasm with phagocytosed material that often obscures the nuclear details.

Problem 1.3

A.	2. Chordoma
B.	3. Liposarcoma
C.	5. HPV infection

D.	1. Adenocarcinoma of stomach
E.	4. Chondrosarcoma

Discussion

Chordoma: The Figure 1.3A shows physaliferous cells of chordoma. The cells contain abundant bubbly cytoplasm with centrally placed round nuclei.

Liposarcoma: The Figure 1.3B shows lipoblast of liposarcoma. The cells show vacuolated cytoplasm with central nuclei. The vacuoles often indent the nuclear margin.

HPV infection: The Figure 1.3C shows classical koilocytes of low grade squamous intraepithelial lesion (LSIL) case. The cells show enlarged central nuclei with perinuclear halo. The cytoplasmic margin of the halo is thick. Koilocytes is seen in HPV infection.

Adenocarcinoma of stomach: The Figure 1.3D shows signet ring cell of adenocarcinoma of stomach. The cell shows abundant vacuolated cytoplasm with eccentric enlarged nuclei.

Chondrosarcoma: The Figure 1.3E shows chondrocytes of chondrosarcoma. The cell is large with abundant reddish blue cytoplasm and enlarged nucleus.

Problem 1.4

A.	3. Urothelial cell carcinoma
B.	4. Rhabdomyosarcoma
C.	1. Hodgkin lymphoma
D.	2. Squamous cell carcinoma

Discussion

Urothelial cell carcinoma: Figure 1.4A shows cercariform cells. It is seen in urothelial cell carcinoma. The particular word "cercaria" is derived from Greek word "kerkos" that means tail. The cell has a globular nucleus in one end with long cytoplasmic tail like process. This is not an artefact due to the smear preparation. These cells are not specific for urothelial cell carcinoma; however the presence of more than 20 cercariform cells in a single case indicates urothelial cell carcinoma.

Rhabdomyosarcoma: The (Fig. 1.4B) strap cell is noted in rhabdomyosarcoma. The cell has eccentric nuclei with abundant eosinophilic cytoplasm with striation that can be seen in careful examination.

Hodgkin lymphoma: The Figure 1.4C shows classical Reed Sternberg (R-S) cell. The cell has two enlarged nuclei which are mirror image of each other. Each nucleus contains a large prominent nucleoli. Many cells may simulate R-S cells such as cells in melanoma, nasopharyngeal carcinoma, anaplastic large cell lymphoma, etc.

Squamous cell carcinoma: The Figure 1.4D shows tadpole cells which is seen in squamous cell carcinoma. The cell is elongated with a hyperchromatic nucleus in one end and elongated orangeophilic cytoplasm in the other end.

Further Reading

1. Dey P, Amir T, Jogai S, Al Jussar A. Fine needle aspiration cytology of metastatic transitional cell carcinoma. Diagn Cytopathol. 2005;32:226-8.
2. Hida CA, Gupta PK. Cercariform cells: are they specific for transitional cell carcinoma? Cancer. 1999;87:69-74.

Problem 1.5

A.	3. Giant cell tumor of bone
B.	4. Pleomorphic sarcoma with tumor giant cell
C.	2. Megakaryocyte in normal bone marrow
D.	1. Fat necrosis of breast

Discussion

Giant cell tumor of bone: Figure 1.5A shows multiple multinucleated giant cells in a case of giant cell tumor of bone. The giant cell contain 30 to 50 round monomorphic nuclei.

Pleomorphic sarcoma with tumor giant cell: Figure 1.5B shows multinucleated tumor giant cells in a case of pleomorphic sarcoma. The nuclei are 5 to 15 in number and show considerable pleomorphism.

Megakaryocyte in normal bone marrow: Figure 1.5C shows a megakaryocyte in the bone marrow. The nuclei are fused together. Background normal marrow elements help to recognize the megakaryocyte. At times platelets may be attached with the megakaryocyte.

Fat necrosis of breast: Figure 1.5D shows the foreign body giant cell in a fat necrosis of breast. The nuclei of the foreign body giant cells are usually variable form 5 to 20. The cytoplasm of the cell often contains foreign body material.

Problem 1.6

A.	3. Microfollicles
B.	4. Papillae
C.	1. Rosettes
D.	2. Glomeruloid body
E.	5. Vertebroid appearance

Discussion

Microfollicles: Figure 1.6A shows microfollicular arrangement of the cells. This is characterized by the circular arrangement of 12 to 15 thyroid follicular cells around a central clear space or colloid.

Papillae: Figure 1.6B shows papillae. These are finger like arrangement of cells. The peripheral layer of cells is often placed horizontally and the long axis of these cells is perpendicular to the long axis of the underlying cells. Papillae may be avascular or they may show central core of vessels.

Rosettes: The cells in the rosettes are arranged around central fibrillary structures. Cells are not attached with any basement membrane. Rosettes like arrangement are often seen in neuroblastoma, peripheral neuroectodermal tumor, and retinoblastoma.

Glomeruloid body: This is a tight cluster of cells resembling glomeruloid structure of kidney. Glomeruloid body is seen in Wilms' tumor.

Vertebroid appearance: This arrangement is characterized by row of single cells that compress each other. This is commonly seen in small cell carcinoma and lobular carcinoma.

Problem 1.7

A.	5. Synovial fluid
B.	4. Mucinous material
C.	3. Chondromyxoid material
D.	2. Necrosis
E.	1. Calcium crystal

Discussion

Synovial fluid (Fig. 1.7A): Mossy proteinaceous background is typical for synovial fluid.

Mucinous material (Fig. 1.7B): This is readily recognized as bluish material in May Grunwald Giemsa stain.

Chondromyxoid material (Fig. 1.7C): It is seen as deep red homogenous material.

Necrosis (Fig. 1.7D): It is characterized by bluish granular material with intermingled nuclear debris.

Calcium crystal (Fig. 1.7E): This is seen as variable sized dark blackish acellular material. This may be mistaken as collection of bacterial organisms. However the calcified material is homogenous and bacteria are granular.

Problem 1.8

A.	2. Crystals in gout
B.	4. Melanin containing cells in melanoma
C.	1. Psammoma body
D.	3. Melanin containing cells in dermatopathic lymphadenitis

Discussion

Crystals in gout (Fig. 1.8A): There are abundant crystals in the background. In gout these crystal are monosodium urate monohydrate crystals.

Melanin containing cells in melanoma (Fig. 1.8B): The figure shows abundant melanin containing pigments within the tumor cells in a case of melanoma. These cells are HMB 45 positive.

Psammoma body (Fig. 1.8C): The concentric laminated calcified material is known as psammoma body. This is commonly seen in papillary carcinoma of thyroid, meningioma, and serous cyst adenocarcinoma of ovary.

Melanin containing cells in dermatopathic lymphadenitis (Fig. 1.8D): The melanin containing benign cells are often seen in dermatopathic lymphadenitis. This is a benign condition and is usually seen in the drainage lymph node of dermatitis.

Further Reading

1. Johannessen JV, Sobrinho-Simões M. The origin and significance of thyroid psammoma bodies. Lab Invest. 1980;43(3):287-96.
2. Das DK, Mallik MK, Haji BE, Ahmed MS, Al-Shama'a M, et al. Psammoma body and its precursors in papillary thyroid carcinoma: a study by fine-needle aspiration cytology. Diagn Cytopathol. 2004;31(6):380-6.

Problem 1.9

A.	3. Cell cannibalism
B.	1. Micronucleus
C.	2. Emperipolesis

Discussion

Cell cannibalism (Fig. 1.9A): The cell cannibalism is an active killing of one cell by the other. The cannibalistic cell phagocytoses the victim cell and actively kills it. The cell cannibalism is usually seen in malignancy.

Micronucleus (Fig. 1.9B): This is the nuclear fragment. The micronucleus is always near the main nucleus and not attached with it. The size of the micronucleus is 2 to 4 micron. The excess number of micronucleus indicates chromosomal damage.

Emperipolesis (Fig. 1.9C): In emperipolesis the histiocyte phagocytose large number of lymphocytes. Emperipolesis is commonly seen in Rosai-Dorfman disease.

Further Reading

1. Samanta S, Dey P. Micronucleus and its applications. Diagn Cytopathol. 2012;40(1):84-90.
2. Sharma N, Dey P. Cell cannibalism and cancer. Diagn Cytopathol. 2011;39(3):229-33.

Problem 1.10

A.	2. Irregular nuclear margin
B.	1. Artifactual nuclear vacuolation
C.	3. Intranuclear inclusion

Discussion

Irregular nuclear margin (Fig. 1.10A): The nucleus show markedly irregular margin. This type of irregular margin is usually seen in malignant cells.

Artifactual nuclear vacuolation (Fig. 1.10B): This type of artifact is commonly mistaken as intranuclear inclusion.

Intranuclear inclusion (Fig. 1.10C): Intranuclear inclusion shows round shaped cytoplasmic invagination within the nucleus. The color of the inclusion is same as cytoplasmic color. This is commonly seen in papillary carcinoma of thyroid, bronchoalveolar carcinoma, melanoma, and meningioma.

Further Reading

1. Arora SK, Dey P. Intranuclear peudoinclusions: morphology, pathogenesis, and significance. Diagn Cytopathol. 2012;40(8):741-4.

Problem 1.11

A.	2. Cartwheel
B.	4. Salt and pepper
C.	1. Clumped chromatin
D.	3. Fine chromatin

Discussion

Cartwheel (Fig. 1.11A): This is characterized by peripheral clumped chromatin in a cartwheel like fashion. Plasma cells show cartwheel chromatin.

Salt and pepper (Fig. 1.11B): The chromatin is granular and gives the salt and pepper appearance. This pattern is typically seen in neuroendocrine tumor.

Clumped chromatin (Fig. 1.11C): Coarsely clumped chromatin is usually seen in malignant nucleus.

Fine chromatin (Fig. 1.11D): Fine chromatin is seen mainly in leukemia, adenocarcinoma and many lymphomas.

Problem 1.12

A. Inhibin	4. Sex cord tumor of ovary
B. Mic 2	3. Peripheral neuroectodermal tumor
C. Napsin	1. Adenocarcinoma of lung
D. CK 20	2. Carcinoma of colon

Discussion

Inhibin: It is one of the markers of sex cord tumor of ovary.

Mic 2: Ewing's tumor and peripheral neuroectodermal tumors are all positive for Mic 2.

Napsin: It shows granular cytoplasmic positivity. Napsin is positive in adenocarcinoma of lung.

CK 20: Carcinoma of colon is positive for CK 20. In addition focal CK 20 positivity is also seen in urothelial cell carcinoma, pancreatic and biliary carcinoma.

Problem 1.13

A. Desmin	6. Skeletal muscle
B. Vimentin	4. Any soft tissue tumor
C. CD15	2. Hodgkin's lymphoma
D. ALK	1. Anaplastic large cell lymphoma
E. CD 68	3. Histiocyte
F. CK 7	5. Adenocarcinoma lung

Discussion

Desmin: Desmin is expressed in smooth muscle, skeletal muscle and cardiac muscle. Desmin expression is a reliable marker of rhabdomyosarcoma and leiomyosarcoma.

Vimentin: Vimentin is positive in mesenchymal cells such as fibroblast, endothelial cells and smooth muscle. It is not a specific marker of mesenchymal cells and has also been demonstrated in epithelial neoplasms such as in renal cell carcinoma, lymphomas and mesothelial cells.

CD15: Positive in Hodgkin's lymphoma.

ALK: It is commonly expressed in anaplastic large cell lymphoma. It is also positive in non-small cell lung carcinoma.

CD 68: It is a marker of histiocyte.

CK 7: Adenocarcinomas of lung, breast carcinoma, and ovarian carcinomas are positive for CK 7.

Further Reading

1. Dey P, Togra J, Mitra S. Intermediate filament: structure, function, and applications in cytology. Diagn Cytopathol. 2014;42(7):628-35.

Problem 1.14

A.	2. All lymphoid cells express
B.	1. B cell lymphoma
C.	4. Light chain restriction
D.	3. Lymphoblastic lymphoma

Discussion

CD 45: It is also known as leucocyte common antigen (LCA) and it is expressed in all types of lymphocytes.

CD 19: It is a B cell marker. Flow cytometry (FCM) shows predominant CD 19 expression.

Kappa chain expression: The present FCM graph shows predominant kappa light chain expression. This is also known as light chain restriction. It indicates monoclonality of B cell lymphoma.

Tdt: It is positive in lymphoblastic lymphoma.

Problem 1.15

1.	a. Heterochromatin
2.	c. Euchromatin
3.	e. Nuclear pore
4.	b. Mitochondria
5.	d. Rough endoplasmic reticulum

Discussion

Heterochromatin and euchromatin: Chromatin is the uncoiled condensed chromosome in the interphase nuclei. Chromatin is classified as heterochromatin and euchromatin. Heterochromatin is condensed part of chromatin and is located on the nuclear membrane where genes are usually inactive. Uncoiled lightly stained chromatin is known as euchromatin.

Nuclear pore: In certain places in the nucleus the outer and inner nuclear membrane fuses and make the hole in the nuclear membrane known as nuclear pore. This nuclear pore selectively allows the passage of the macromolecules from the nucleus to cytoplasm and vice versa.

Mitochondria: Mitochondria are 0.5 to 1 micron diameter organelle and act as the power house of the cell. Mitochondria are double membrane-bound structures and consist of outer membrane, inner membrane, intermembranous space, cristae and mitochondrial matrix.

Endoplasmic reticulum: These are tubular and cistern like spaces and vesicular structures folded within the cytoplasm. They are involved with synthesis of secretary protein and lysosomal enzymes.

Problem 1.16

A.	2. Candida
B.	3. Aspergillosis
C.	4. Mucormycosis
D.	1. Nocardia

Discussion

Candida: Multiple spores of candida are seen. This is an opportunistic fungus and is seen in diabetes or other immunocompromised condition. The candida spore is 3 to 4 micron in size and is easily recognizable.

Aspergillosis: Thin slender fungi with acute angle branching is characteristic of aspergillus.

Mucormycosis: Mucormycosis shows non-septate broad fungal hyphae with wide angle branching.

Nocardia: Thin slender nocardia is demonstrated by modified Ziehl-Neelsen stain.

Problem 1.17

A.	3. Cryptococci
B.	4. Histoplasma
C.	1. *Trichomonas vaginalis*
D.	2. Leishmania donovani body

Discussion

Cryptococci: Small round predominantly intracellular organism 5–8 μ in diameter. This is stained by PAS or India ink stain.

Histoplasma: Small round 2–5 μ in diameter, narrow budding, usually remains inside the macrophages.

***Trichomonas vaginalis*:** It is well visualized in Papanicolaou's stained cervical smear as 15–30 μ long, round-to-oval "pear-shaped" organisms with an indistinct eccentric vesicular nucleus and red cytoplasmic granules.

Leishmania donovani body (LD body): These are amastigote form of leishmania and 3–5 μ in diameter. It consists of a kinetoplast and basal body. LD body usually remains within the histiocytes.

Problem 1.18

A.	2. Hydatid
B.	1. Microfilaria
C.	4. Cysticercosis
D.	3. Actinomycosis

Discussion

Hydatid: The smear shows characteristic hooklets of hydatid cyst. The disease is caused by *Echinococcus* granulosis. The smear from hydatid cyst aspirate may contain laminated membrane, scolex or hooklet.

Microfilaria: The smear shows microfilaria of *Wuchereria bancrofti.*

Cysticercosis: The microphotograph shows tegumentum layer of cysticercosis. This infection is caused by *Taenia solium.*

Actinomycosis: The slender filamentous organism arranged as sun ray like fashion.

CHAPTER 2

Cervical Cytology

Introduction

In 2001 the Bethesda system of reporting cytology smear was revised and the reporting format is divided into three parts (Box 2.1):

(a) Specimen adequacy, (b) General categorization and (c) Terminologies.

Box 2.1: The Bethesda system 2001[1]

Specimen type
- Conventional or liquid based cytology
- Specimen adequacy

Satisfactory for evaluation
- Adequate number of squamous cells:
  - Conventional smear: 8000 to 12000 squamous cells
  - Liquid based cytology: 5000 squamous cells
- Mention the presence or absence of endocervical cells or transformation zone components: at least 10 endocervical or metaplastic cells should be present, no need for cell clusters
- Any other quality elements such as drying artifact, obscured by blood or inflammation, etc. Partially obscured: 50 to 75% epithelial cells cannot be visualized, unsatisfactory: when more than 75% epithelial cells cannot be visualized

Unsatisfactory for evaluation
- Specimen rejected: Specimen rejected due to broken slide or mislabelled specimen
- Fully evaluated unsatisfactory specimen: Specimen processed but unsatisfactory because:
  - Insufficient squamous cells
  - More than 75% of smear is obscured by blood or inflammatory cells

General Categorization (Optional)
- Negative for intraepithelial lesion or malignancy
- Epithelial cell abnormality

Interpretation/Result

Negative for intraepithelial lesion or malignancy

Organisms
- *Trichomonas*
- Bacterial vaginosis
- Fungal morphology consistent with *Candida*
- Bacterial morphology consistent with actinomycosis
- Cellular morphology consistent with herpes simplex virus

Contd...

Contd...

Other non-neoplastic findings
- Reactive cellular changes associated with
  - Inflammation
  - Intrauterine contraceptive device
  - Radiation
- Glandular cell status post-hysterectomy
- Atrophy

Others

Epithelial cell abnormality

Squamous cell
- Atypical squamous cell (ASC)
  - Of undetermined significance (ASC-US)
  - Cannot exclude HSIL (ASC-H)
- Low-grade squamous intraepithelial lesion (LSIL)
- High-grade squamous intraepithelial lesion (HSIL)
- Squamous cell carcinoma

Glandular cell
- Atypical
  - Endocervical in origin
  - Endometrial in origin
  - Glandular cell, not otherwise specified
- Atypical
  - Endocervical, favor neoplastic
  - Glandular, not otherwise specified
- Endocervical adenocarcinoma in situ
- Adenocarcinoma
  - Endocervical
  - Endometrial
  - Extrauterine
  - Not otherwise specified
- Other malignant neoplasm

Explanation

Specimen adequacy:

Adequate number of cells

Conventional smear: 8000 to 12000 squamous cells

Liquid based cytology: 5000 squamous cells.

Only the well visualized and well preserved squamous cells should be taken into consideration. The approximate number of cells should be estimated only and the cytologists are not expected to count individual cells. TBS 2001 published the reference images in their website to help in the estimation of the number of cells on conventional smear and liquid based preparations. Squamous metaplastic cells in the smear can be considered as squamous cells during the estimation of the cell number. The minimum number of squamous cells is required less in case of liquid based cytology as this technique represents homogenous population of cell. A minimum number of 10 microscopic filed at 40 X, should be assessed for the cellularity.

Squamous Cell Abnormality

Atypical Squamous Cells of Undetermined Significance (ASCUS)

Criteria

- Nuclear atypia in the squamous cells: The nuclear area of the atypical cells is 2.5 to 3 times more than the nucleus of the normal intermediate cell
- Mildly increased nucleo-cytoplasmic ratio
- Nuclear hyperchromasia is minimal
- Regular margin with fine granular chromatin.

Atypical Squamous Cell High-grade SIL cannot be Excluded (ASC-H)

Criteria

- Single and loose cohesive group of cells
- Immature small squamous cells
- Nucleus: Mild to moderate enlargement, hyperchromasia and pleomorphism
- Increased nucleocytoplasmic ratio.

Low-grade Squamous Intraepithelial Lesion (LSIL)

The cytological diagnostic criteria of LSIL are:

- Single or sheets of superficial and intermediate cells.
- Large cells with abundant cytoplasm. Cytoplasmic differentiation is evident.
- Well defined cytoplasmic margin of the atypical cells.
- Enlarged nucleus: Nuclear area is more than three times of that a normal intermediate cell nucleus.
- Mildly increased nucleocytoplasmic (N/C) ratio.
- Mild hyperchromatic nuclei.
- Variably irregular nuclear margin.
- Chromatin is usually evenly distributed coarsely granular.
- Nucleoli are commonly absent or inconspicuous.

Koilocytosis: Distinct perinuclear cytoplasmic vacuoles with sharp condensed thick cytoplasmic margin. Nuclei are enlarged. The perinuclear halos along with the nuclear abnormalities are essential for the qualification of koilocytosis to diagnose LSIL. The presence of koilocytosis is the characteristic feature of LSIL but it is not an essential feature to interpret LSIL.

High-grade Squamous Intraepithelial Lesion (HSIL)

HSIL in Bethesda system encompasses cervical intraepithelial neoplasia (CIN 2) (moderate dysplasia) and CIN 3 (severe dysplasia/carcinoma in situ). Approximately 0.5% of all cervical smears show HSIL. About 97% HSIL cases are associated with high-risk HPV infection. The diagnostic criteria of HSIL include:

- Single, sheets or syncytial aggregates of cells
- Parabasal or basal types of cells are commonly seen. These cells are smaller and less mature
- Cytoplasm is dense or lacy and orangeophilic in keratinized variant
- Enlarged nuclei with marked increased N/C ratio
- Usually coarsely granular chromatin, but it may be fine granular
- Marked irregular nuclear membrane with frequent indentation

- Nucleoli are commonly absent. However, the dysplastic cells may show prominent nucleoli if the endocervical glands are involved by the dysplastic cells.

Squamous Cell Carcinoma

The criteria of squamous cell carcinoma of cervix include:

- Scattered and small clusters of tumor cells
- Moderately pleomorphic cell with high N/C ratio
- Moderate to markedly pleomorphic nuclei
- Moderately hyperchromatic nuclei
- Irregular nuclear membrane
- Irregular coarse chromatin
- Occasionally macronucleoli may be seen
- Cytoplasm: Dense orangeophilic
- Tadpole cells: Elongated cell with large abnormal nuclei in one corner
- Fibre cell: Spindle shaped hyperchromatic nuclei
- Tumor diathesis: Dirty background with necrotic tissue fragments and RBCs.

Tumor diathesis is not always obvious in case of liquid based cytology (LBC) preparation.

Glandular Cell Abnormality

Atypical Glandular Cells

The terminology of atypical glandular cell should be more precisely defined as:

- Endocervical
- Endometrial
- Glandular cell, not otherwise specified.

The criteria of diagnosis of atypical endocervical cells are:

- Cells are in sheets or small overcrowded groups
- Cytoplasm: Abundant, vacuolated with distinct cell border
- Nucleus
    - Mild enlargement: Three to five times the area of normal endocervical nuclei
    - Mildly pleomorphic
    - Hyperchromatic
    - Small nucleoli
- Rare mitotic figures.

The criteria of the atypical endometrial cells are:

- Cohesive clusters of small groups of endometrial cells
- Mildly enlarged nuclei compared to normal endometrial cells
- Mild hyperchromasia
- Small nucleoli, more prominent in LBC (liquid based cytology) preparation
- Scanty cytoplasm with ill-defined margin.

Atypical Glandular Cells (AGC) Favors Neoplastic

- Endocervical, favor neoplastic
- Glandular, not otherwise specified.

The cases of atypical endocervical cells, favor neoplastic show cellular morphology which is abnormal but fall short of unequivocal diagnosis of adenocarcinoma in situ or invasive carcinoma.

The criteria of AGC, endocervical favor neoplastic are:

- Cells are in sheets or small overcrowded groups
- Rarely rosetting or feathering appearance of the cell cluster
- Cytoplasm: Abundant and vacuolated
- Nucleus: Enlarged, hyperchromatic, pleomorphic with irregular margin
- N/C ratio increased
- Rare mitotic figures.

Endocervical Adenocarcinoma

The diagnostic criteria of endocervical adenocarcinoma are:

- High cellularity: Abundant atypical cells
- Columnar to cuboidal in appearance
- Single cells, loose clusters of cells, three dimensional tight clusters and syncytial aggregates
- Well defined scalloped borders of the cluster
- Nucleus
  - Large and round
  - Pleomorphic
  - Irregular coarse chromatin
  - Parachromatin clearing
  - Irregular nuclear membrane
  - Prominent nucleoli, macronucleoli may be present
- Cytoplasm: Abundant and finely vacuolated
- Tumor diathesis present.

Endometrial Adenocarcinoma

The following criteria should be present:

- Tight clusters or small groups of cells
- Pleomorphic nuclei
- Nuclear crowding indicating loss of polarity
- Nuclear hyperchromasia, and irregular clumped chromatin
- Multiple prominent nucleoli depending on grade of the tumor
- Scanty vacuolated cytoplasm
- Intracytoplasmic polymorphs
- Tumor diathesis may be present.

Reference

1. Solomon D, Davey D, Kurman R, et al. for the Forum Group Members and the Bethesda 2001 Workshop. The 2001 Bethesda System—terminology for reporting results of cervical cytology. JAMA, 2002;287:2114-9.

QUESTIONS

Problem 2.1

Match the following (Fig. 2.1)

A.	1. Metaplastic cells
B.	2. Superficial squamous cells
C.	3. Intermediate cell
D.	4. Endometrial cells
E.	5. Endocervical cells
F.	6. Parabasal and basal cell

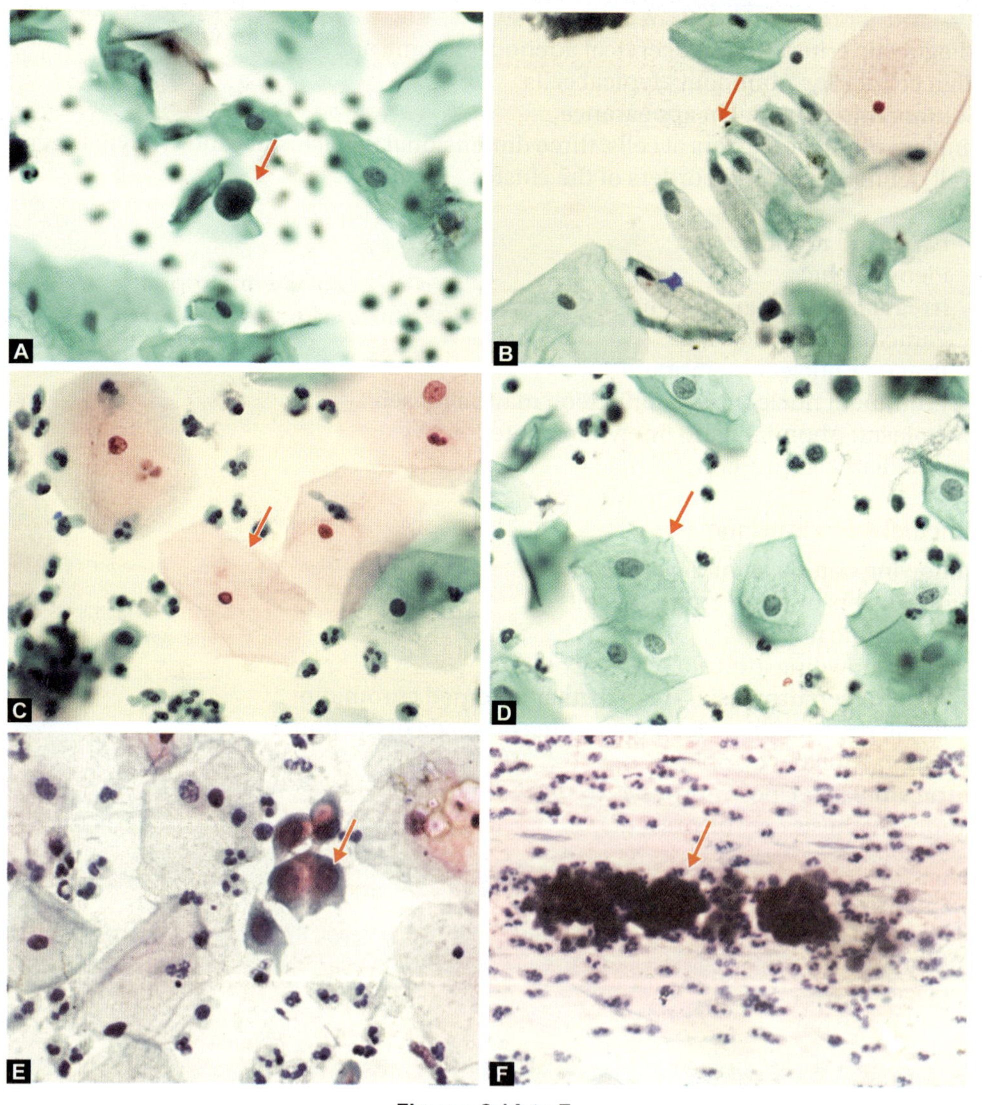

Figures 2.1A to F

Problem 2.2

A 41-year-old female with routine cervical smear screening. How do you evaluate the smear (Fig. 2.2) in Bethesda system of reporting?

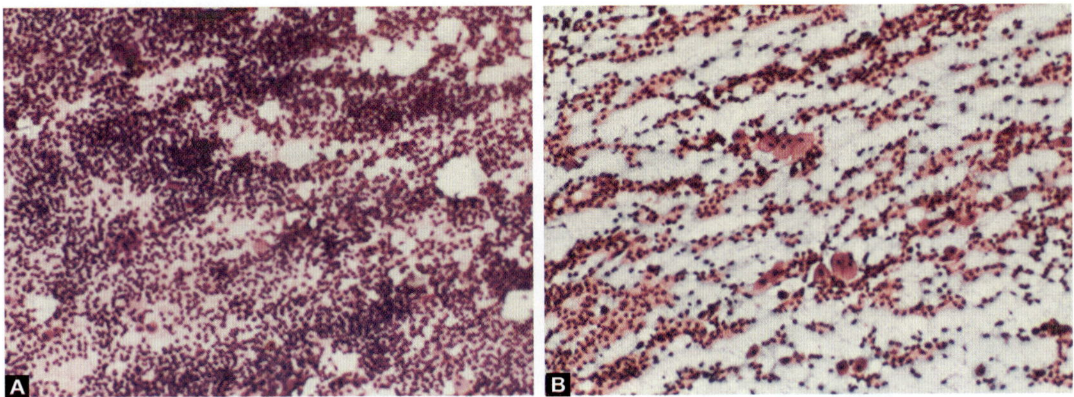

Figures 2.2A and B

Problem 2.3

A 37-year-old female with routine cervical smear screening. The microphotographs of the SurePath preparation are shown in the Figure 2.3. How do you interpret according to Bethesda system?

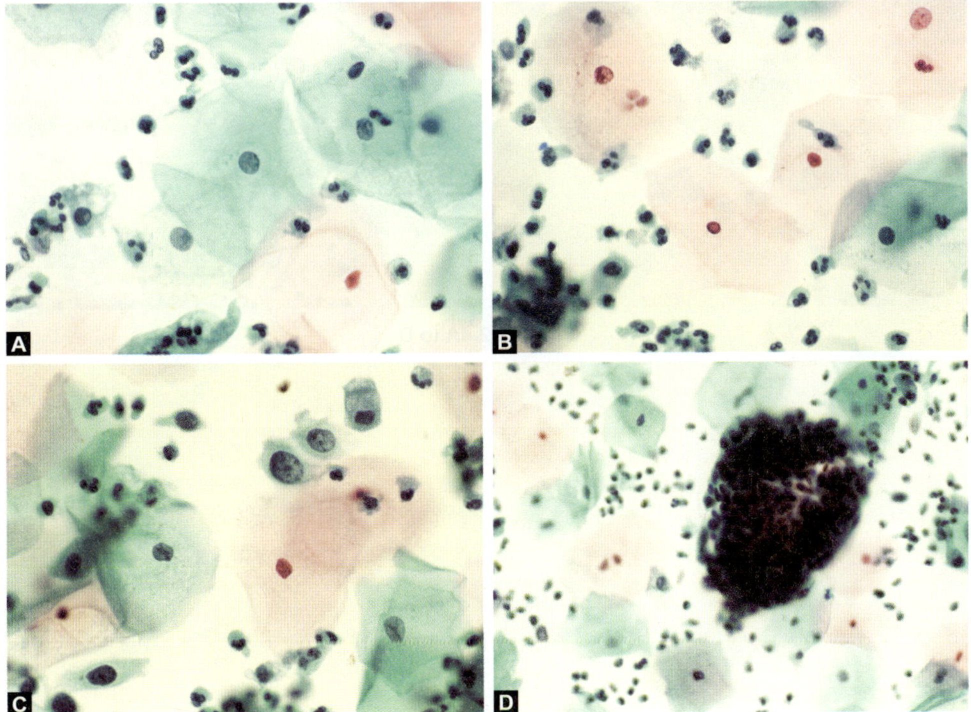

Figures 2.3A to D

Problem 2.4

Routine cervical smear in a 41-year-old female. The most likely interpretation from the SurePath smear (Fig. 2.4) is:

A. Atrophic smear
B. Inflammatory smear
C. Atypical squamous cells of undetermined significance (ASCUS)
D. *Trichomonas vaginalis*

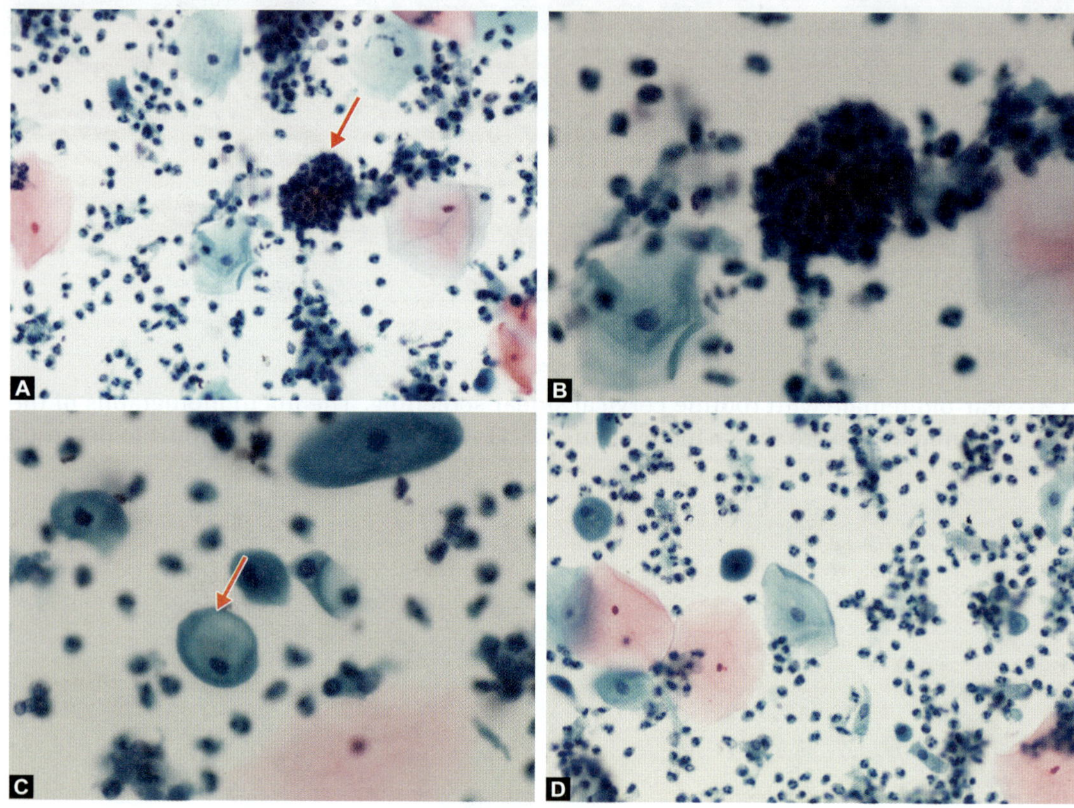

Figures 2.4A to D

Problem 2.5

Routine cervical smear in a 49-year-old female. What is your interpretation from the microphotograph (Fig. 2.5) of SurePath preparation?

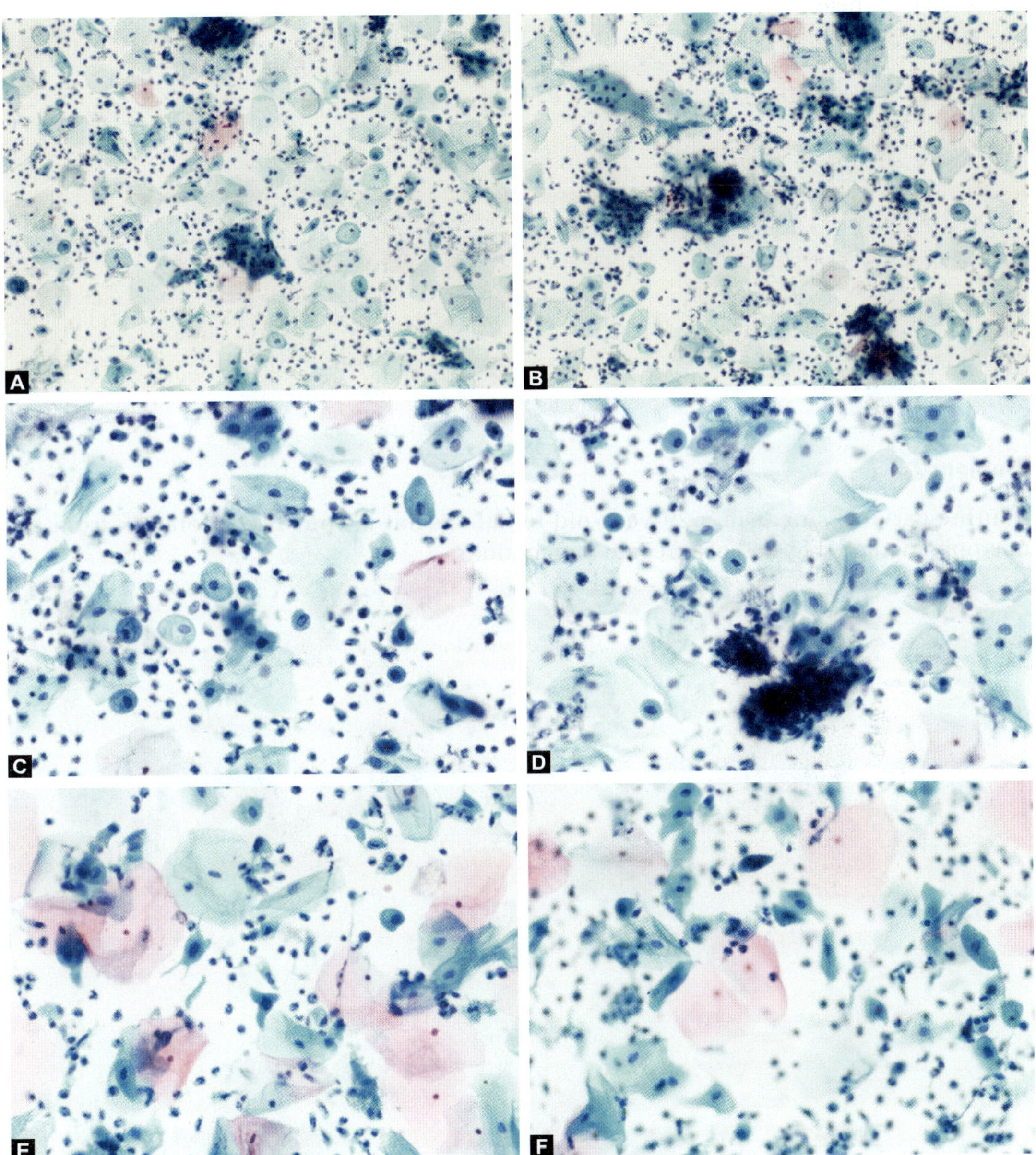

Figures 2.5A to F

Problem 2.6

Routine cervical smear in a 34-year-old female. What is your interpretation from the microphotographs (Fig. 2.6) of SurePath preparation?

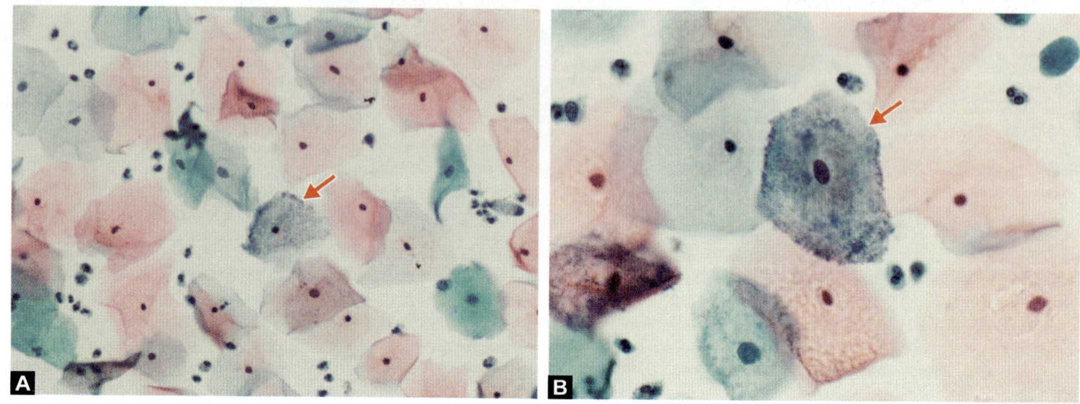

Figures 2.6A and B

Problem 2.7

Routine cervical smear in a 29-year-old female. What is your interpretation from the microphotograph (Fig. 2.7) of SurePath preparation?

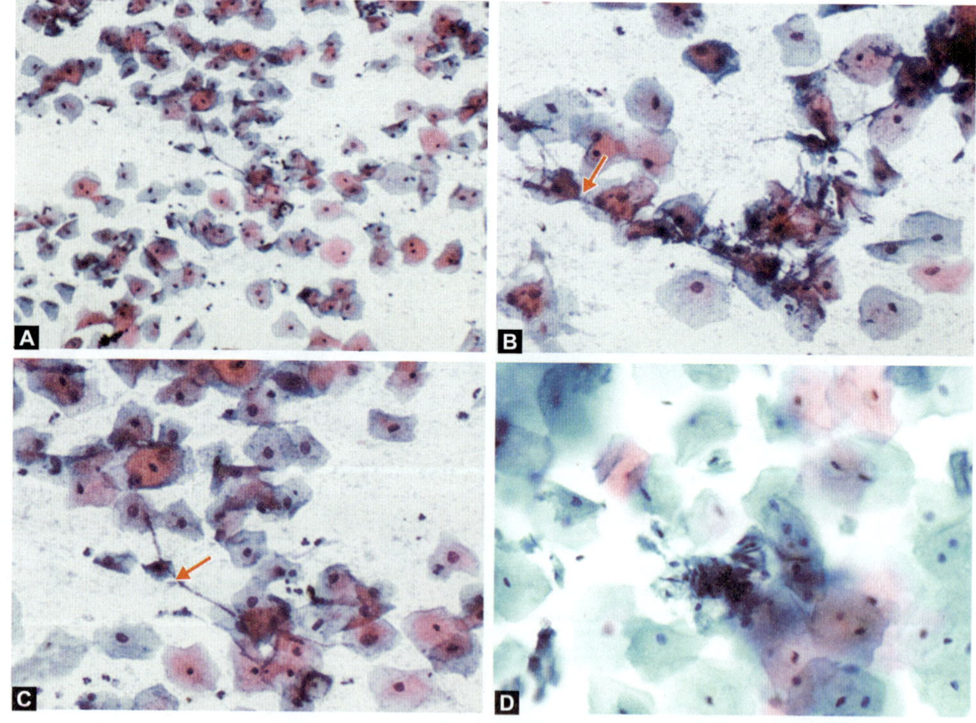

Figures 2.7A to D

Problem 2.8

Routine cervical smear in a 31-year-old female. The most likely diagnosis from the Figure 2.8 is:

A. *Trichomonas vaginalis*
B. *Leptothrix*
C. *Candida*
D. Bacterial vaginosis

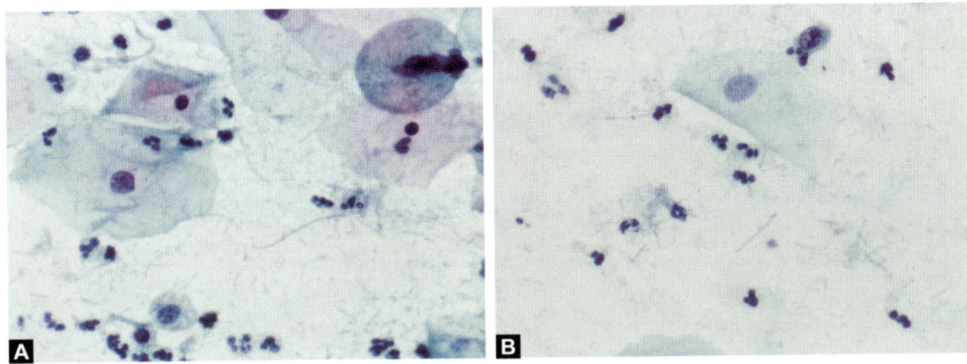

Figures 2.8A and B

Problem 2.9

Routine cervical smear in a 35-year-old female patient. The most likely diagnosis from Figure 2.9 is:

A. Normal cervical smear
B. Atrophic smears
C. *Candida*
D. *Trichomonas vaginalis*

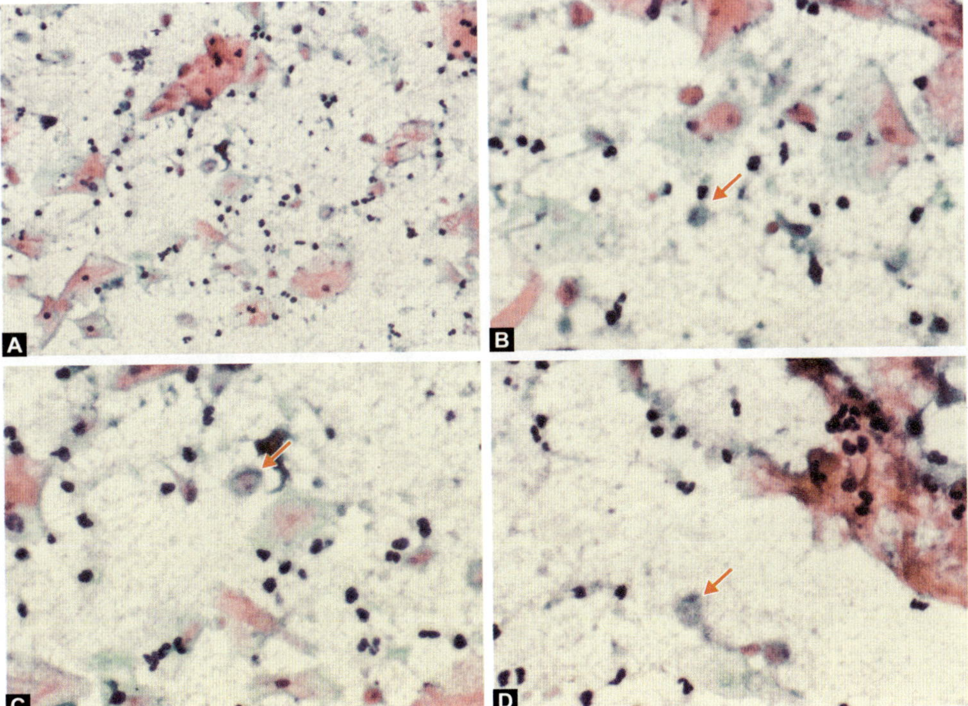

Figures 2.9A to D

Problem 2.10

Routine cervical smear in a 41-year-old female. What is your interpretation from the microphotograph (Fig. 2.10) of SurePath preparation?

A. Atypical squamous cells of undetermined significance (ASCUS)
B. Atypical squamous cells cannot exclude HSIL (ASC-H)
C. Low-grade squamous intraepithelial lesion (LSIL)
D. High-grade squamous intraepithelial lesion (HSIL)

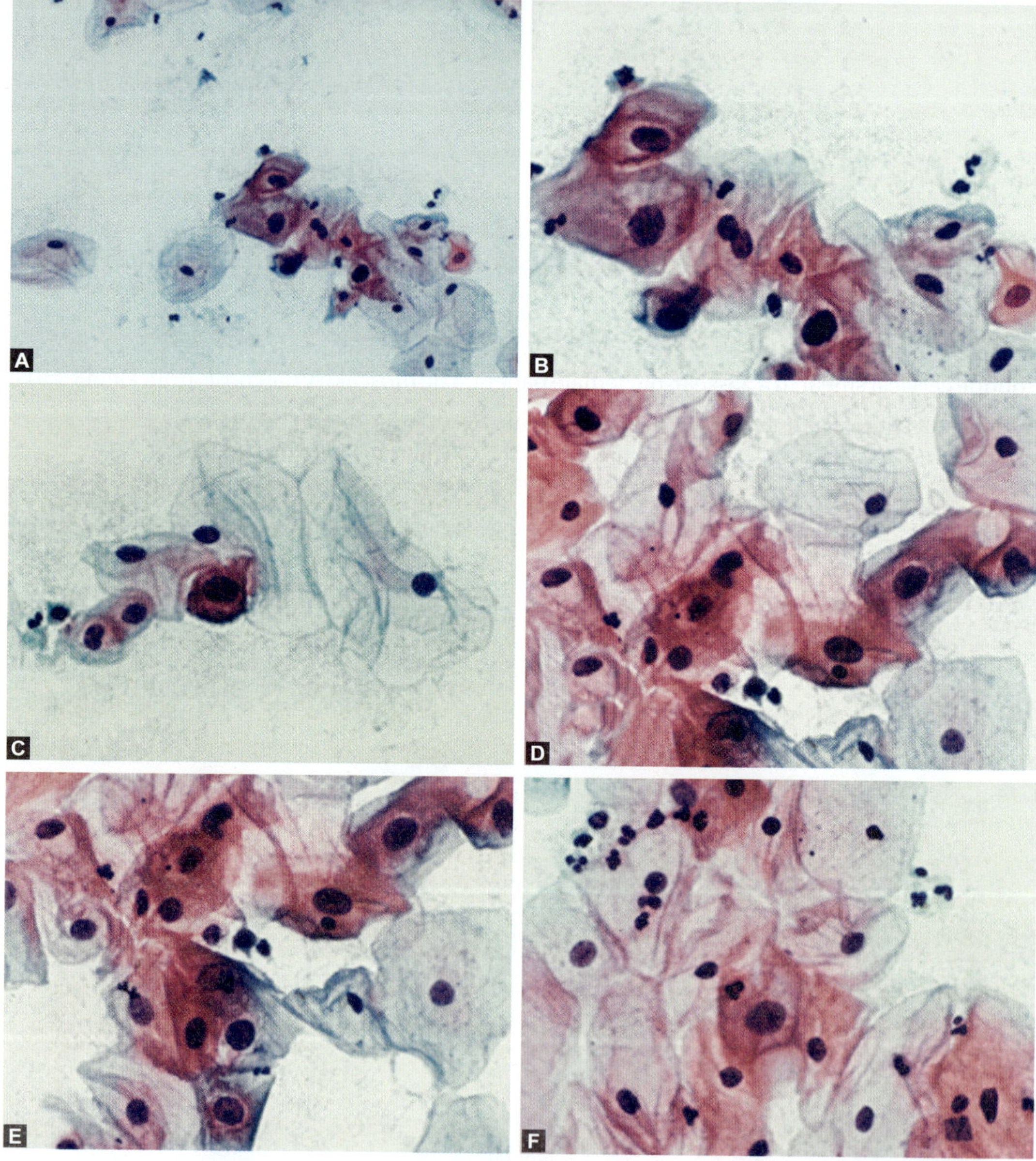

Figures 2.10A to F

Problem 2.11

Routine cervical smear in a 41-year-old female. What is your interpretation from the Figure 2.11 of SurePath preparation?

A. ASCUS
B. LSIL
C. ASC-H
D. HSIL

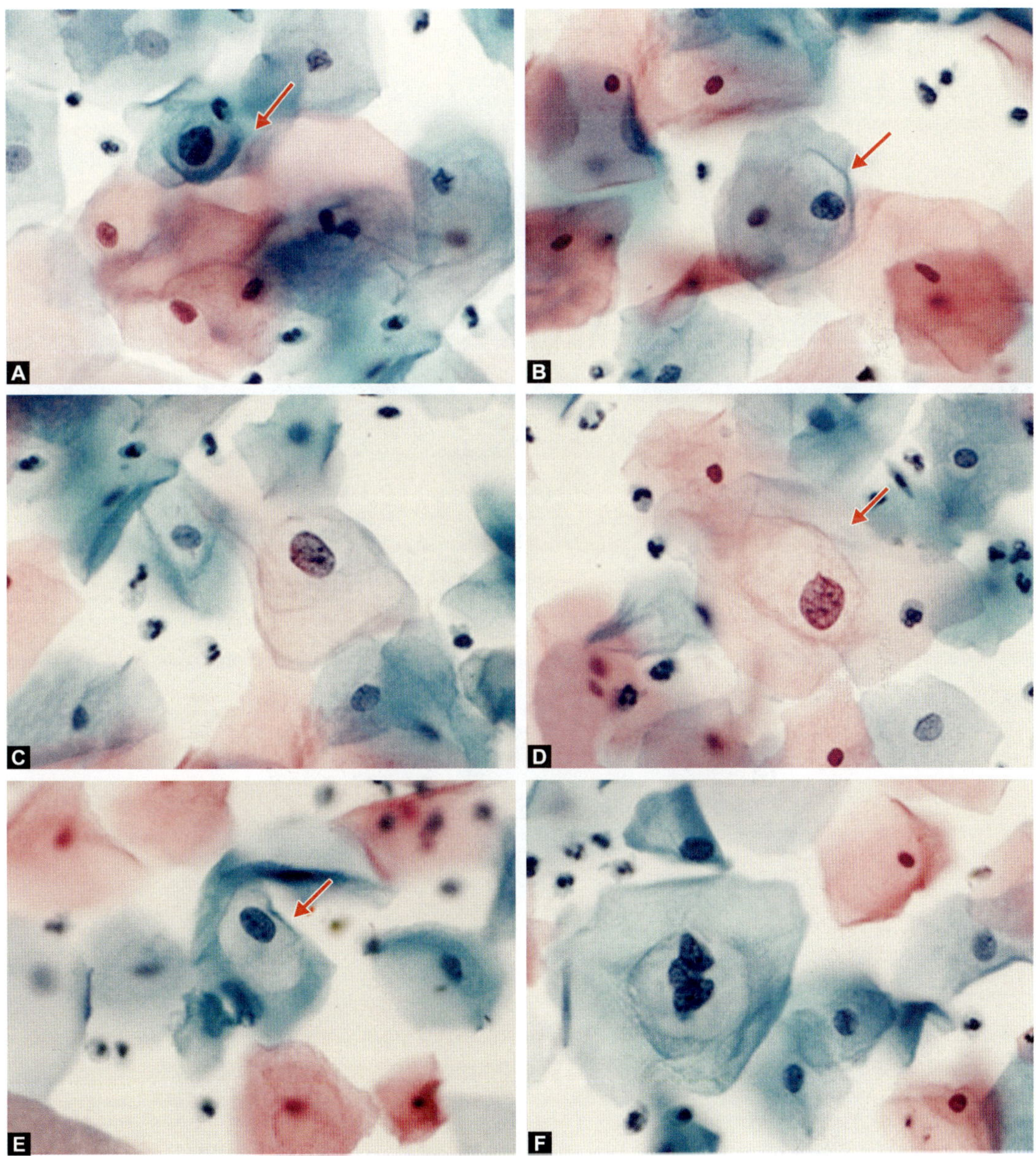

Figures 2.11A to F

Problem 2.12

Match the following (Fig. 2.12)

A.	1. True koilocytotic cell
B.	2. Koilocyte like cell in inflammation
C.	3. Koilocyte like cell due to glycogen

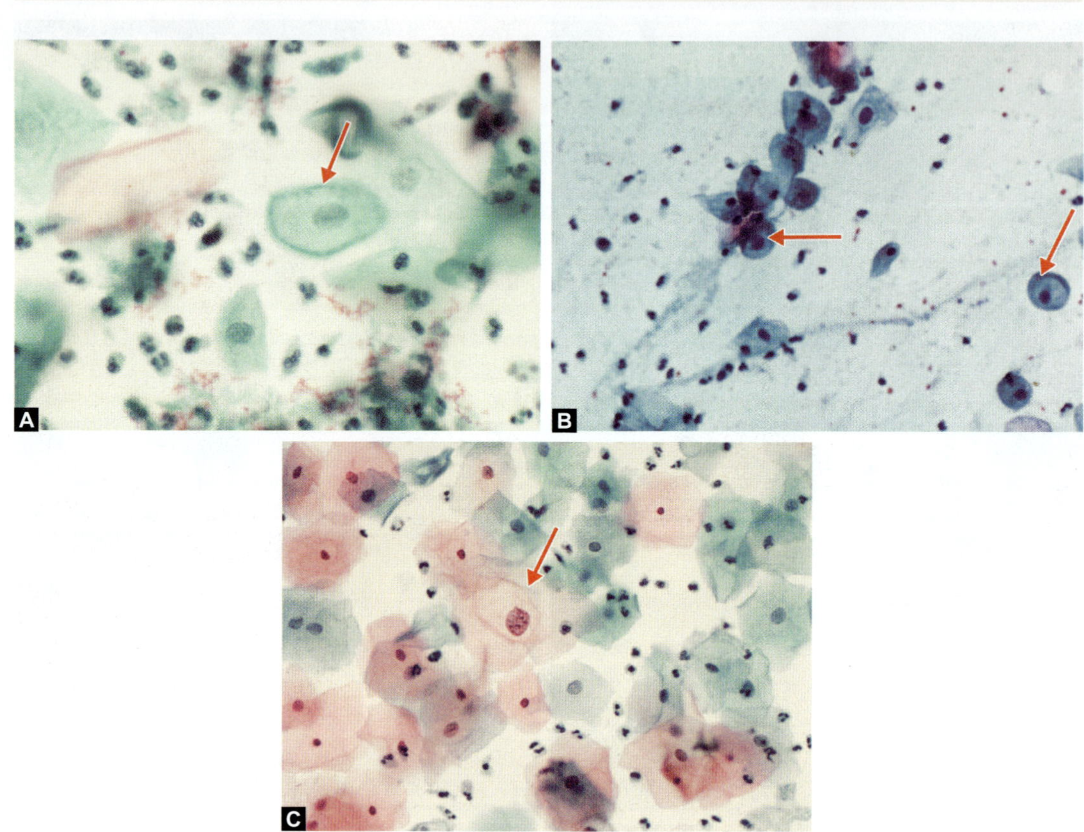

Figures 2.12A to C

Problem 2.13

Routine cervical smear in a 36-year-old female. What is your interpretation from the microphotograph of (Fig. 2.13) SurePath smear?

A. ASCUS
B. LSIL
C. ASC-H
D. HSIL

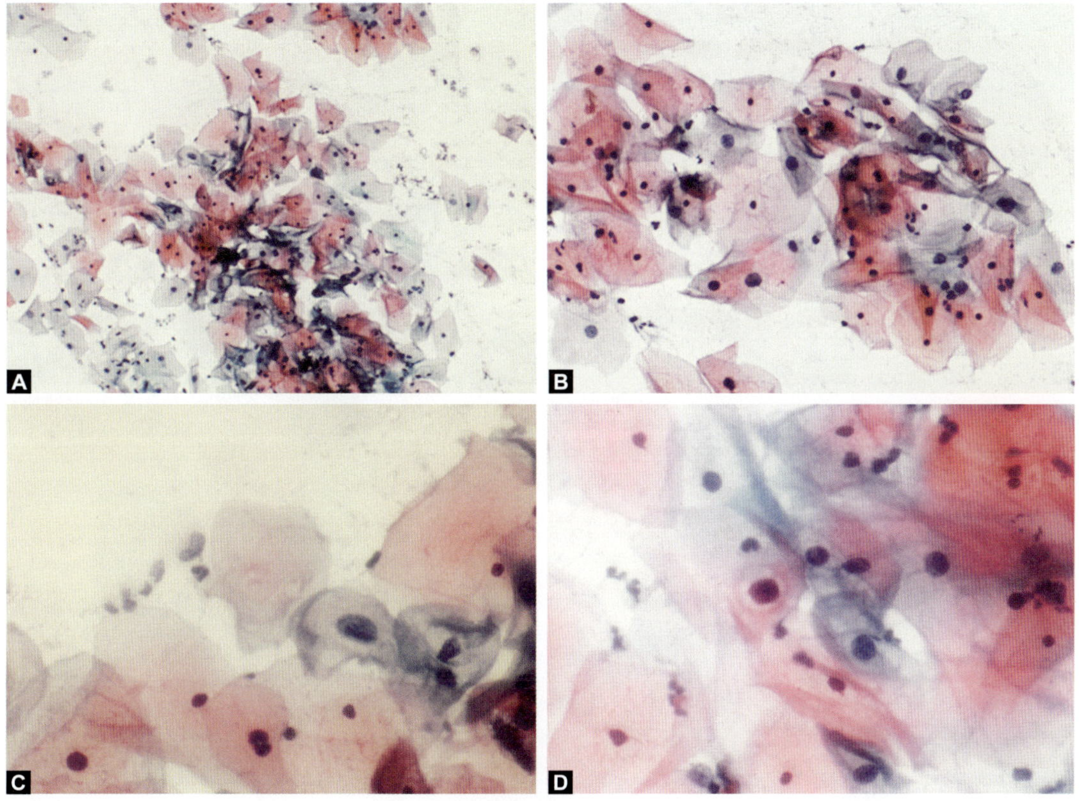

Figures 2.13A to D

Problem 2.14

Routine cervical smear in a 39-year-old male. What will be the interpretation according to Bethesda system of reporting from the Figure 2.14?

A. ASCUS
B. LSIL
C. ASC-H
D. HSIL

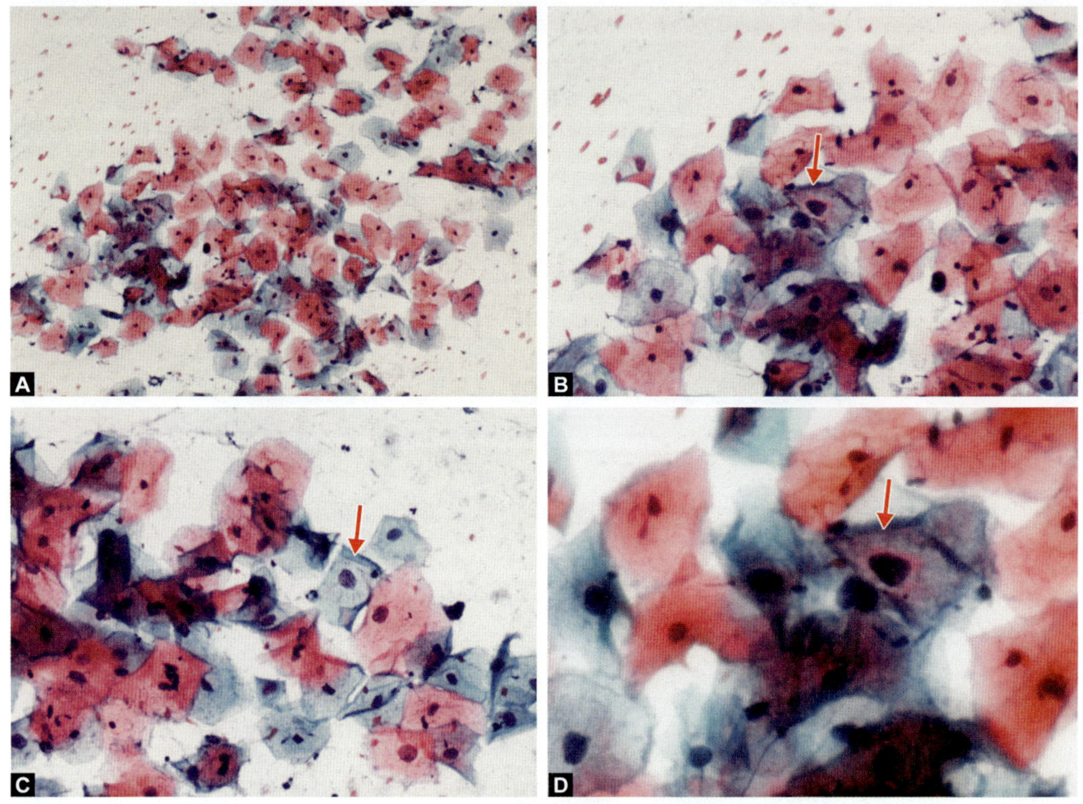

Figures 2.14A to D

Problem 2.15

Routine cervical smear in a 29-year-old female. What is your interpretation from the Figure 2.15 of SurePath preparation?

A. ASCUS
B. LSIL
C. ASC-H
D. HSIL

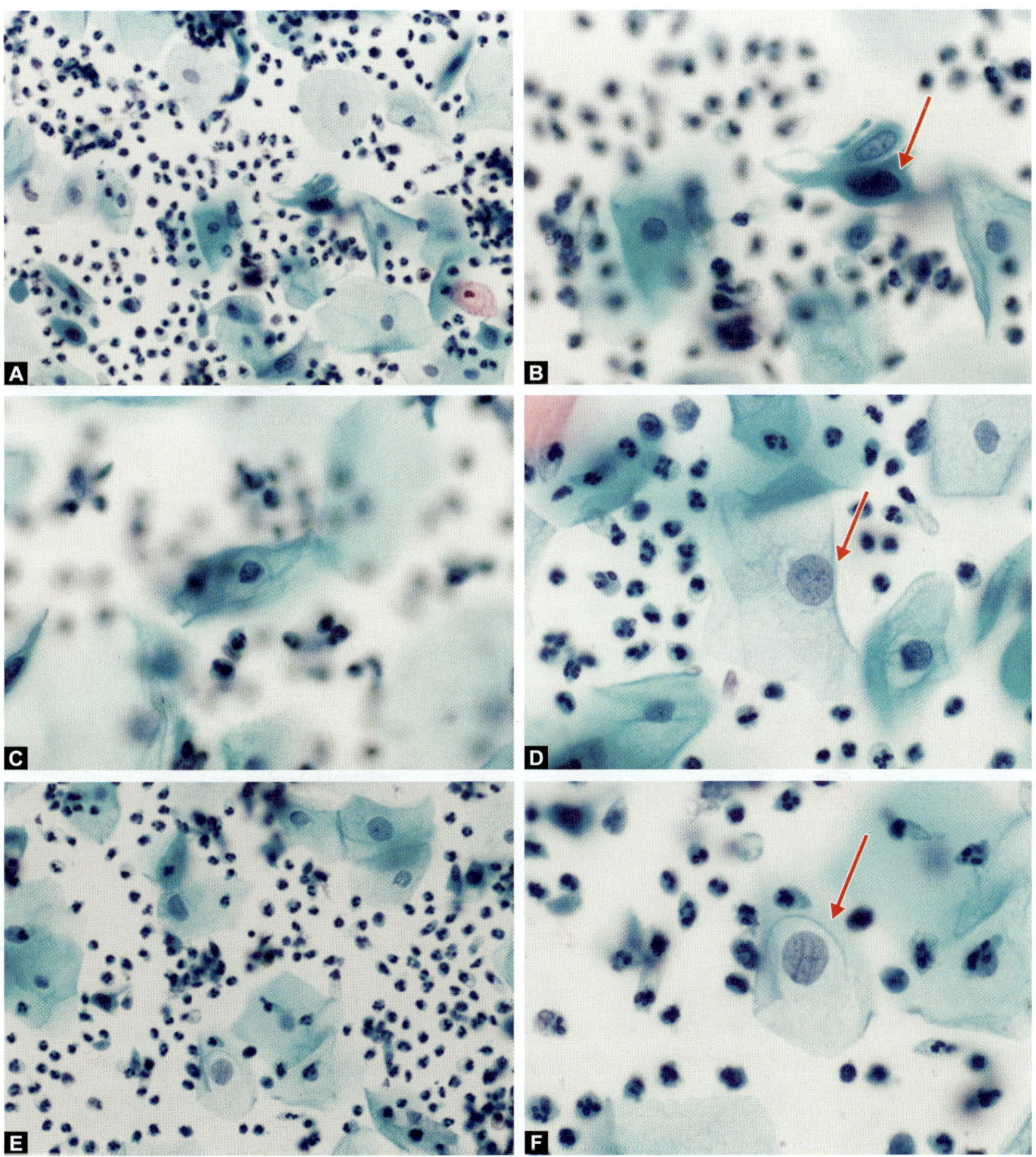

Figures 2.15A to F

Problem 2.16

Routine cervical smear in a 34-year-old female. What is your interpretation from the microphotographs (Fig. 2.16) of SurePath preparation?

A. ASCUS
B. ASC-H
C. LSIL
D. HSIL
E. Squamous cell carcinoma

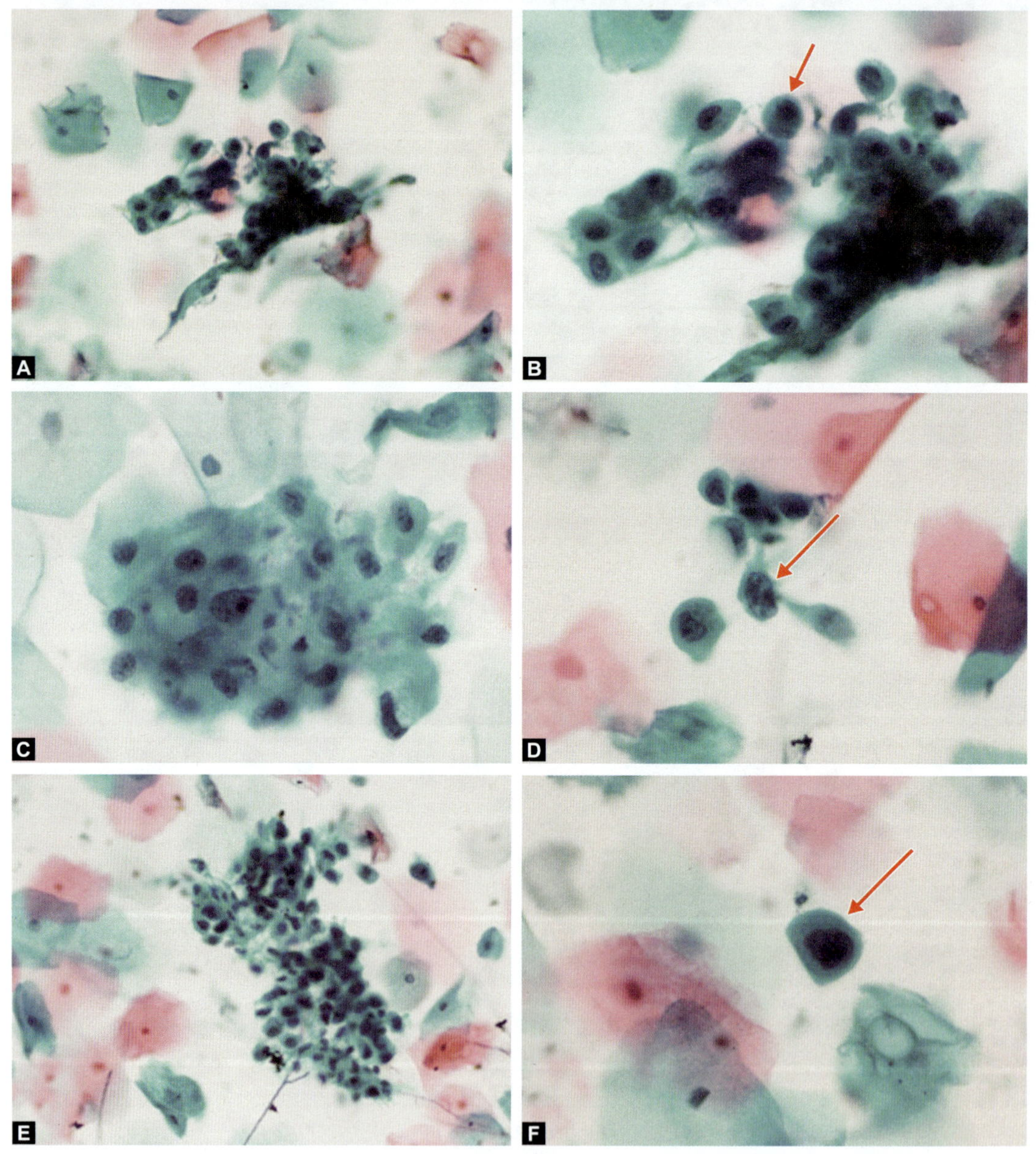

Figures 2.16A to F

Problem 2.17

Routine cervical smear in a 39-year-old female. What is your interpretation from the microphotograph (Fig. 2.17) of SurePath preparation?

A. ASCUS
B. ASC-H
C. LSIL
D. HSIL

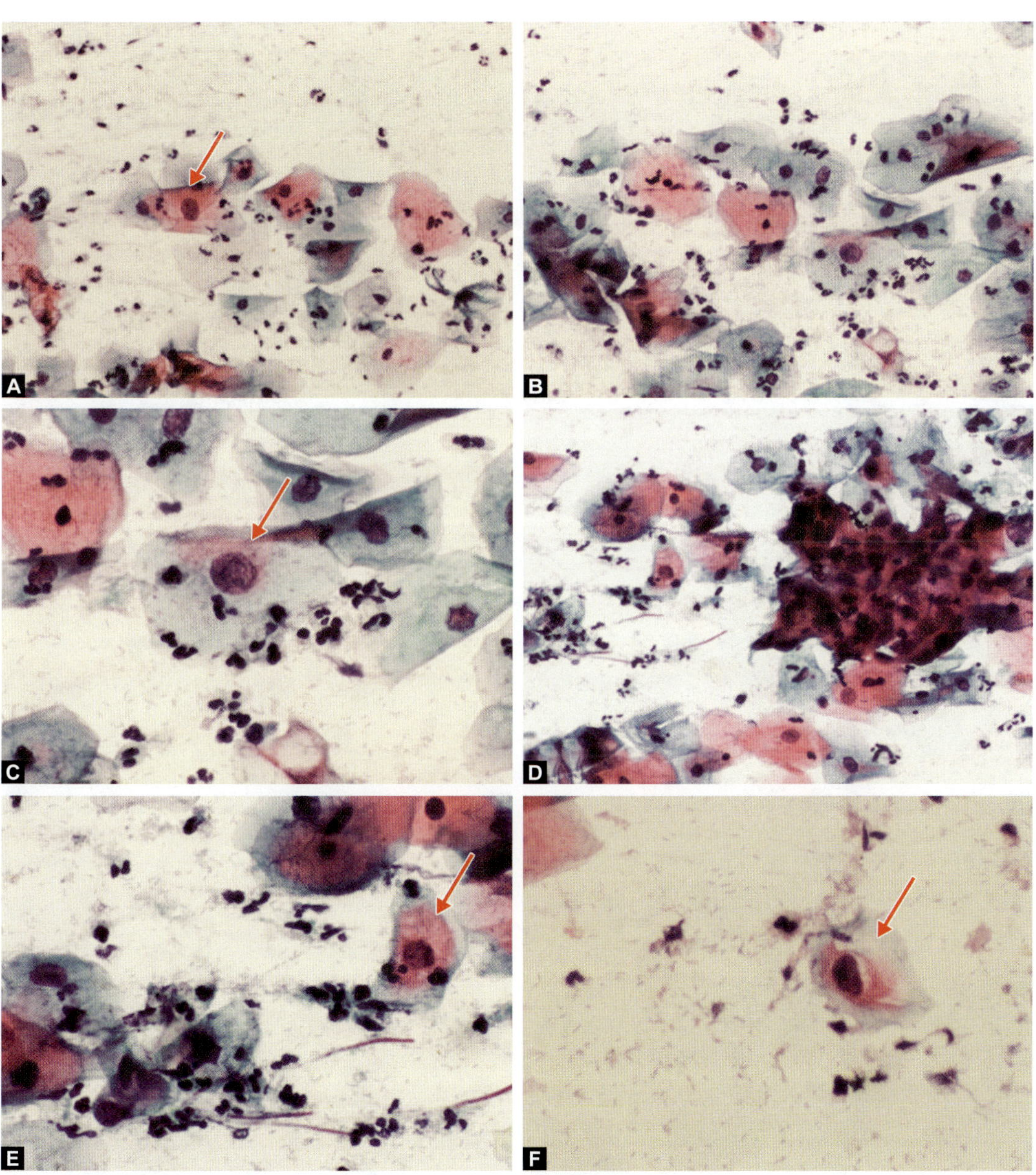

Figures 2.17A to F

Problem 2.18

Routine cervical smear in a 33-year-old female. Interpret the smear by Bethesda system of reporting from the microphotographs of (Fig. 2.18) SurePath preparation.

A. ASCUS
B. LSIL
C. ASC-H
D. HSIL
E. Squamous cell carcinoma

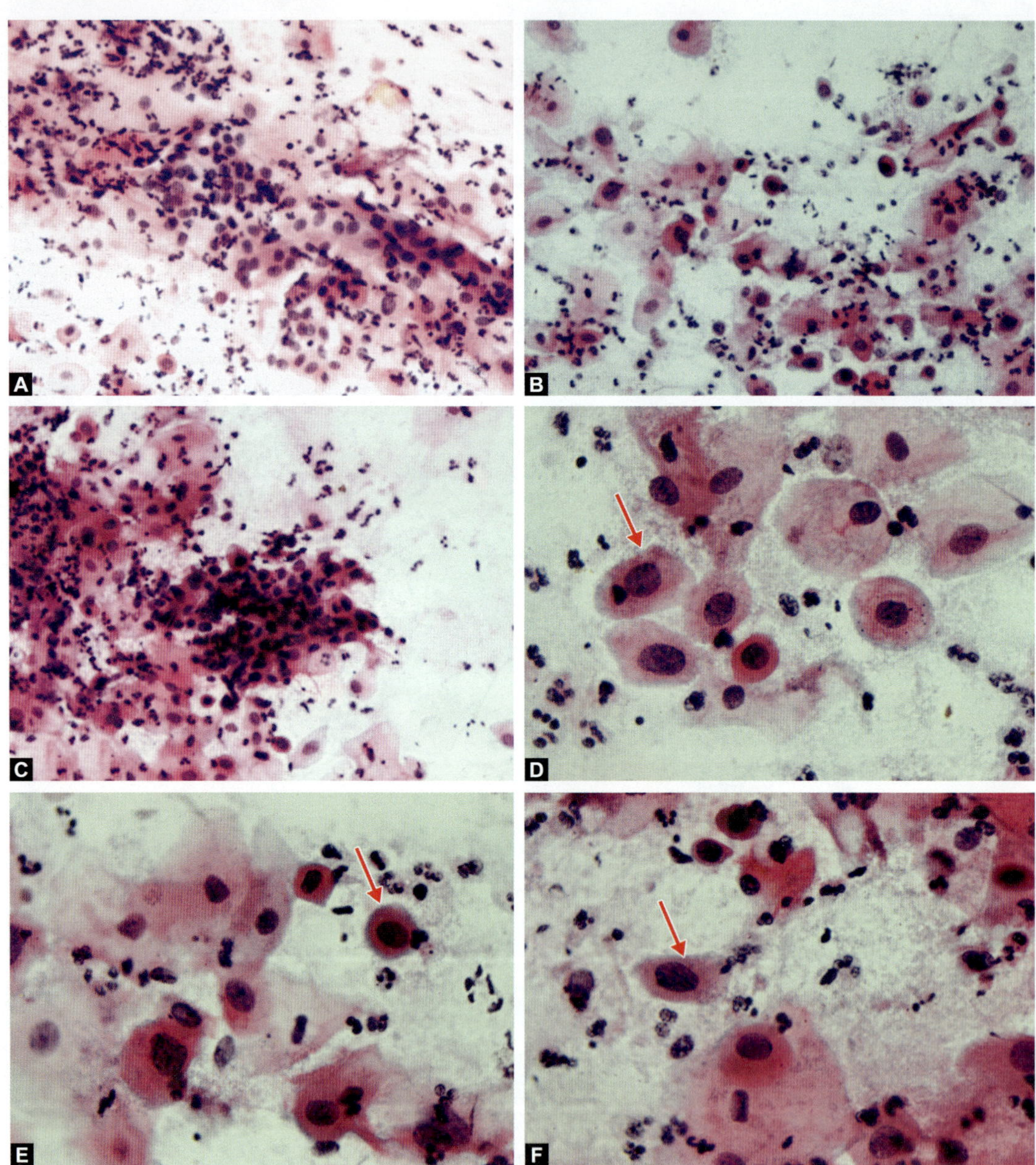

Figures 2.18A to F

Problem 2.19

Routine cervical smear in a 32-year-old female. What is your interpretation from the Figure 2.19 of SurePath smear?

A. ASCUS
B. ASC H
C. LSIL
D. HSIL

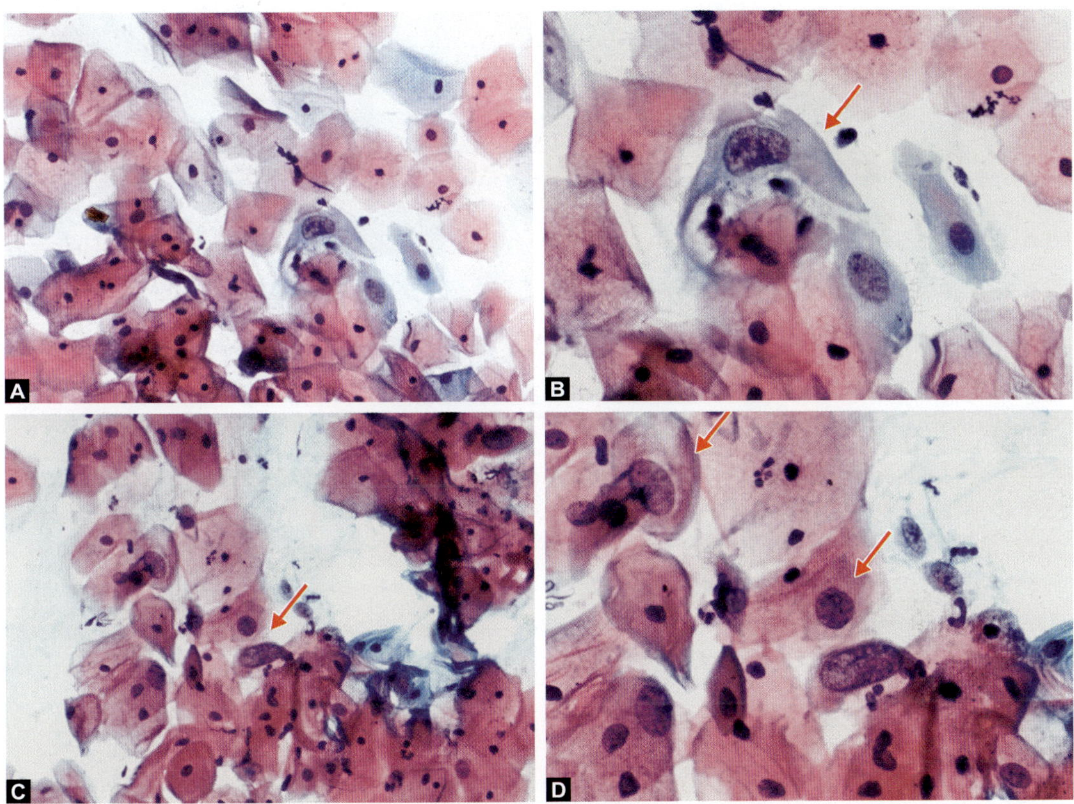

Figures 2.19A to D

Problem 2.20

Routine cervical smear in a 37-year-old female. What is your interpretation from the Figure 2.20 of SurePath preparation?

A. ASCUS
B. ASC H
C. LSIL
D. HSIL
E. Squamous cell carcinoma

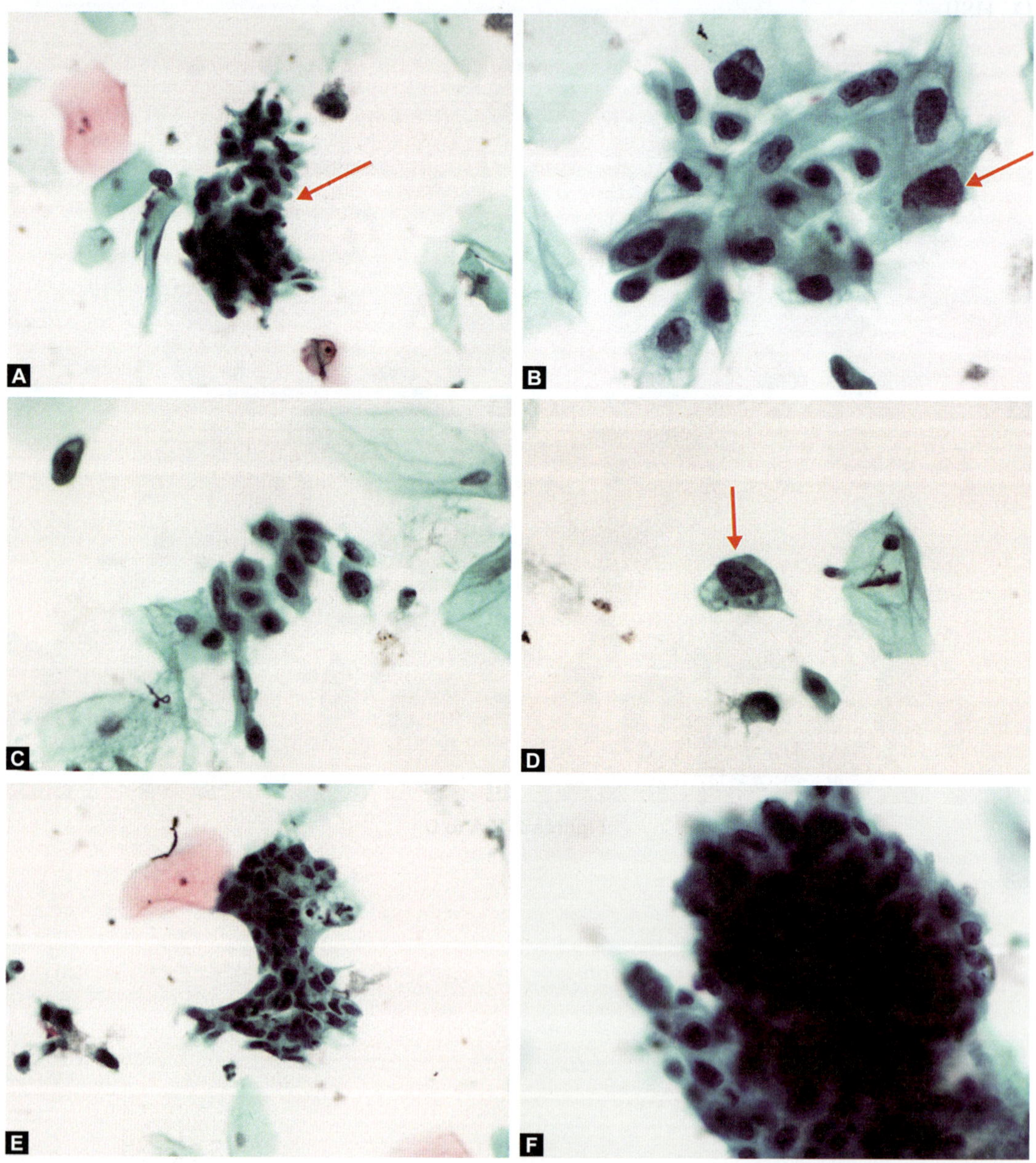

Figures 2.20A to F

Problem 2.21

Routine cervical smear in a 51-year-old female. What is your interpretation from the microphotographs (Fig. 2.21A to H) of SurePath smear?

A. ASCUS
B. ASC-H
C. LSIL
D. HSIL
E. Squamous cell carcinoma

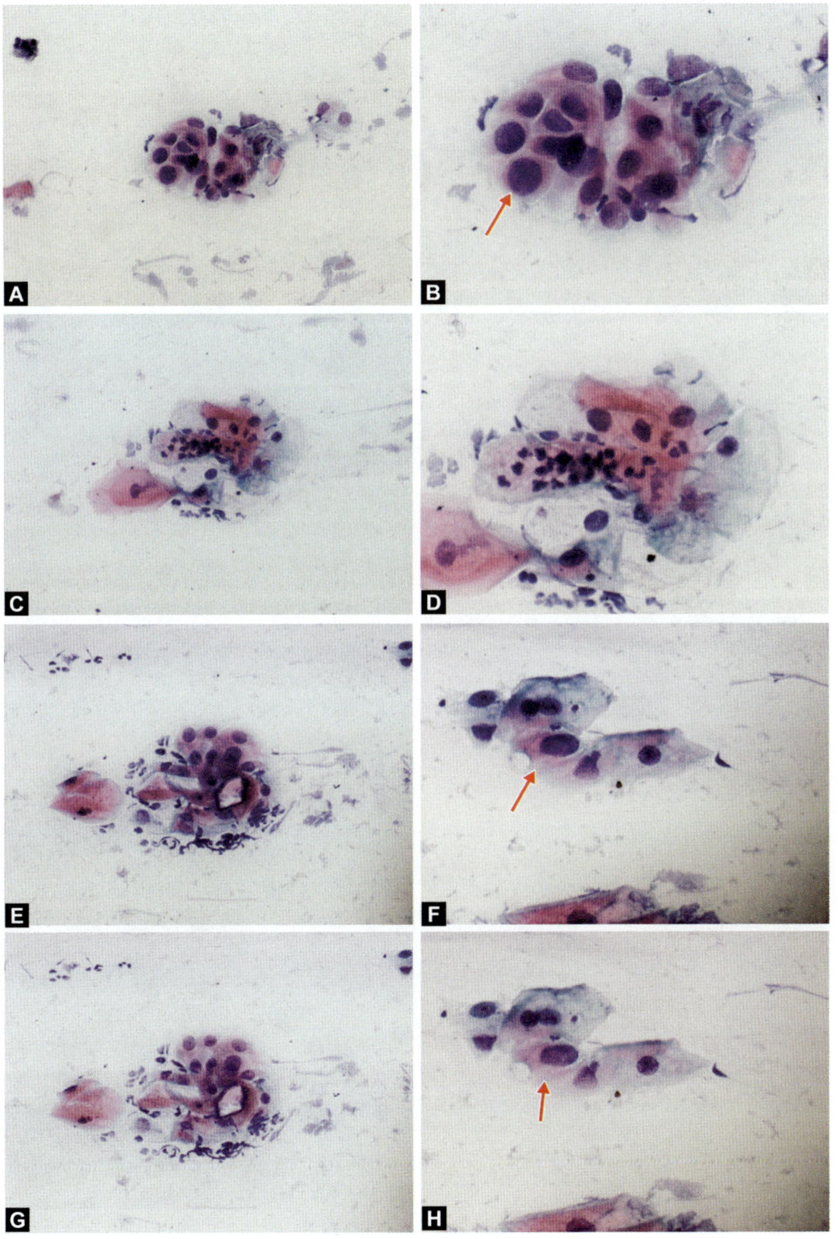

Figures 2.21A to H

Problem 2.22

Routine cervical smear in a 50-year-old female. What is your interpretation from the microphotographs (Figs 2.22A to H) of conventional cervical cytology preparation?

A. ASC-H
B. HSIL
C. Squamous cell carcinoma
D. Atrophic smear

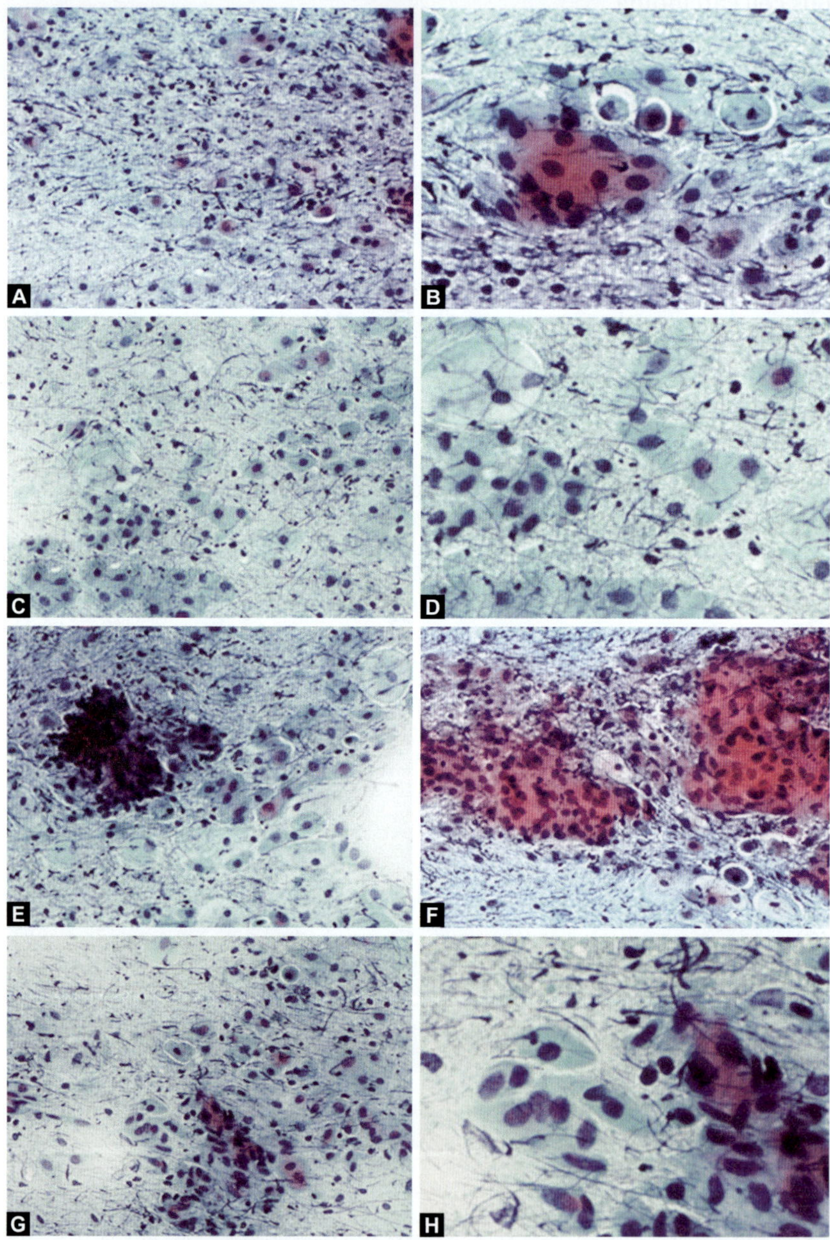

Figures 2.22A to H

Problem 2.23

Routine cervical smear in a 39-year-old female. What is your interpretation from the Figure 2.23 of SurePath preparation?

A. ASCUS
B. ASC-H
C. LSIL
D. HSIL

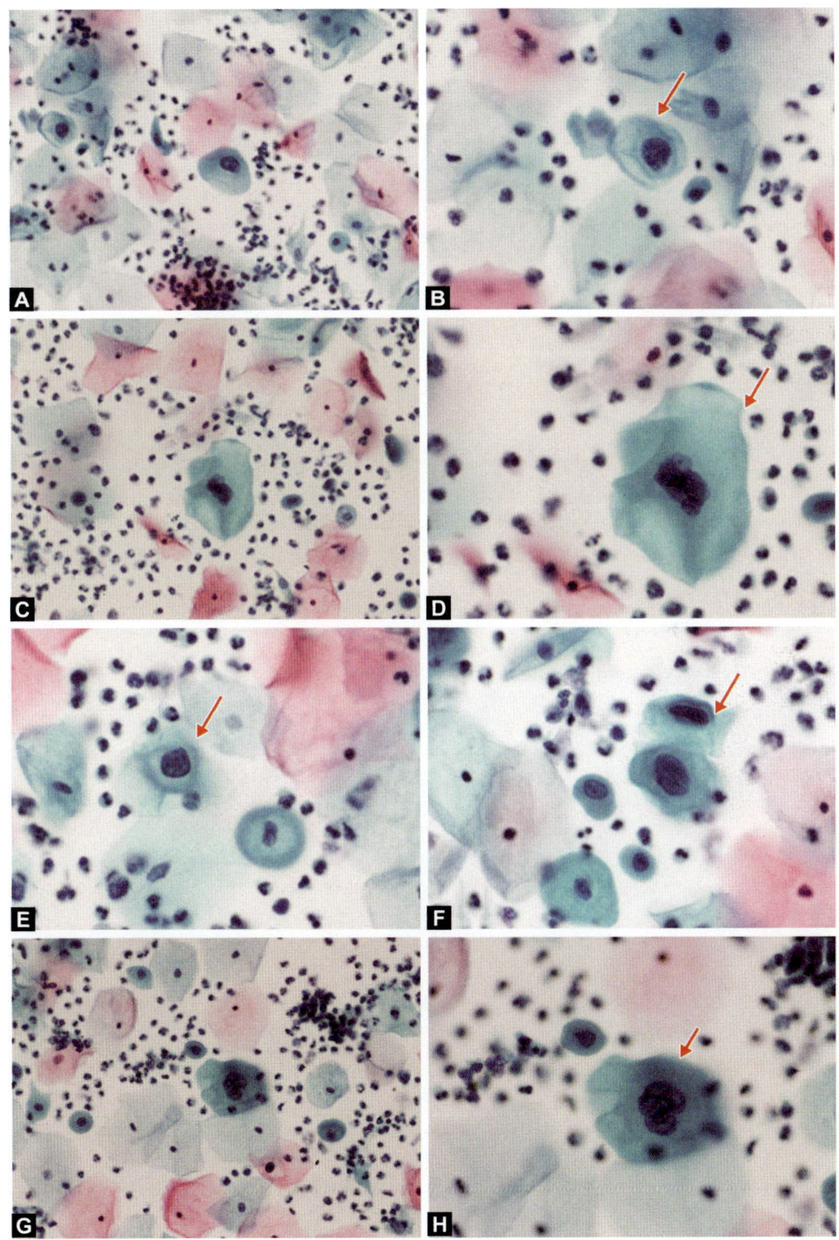

Figures 2.23A to H

Problem 2.24

Routine cervical smear in a 48-year-old female. The most possible interpretation from the Figure 2.24 of SurePath preparation is:

A. Atrophic smear
B. HSIL
C. ASC-H
D. Squamous cell carcinoma

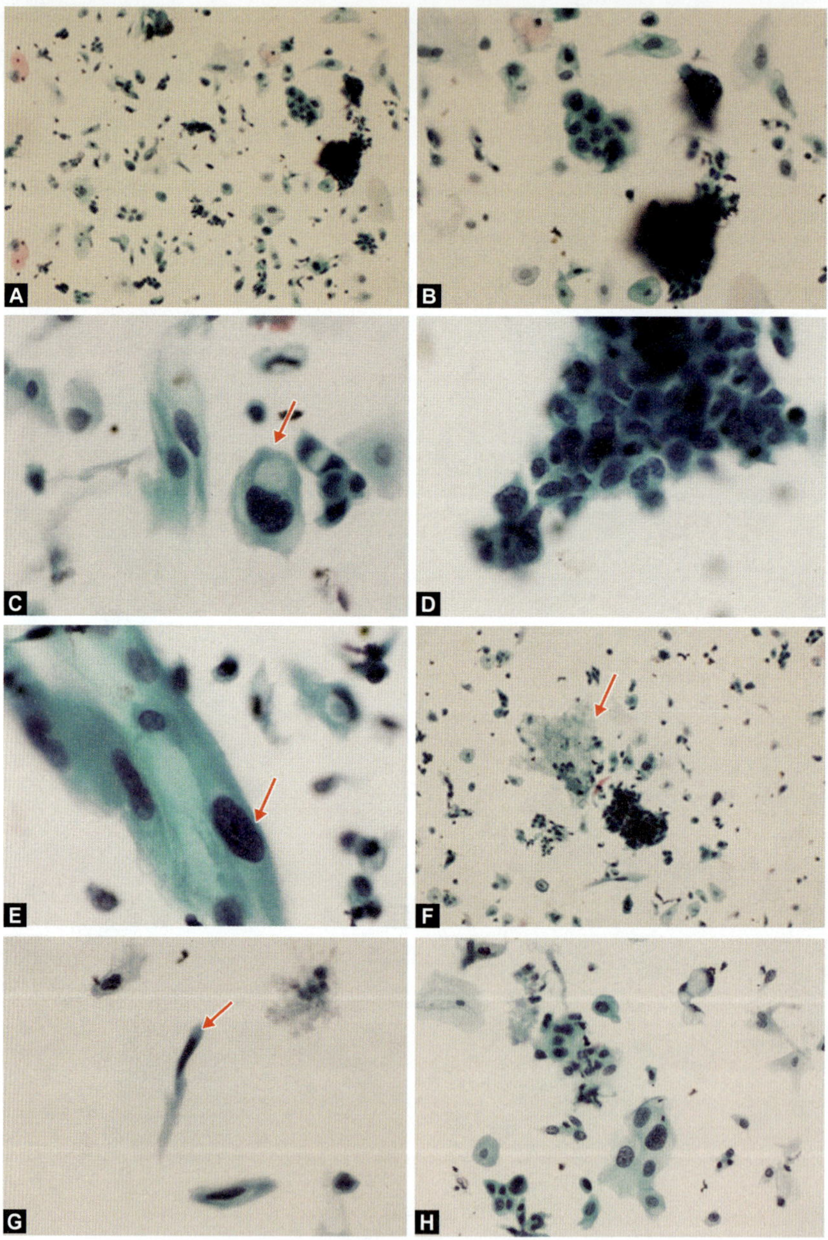

Figures 2.24A to H

Problem 2.25

Routine cervical smear in a 39-year-old female. The most likely diagnosis from the microphotographs of the Surepath preparation of Figure 2.25 is:

A. ASCUS B. ASC-H
C. HSIL D. LSIL

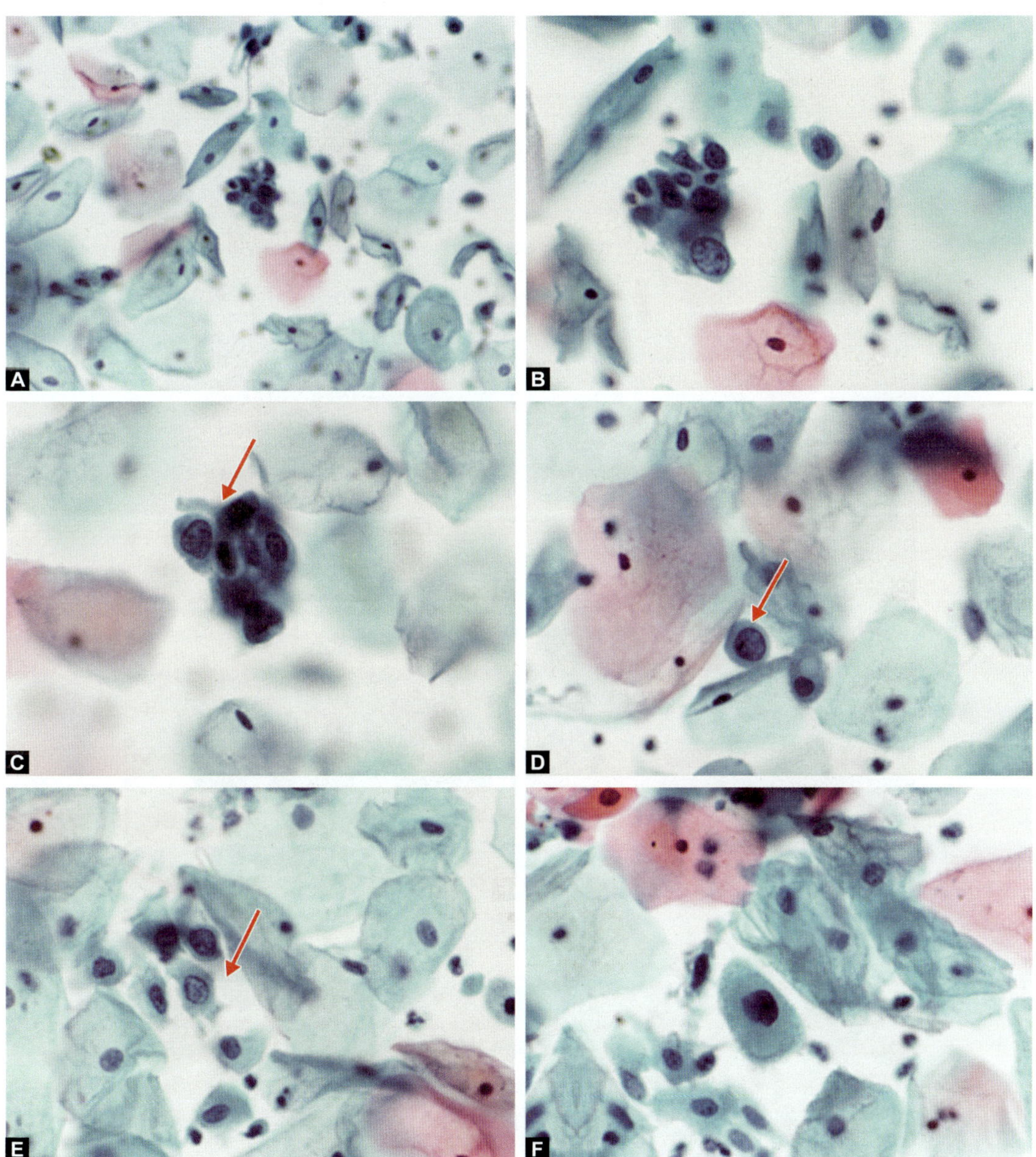

Figures 2.25A to F

Problem 2.26

A 61-year-old female presented with blood stained discharge for 15 days. Routines cervical smear examination was done. The most likely diagnosis form Figure 2.26 is:

A. LSIL
B. HSIL
C. AGCUS
D. Squamous cell carcinoma

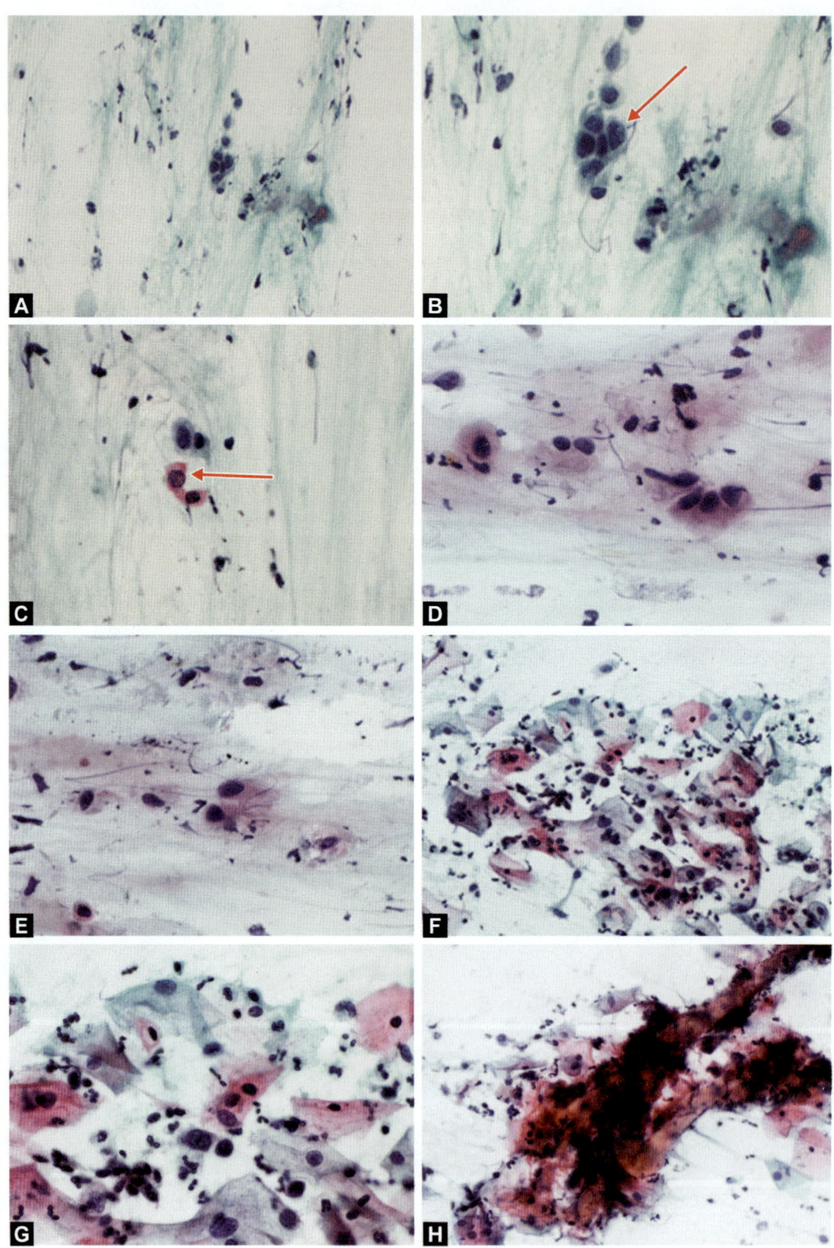

Figures 2.26A to H

Problem 2.27

A 34-year-old female was undergone colposcopic guided biopsy and was diagnosed as CIN3. Now repeat cervical smear examinations was done after 6 months. The most likely diagnosis from Figure 2.27 is:

A. LSIL
B. HSIL
C. ASCUS
D. Tubal metaplasia

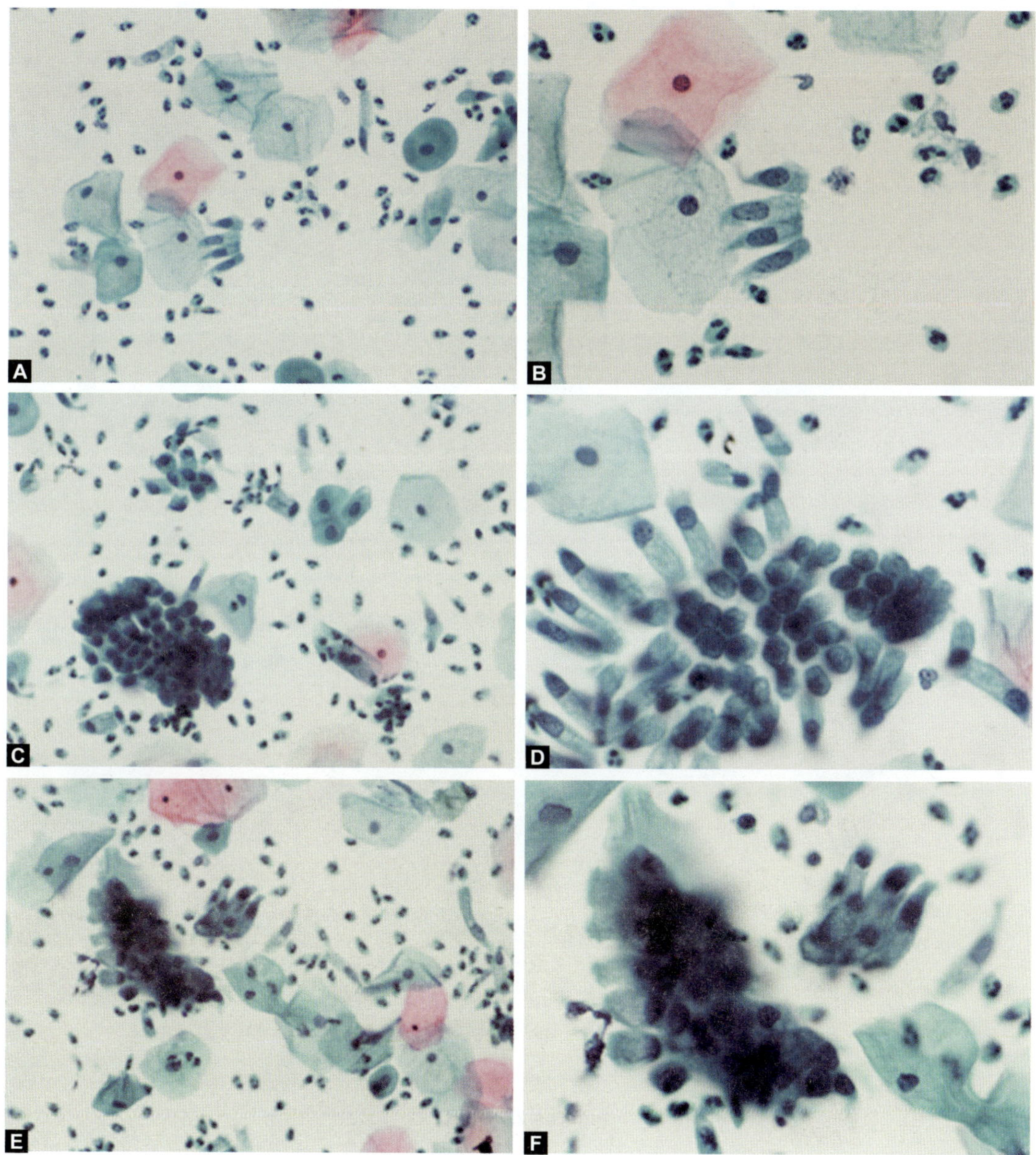

Figures 2.27A to F

Problem 2.28

Unhealthy cervix in a 43-year-old female. The most likely diagnosis from the Figure 2.28 is.

A. Squamous cell carcinoma
B. HSIL
C. LSIL
D. Atrophic smear

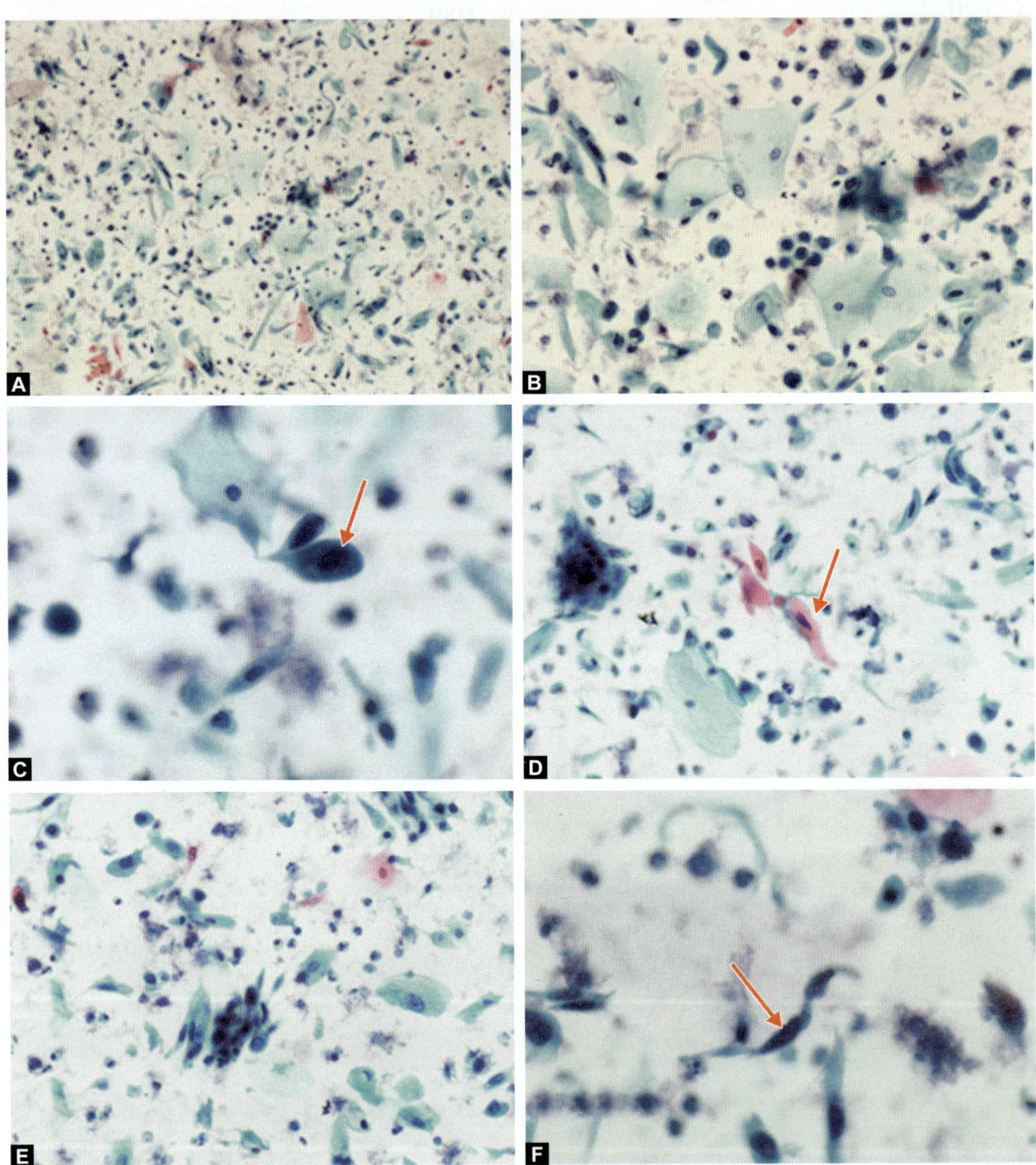

Figures 2.28A to F

Problem 2.29

A 41-year-old female with history of radiation 4 months back for squamous cell carcinoma of cervix. The most likely diagnosis from the Figure 2.29 of SurePath smear is:

A. Radiation effect
B. Squamous cell carcinoma
C. Viral infection
D. HSIL

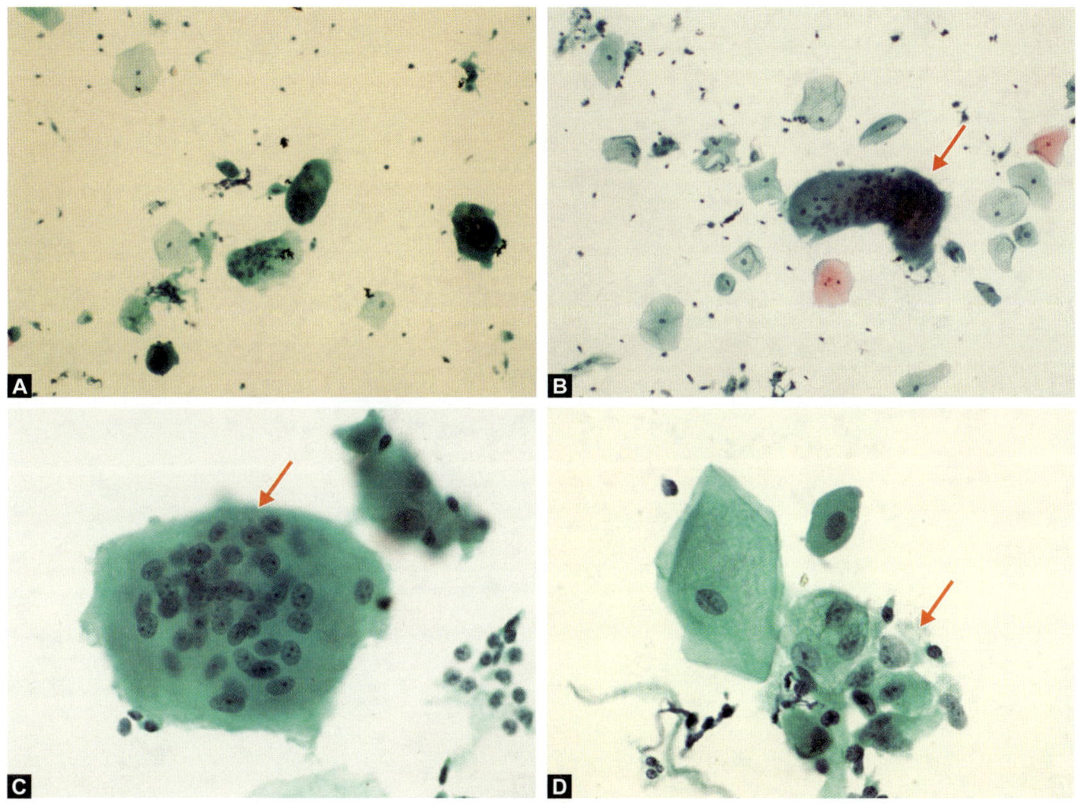

Figures 2.29A to D

Problem 2.30

Routine cervical smear in a 37-year-old female. Interpret the smear by Bethesda system of reporting. The most likely diagnosis from Figure 2.30 is:

A. LSIL
B. ASCUS
C. ASCH
D. HSIL

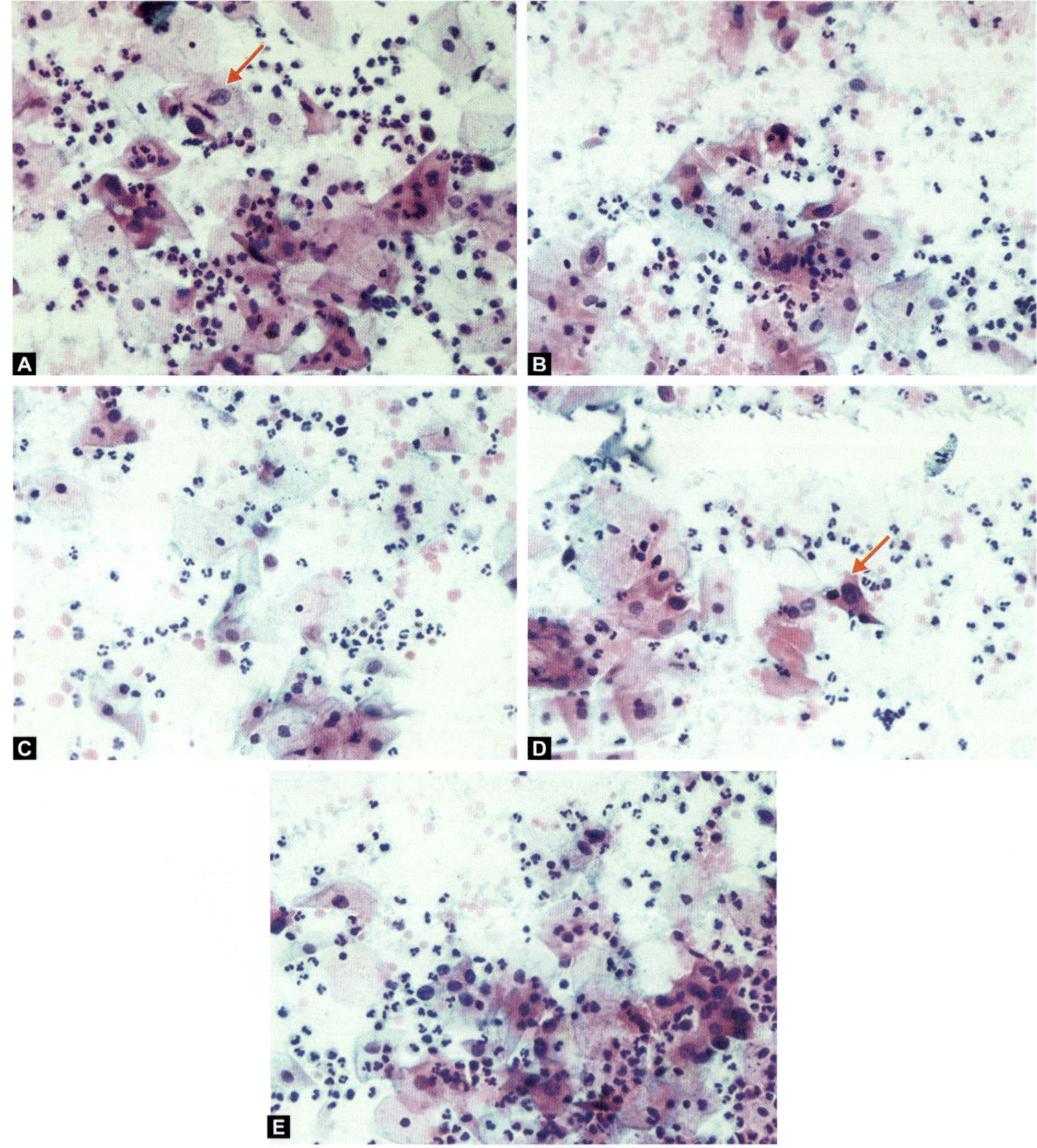

Figures 2.30A to E

Problem 2.31

Routine cervical smear (SurePath, Fig. 2.31) in a 48-year-old female. What is your interpretation?

A. LSIL
B. ASCUS
C. HSIL
D. Atrophic smear

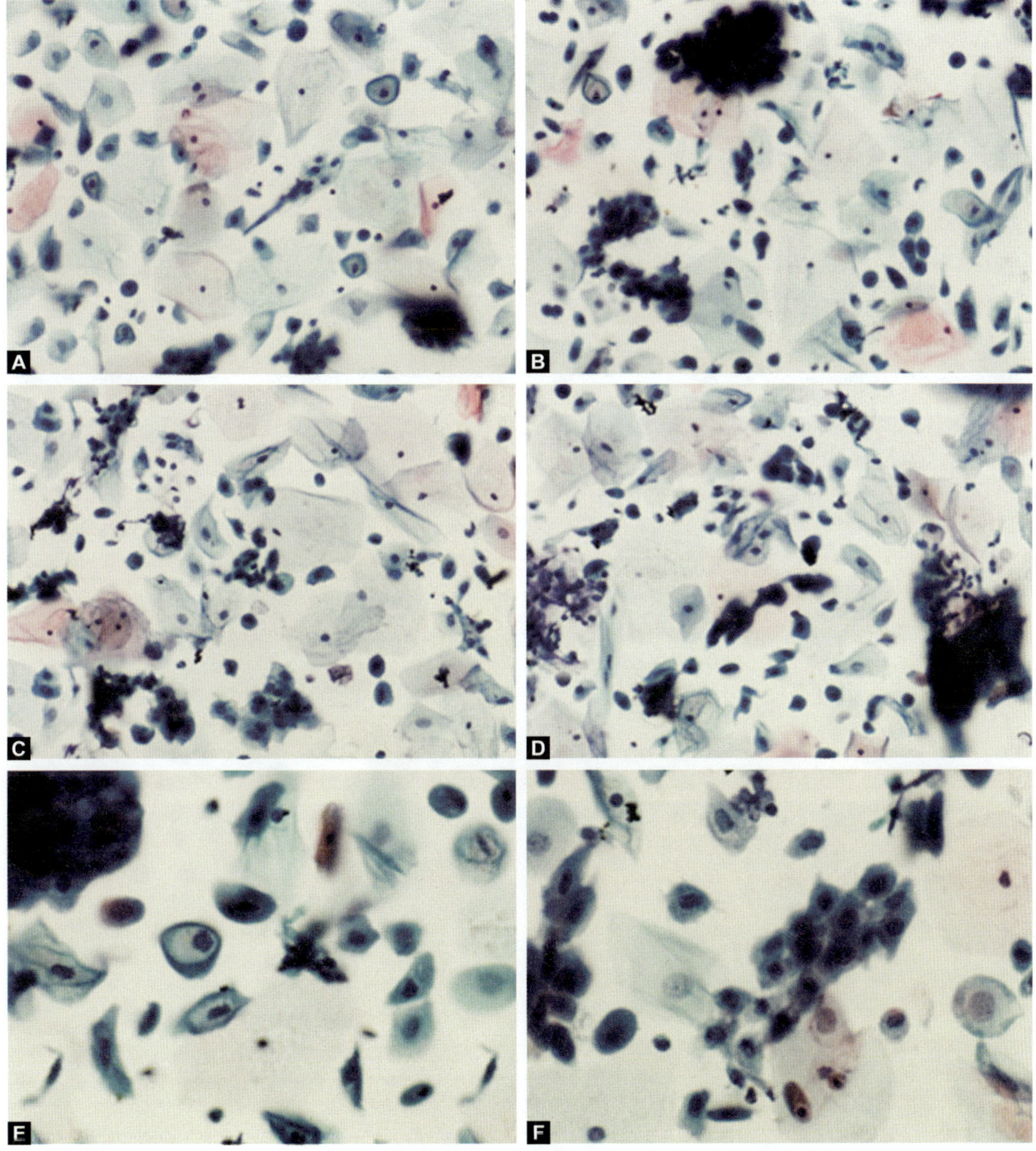

Figures 2.31A to F

Problem 2.32

A 38-year-old female presented with menorrhagia for 3 months. Routine cervical smear examination was done. The most likely diagnosis from the Figure 2.32 of SurePath preparation is:

A. HSIL
B. AGUS
C. Adenocarcinoma
D. Squamous cell carcinoma

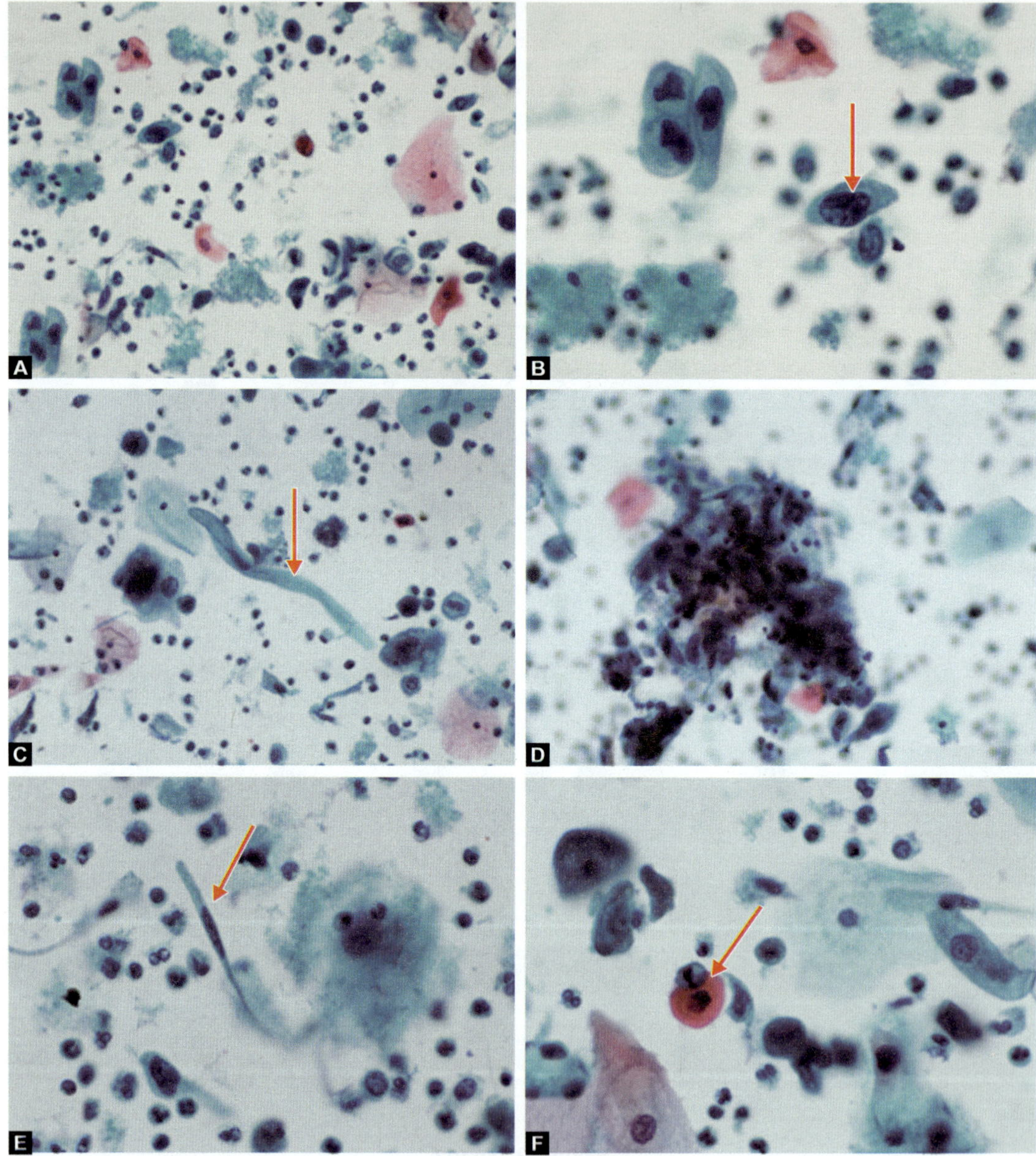

Figures 2.32A to F

Problem 2.33

A 41-year-old female with history of radiation one year back for squamous cell carcinoma of cervix. The most likely diagnosis from the (Fig. 2.33) SurePath smear is:

A. Radiation effect only
B. HSIL
C. Squamous cell carcinoma recurrence with radiation changes
D. Viral infection

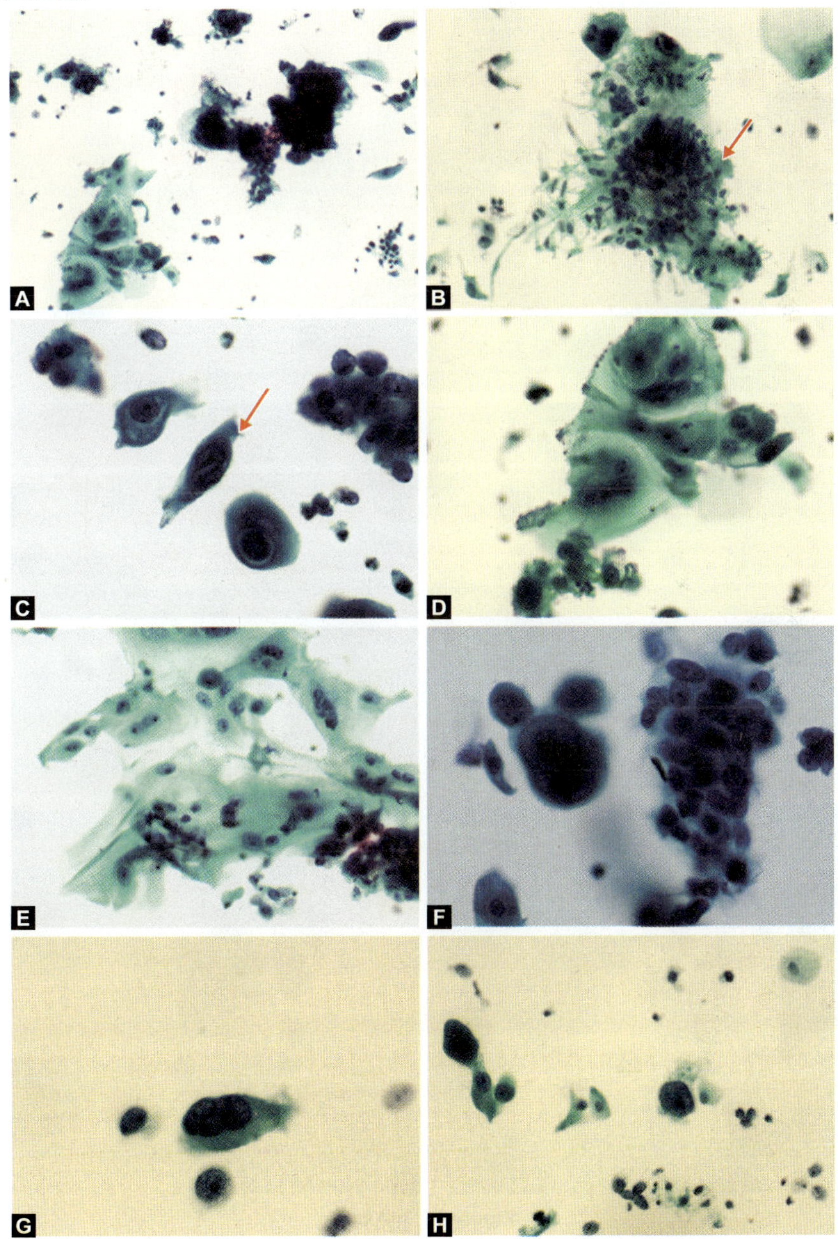

Figures 2.33A to H

Problem 2.34

A 43-year-old female was diagnosed as squamous cell carcinoma 5 year back with history of radiation therapy. Routine follow up cervical smear was done (Fig. 2.34).

The most likely diagnosis is:

A. Radiation induced changes
B. Squamous cell carcinoma: recurrence
C. Radiation induced dysplasia
D. ASCUS

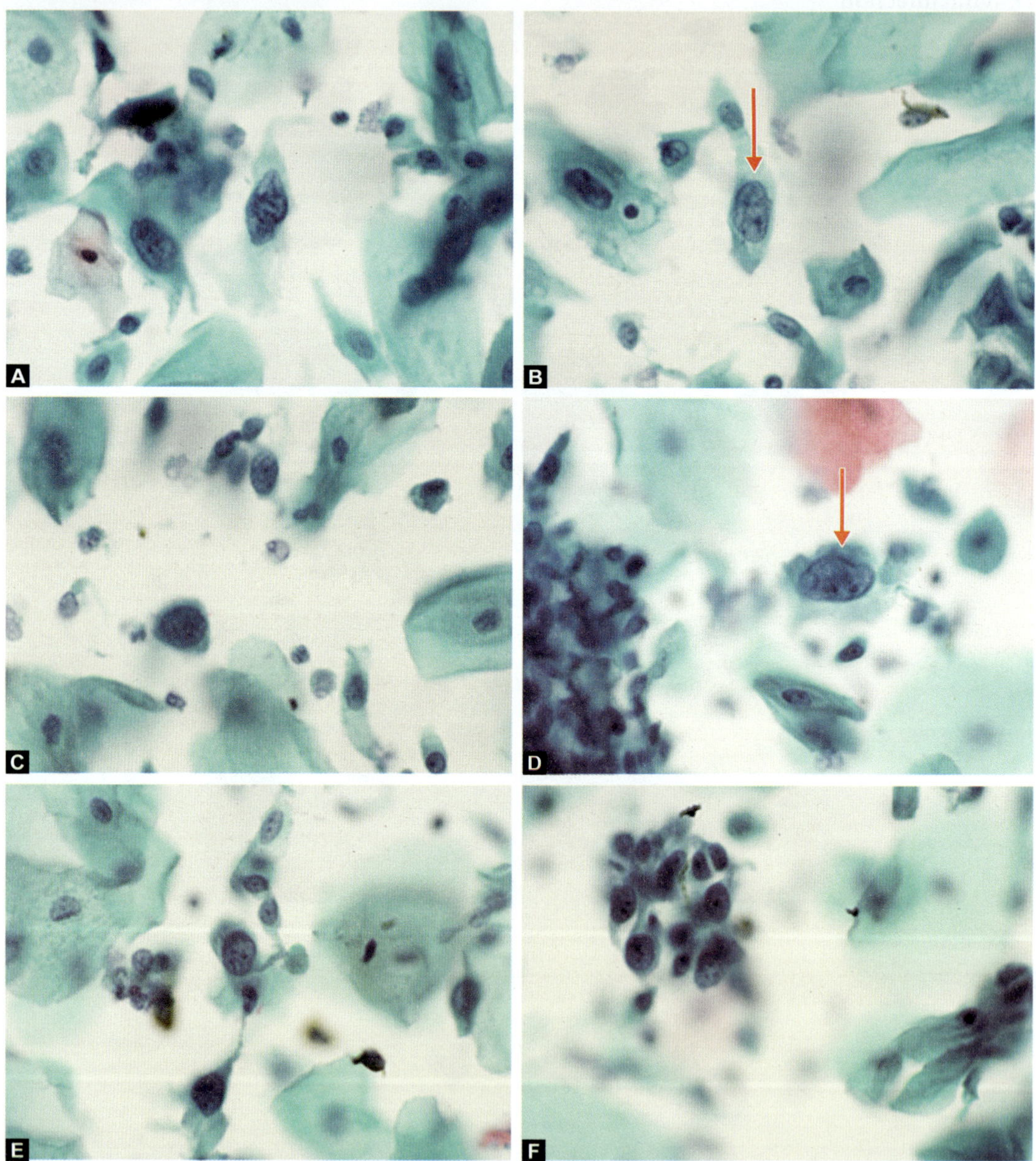

Figures 2.34A to F

Problem 2.35

Routine cervical cytology screening was done in a 51-year-old female. Microphotographs of the conventional cervical smear are shown in Figures 2.35A to F.

1. **What should you advice for the next step after seeing the microphotographs of the smear?**
   A. Colposcopy followed by biopsy
   B. Follow up smear after 6 months
   C. Hysterectomy
   D. Visual inspection under acetic acid
2. **Choose the probable histopathology outcome in this case.**
   G.
   H.
   I.
   J.

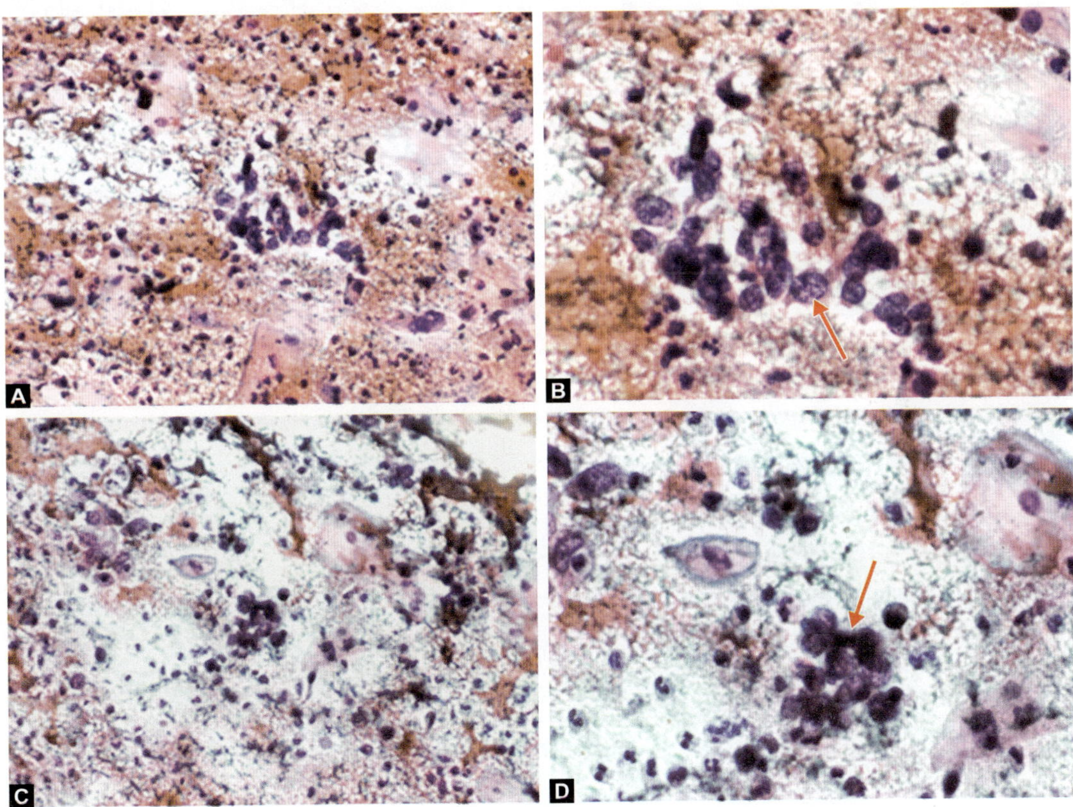

Figures 2.35A to D

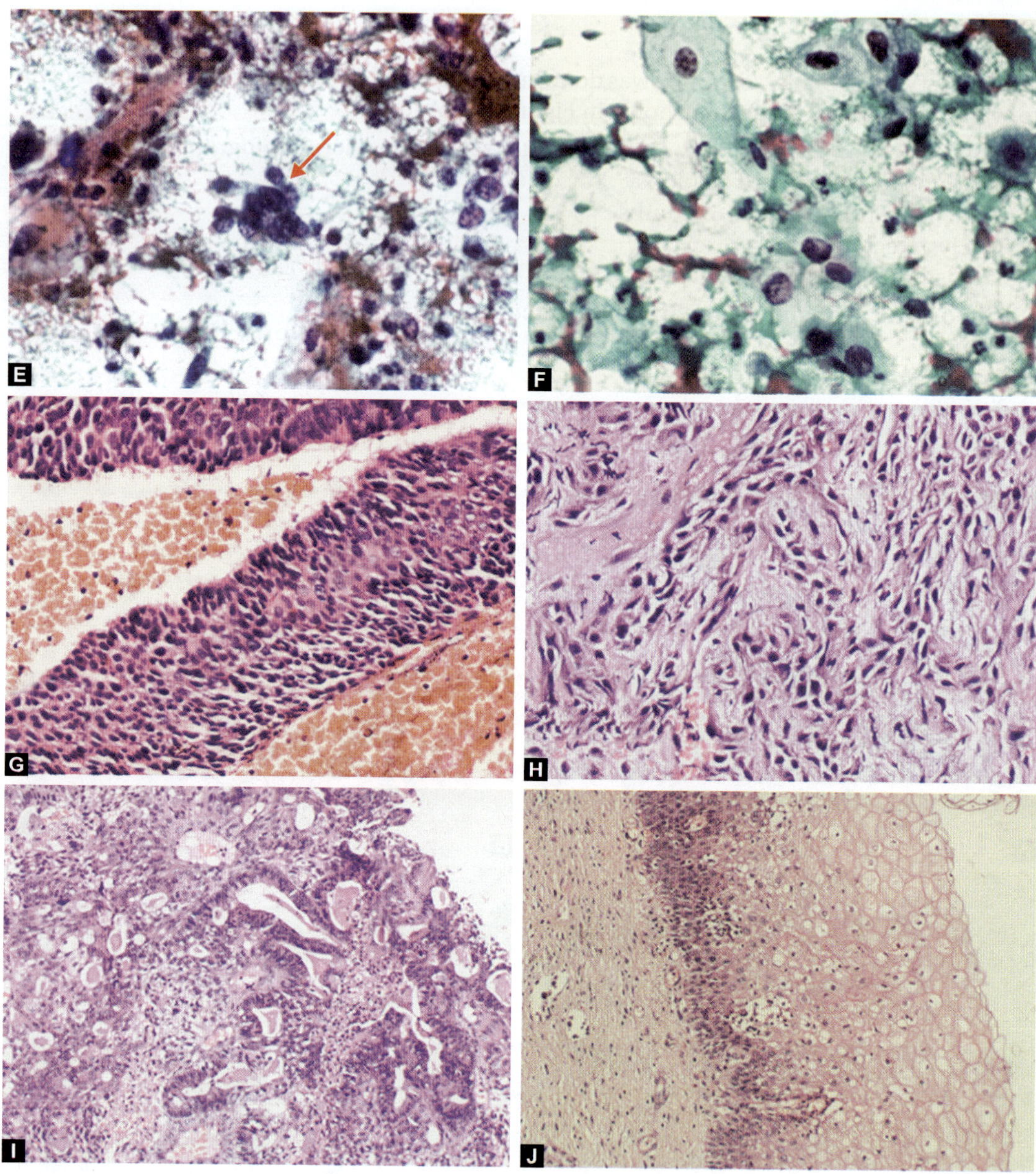

Figures 2.35E to J

Problem 2.36

Match the following (Fig. 2.36)

A.	1. Tubal metaplasia
B.	2. Fibre cells
C.	3. Foreign body giant cells
D.	4. Corn flakes cell

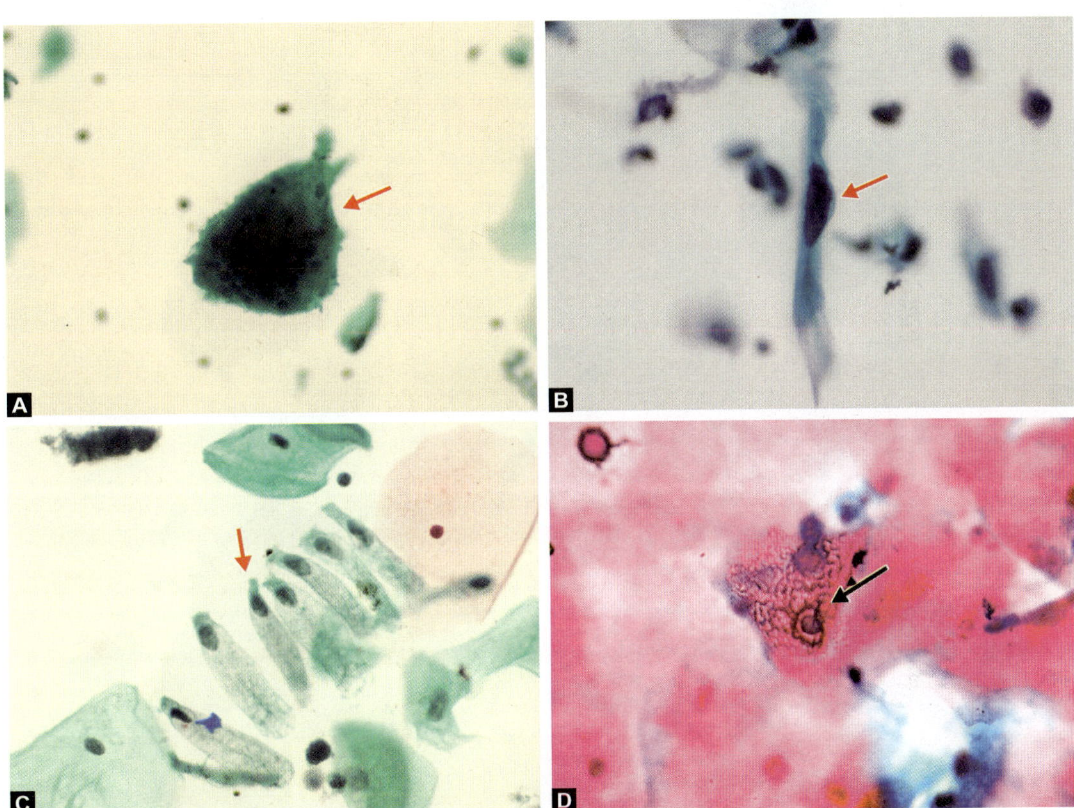

Figures 2.36A to D

Problem 2.37

Match the following (Fig. 2.37)

A.	1. Hyperchromatic crowded group of endocervical adenocarcinoma
B.	2. Hyperchromatic crowded cells in HSIL
C.	3. Hyperchromatic crowded cells of endometrial cells
D.	4. Hyperchromatic crowded cells of benign endocervical cells

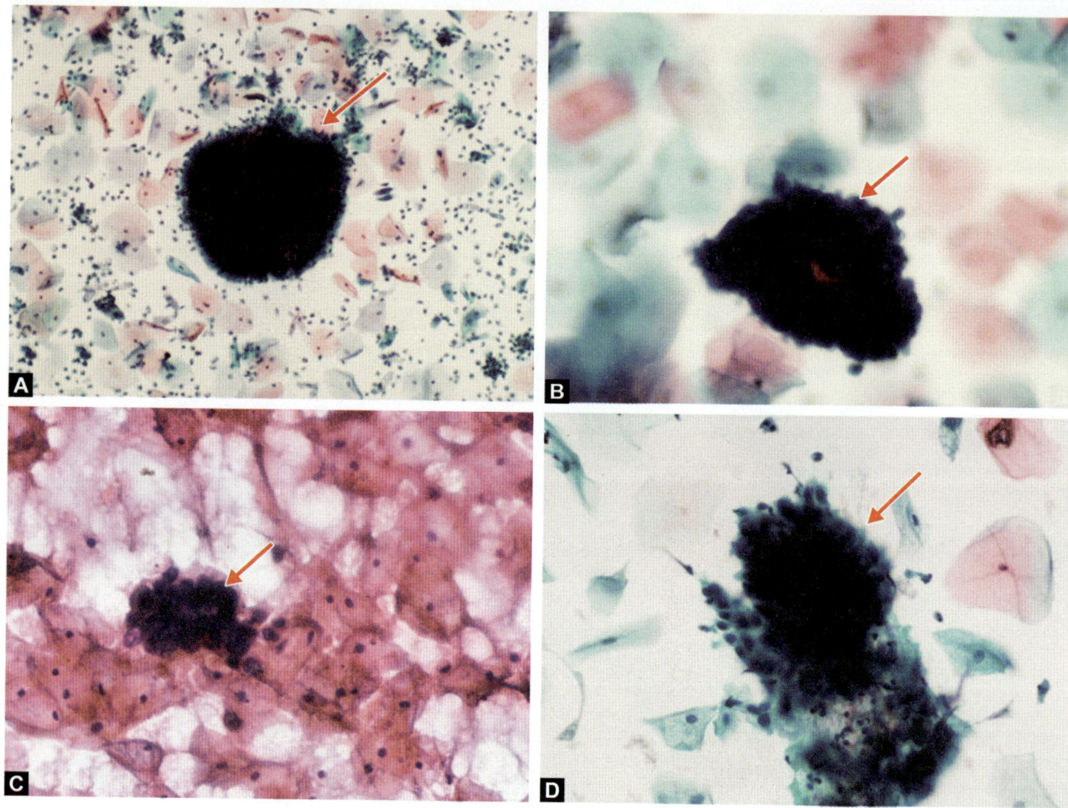

Figures 2.37A to D

Problem 2.38

Match the smear with the findings from Figure 2.38

Figure A	1. Koilocytosis
Figure B	2. Squamous pearl
Figure C	3. Tumor diathesis

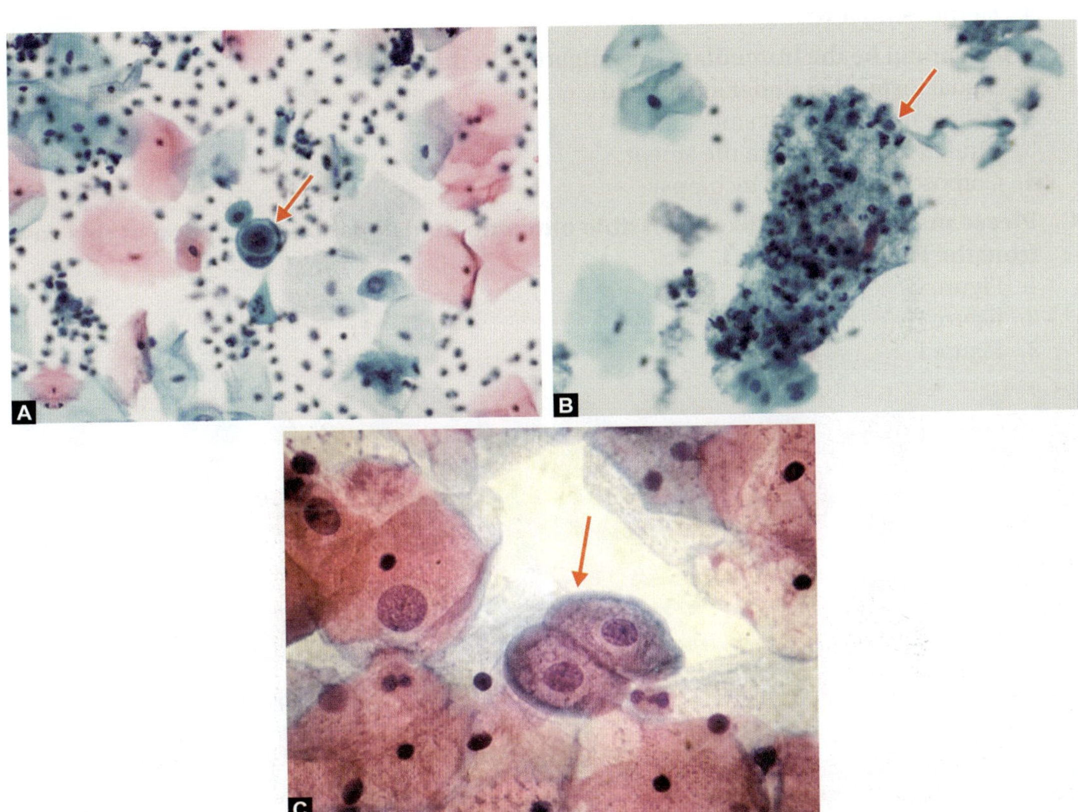

Figures 2.38A to C

Problem 2.39

Conventional cervical smear in a 41-year-old female. Microphotographs of the smear are shown in Figures 2.39A to F.

A. Which virus is related to this smear?

1. HPV 6
2. HPV 11
3. HPV 16
4. Herpes simplex

B. What should be the immediate investigation in this case?

1. Visual inspection under acetic acid
2. Follow up and repeat smear
3. Viral demonstration in the smear
4. Colposcopy followed by biopsy

C. Please match the most likely possible outcome in histopathology section of this case from the Figures 2.39G to I.

1. Figure G
2. Figure H
3. Figure I

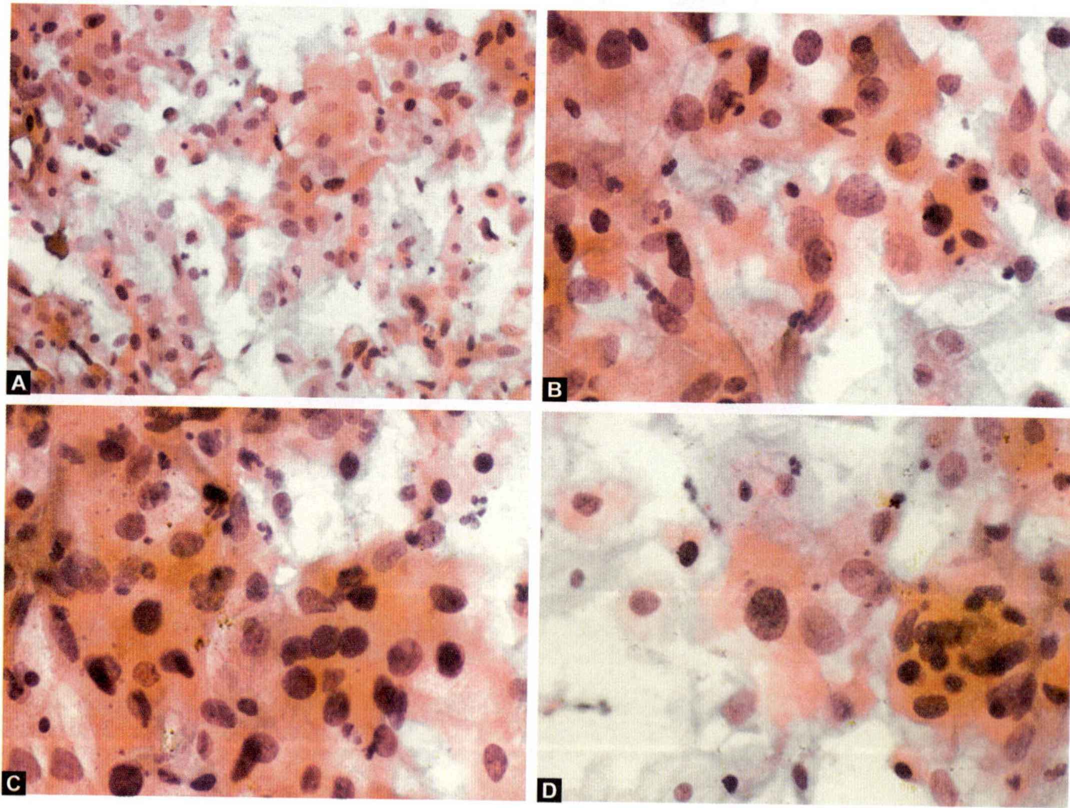

Figures 2.39A to D

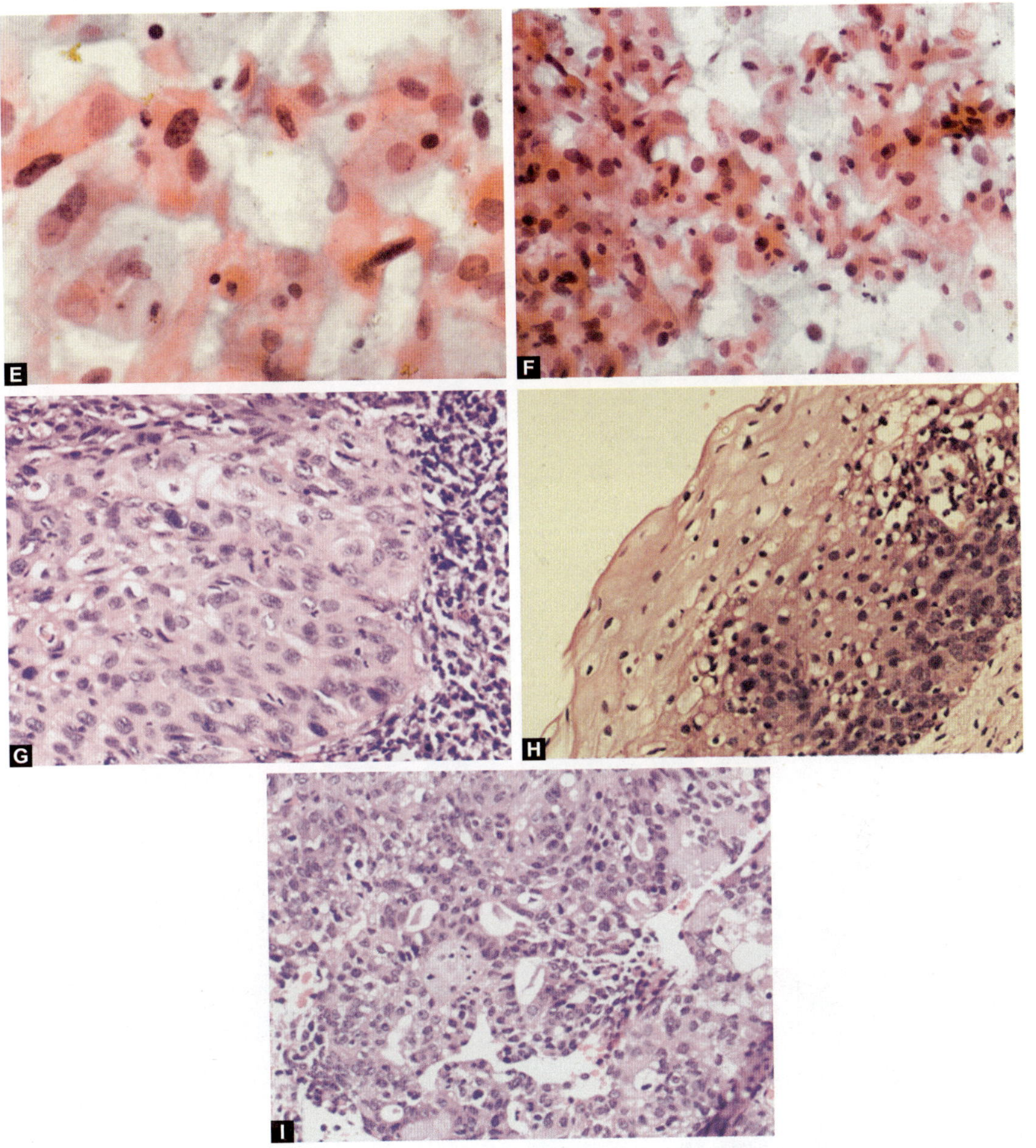

Figures 2.39E to I

Problem 2.40

Match the cervical smear (Fig. 2.40) with the possible histopathology

Figure A	Figure 1
Figure B	Figure 2
Figure C	Figure 3
Figure D	Figure 4

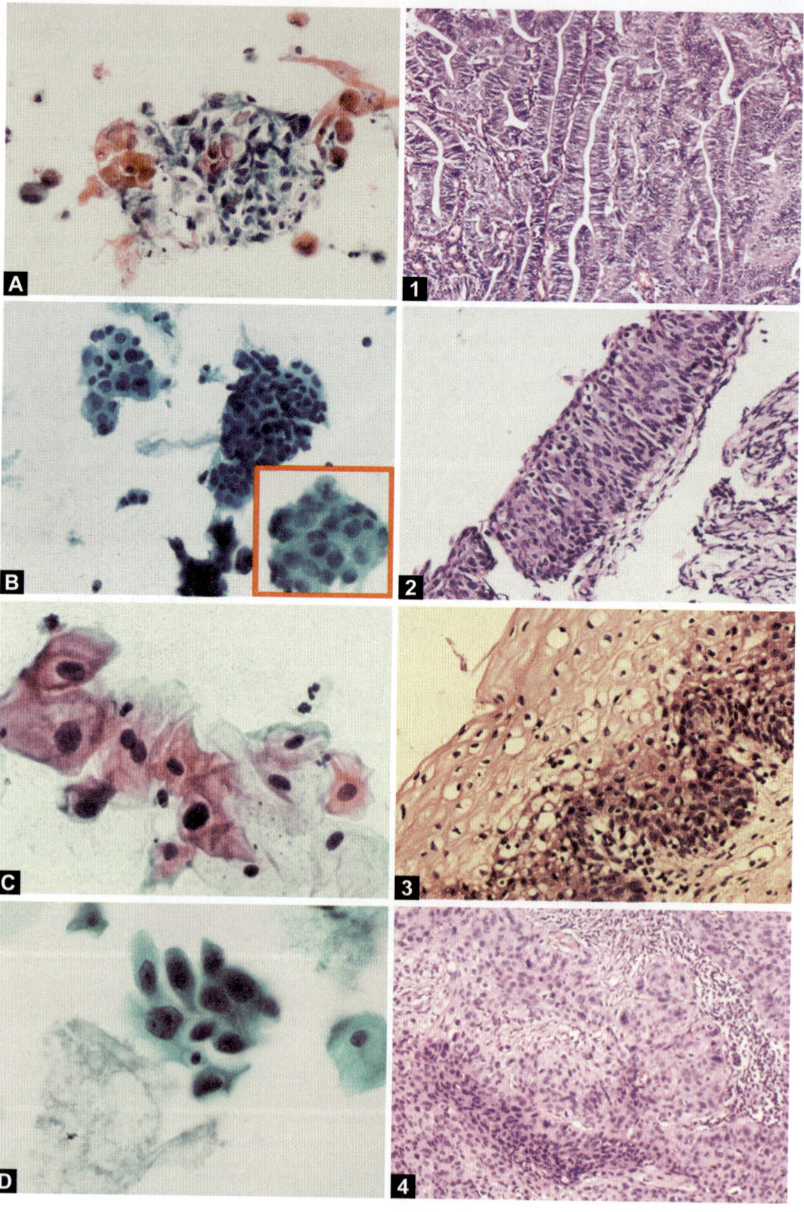

Figures 2.40A to D

Problem 2.41

A 34-year-old female using intrauterine contraceptive device for 2 year. Routine cervical smear is done (Fig. 2.41). What is your interpretation by Bethesda system 2001?

A. Candidal infection
B. Bacterial vaginosis
C. Actinomycosis
D. *Trichomonas vaginalis*

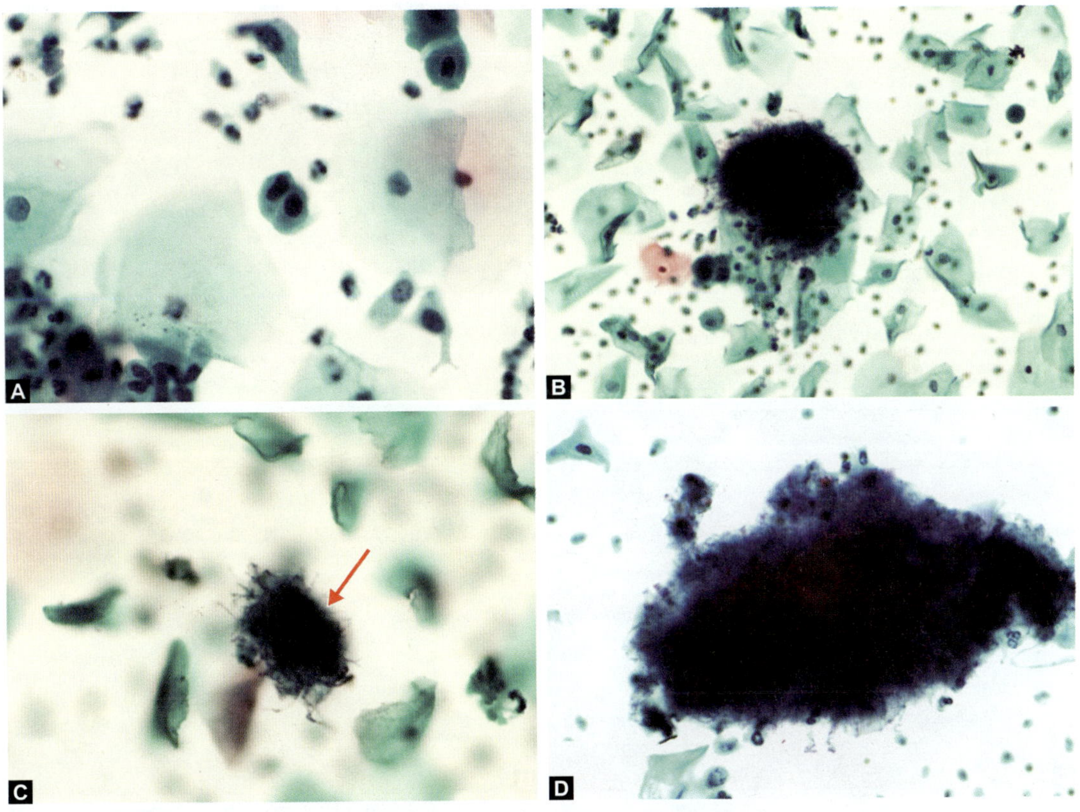

Figures 2.41A to D

Problem 2.42

Routine cervical smear in a 42-year-old female. What is the most likely interpretation from this SurePath smear (Fig. 2.42)?

A. AGCUS
B. ASCUS
C. LSIL
D. HSIL

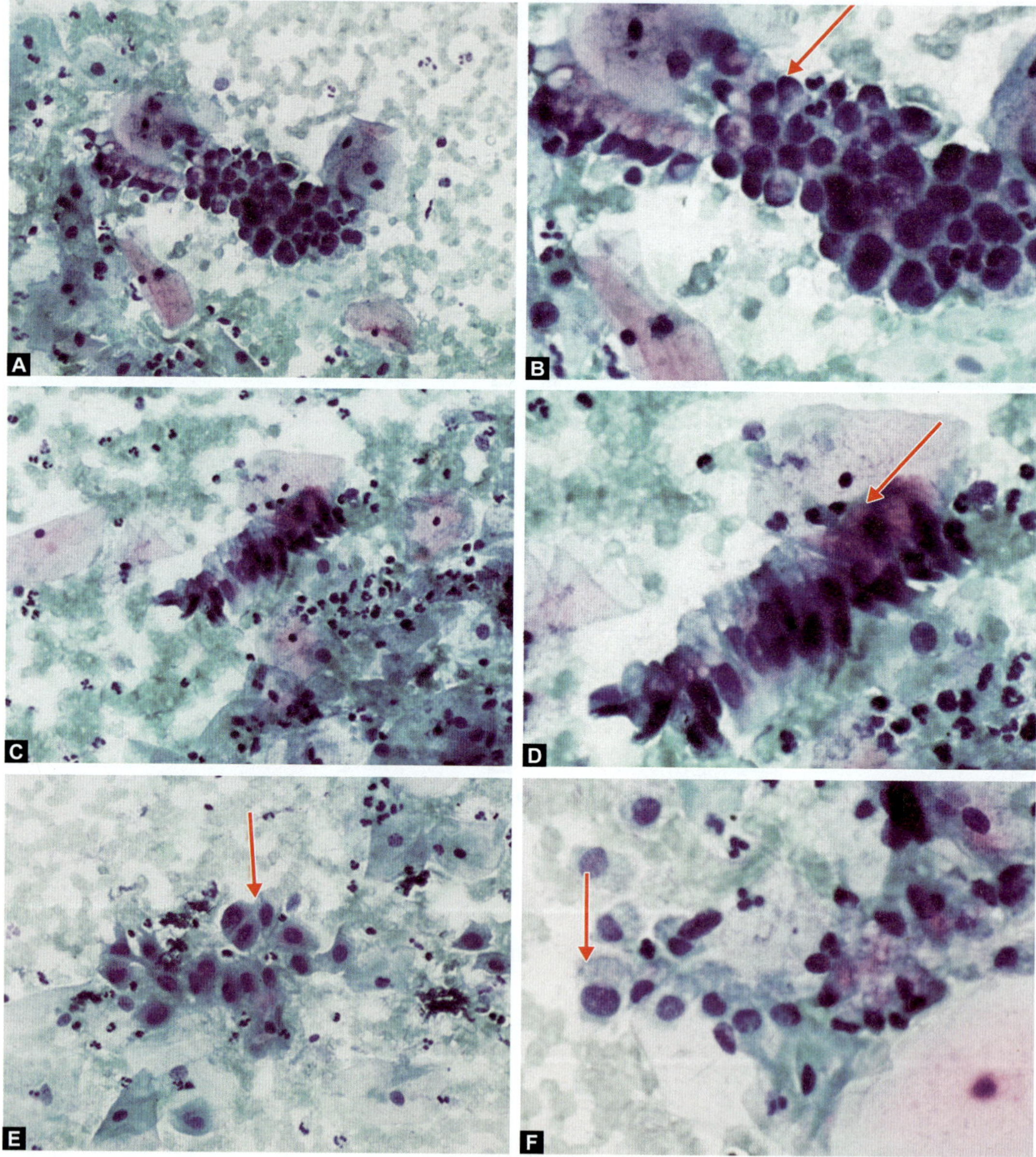

Figures 2.42A to F

Problem 2.43

Routine cervical smear in a 32-year-old female. What is the most likely interpretation from this SurePath smear (Fig. 2.43)?

A. ASCUS
B. AGC, favors neoplastic
C. LSIL
D. HSIL

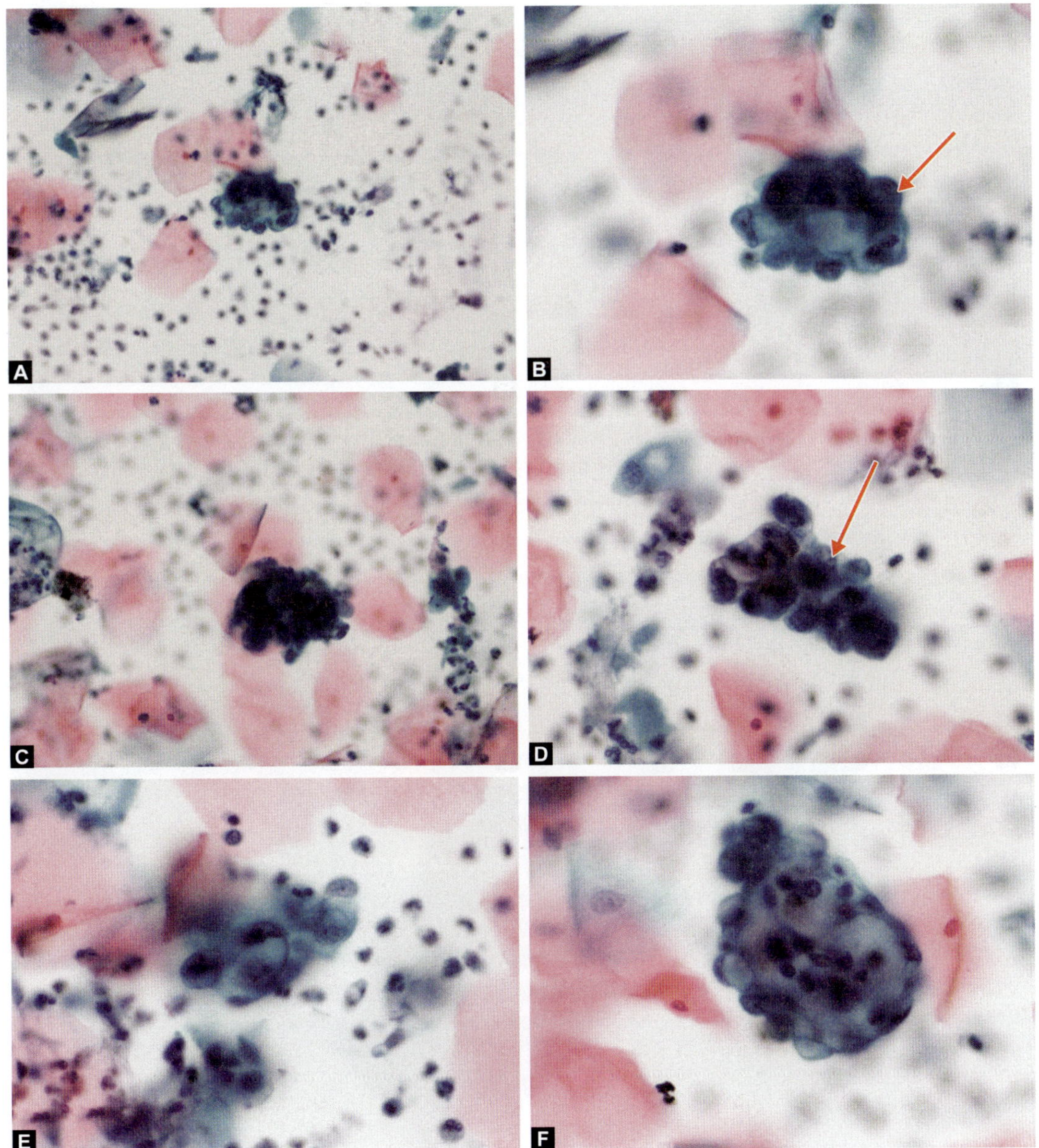

Figures 2.43A to F

Problem 2.44

Routine cervical smear in a 53-year-old female. What is the most likely interpretation from this SurePath smear (Fig. 2.44)?

A. AGC: Possibly endometrial
B. Adenocarcinoma
C. ASCUS
D. LSIL
E. HSIL

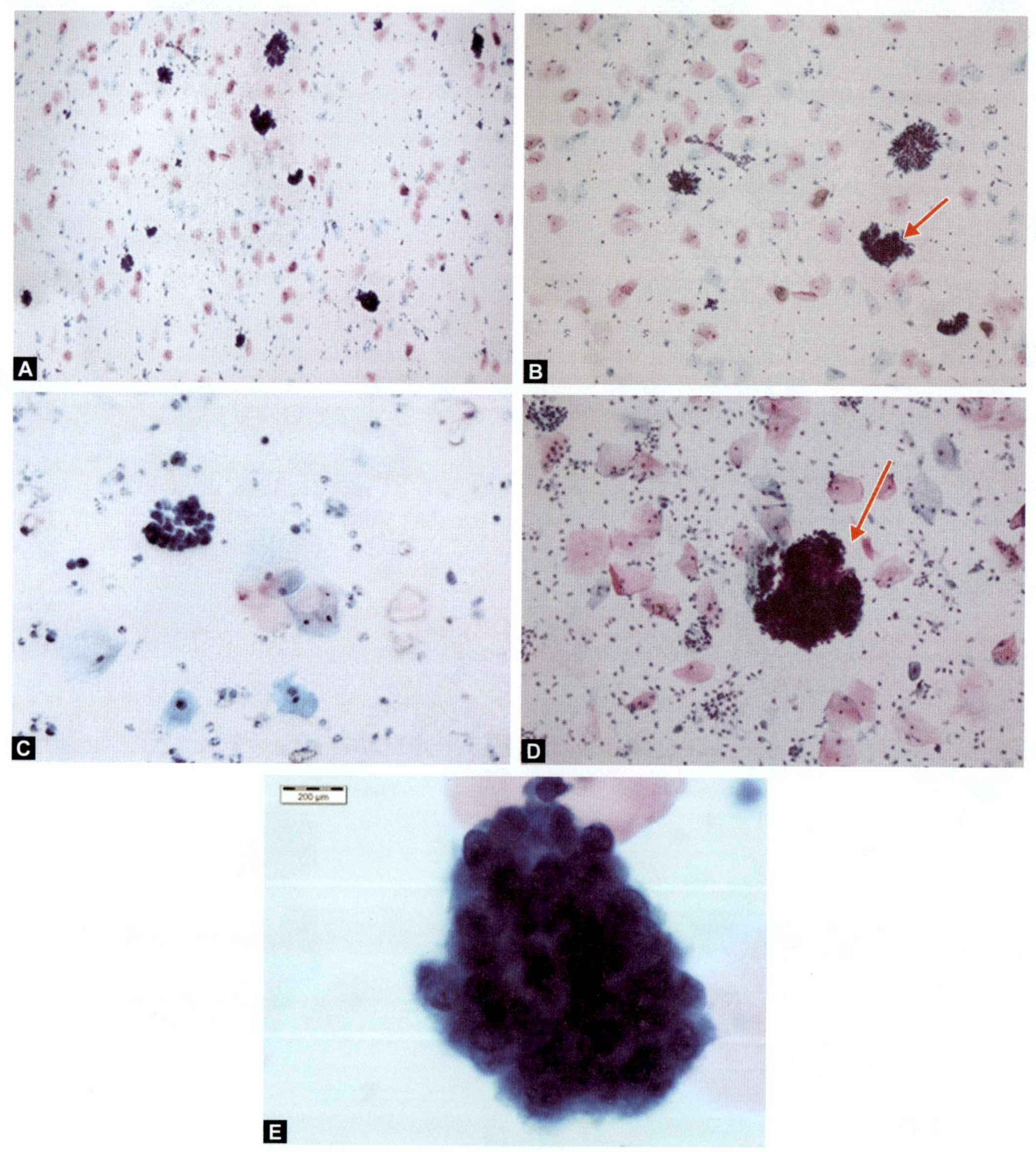

Figures 2.44A to E

Problem 2.45

A 38-year-old female with routine cervical smear screening. What is your diagnosis from the SurePath cervical smear (Fig. 2.45)?

A. Atypical endocervical cells favors neoplasia
B. Atypical endometrial cells favors neoplasia
C. Adenocarcinoma
D. HSIL

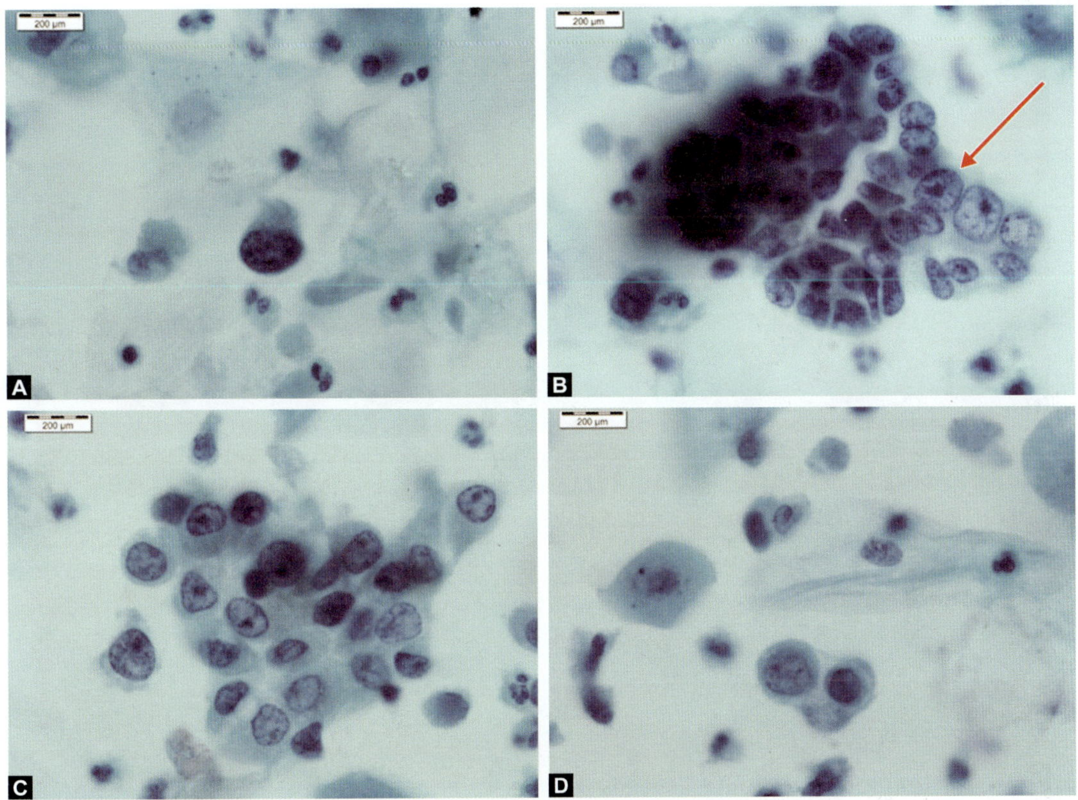

Figures 2.45A to D

Problem 2.46

A 48-year-old female presented with menorrhagia for 3 months. USG uterus showed thickened endometrium. Cervical cytology smear of the patient was done. What is your interpretation from the Figure 2.46?

A. HSIL
B. AGUS
C. Adenocarcinoma possibly endometrial origin
D. ASCUS

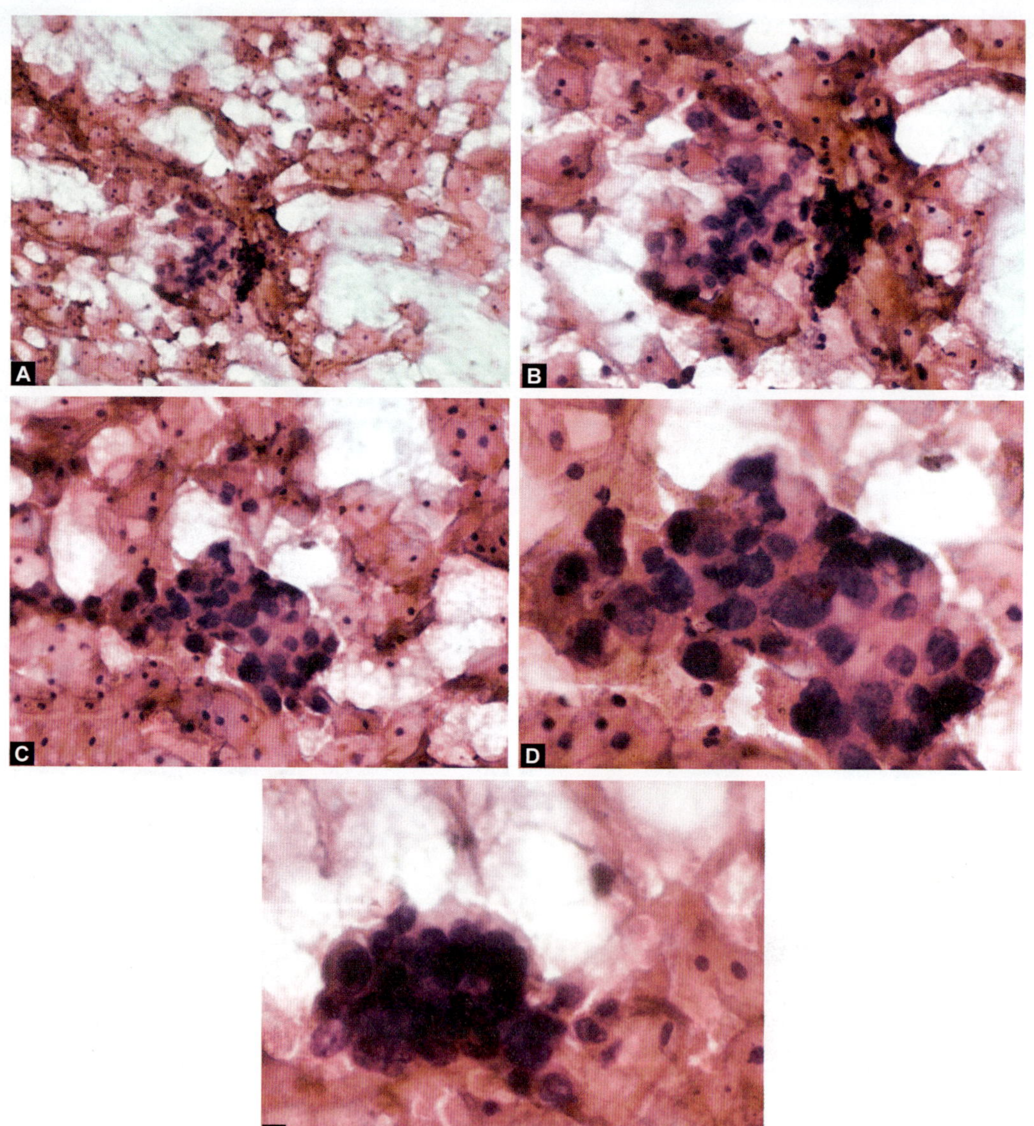

Figures 2.46A to E

Problem 2.47

A 45-year-old female presented with bleeding per vagina for 17 days. What is your interpretation from the Figure 2.47 of cervical cytology (SurePath preparation)?

A. Adenocarcinoma, favors endocervical origin
B. AGCUS
C. HSIL
D. Squamous cell carcinoma

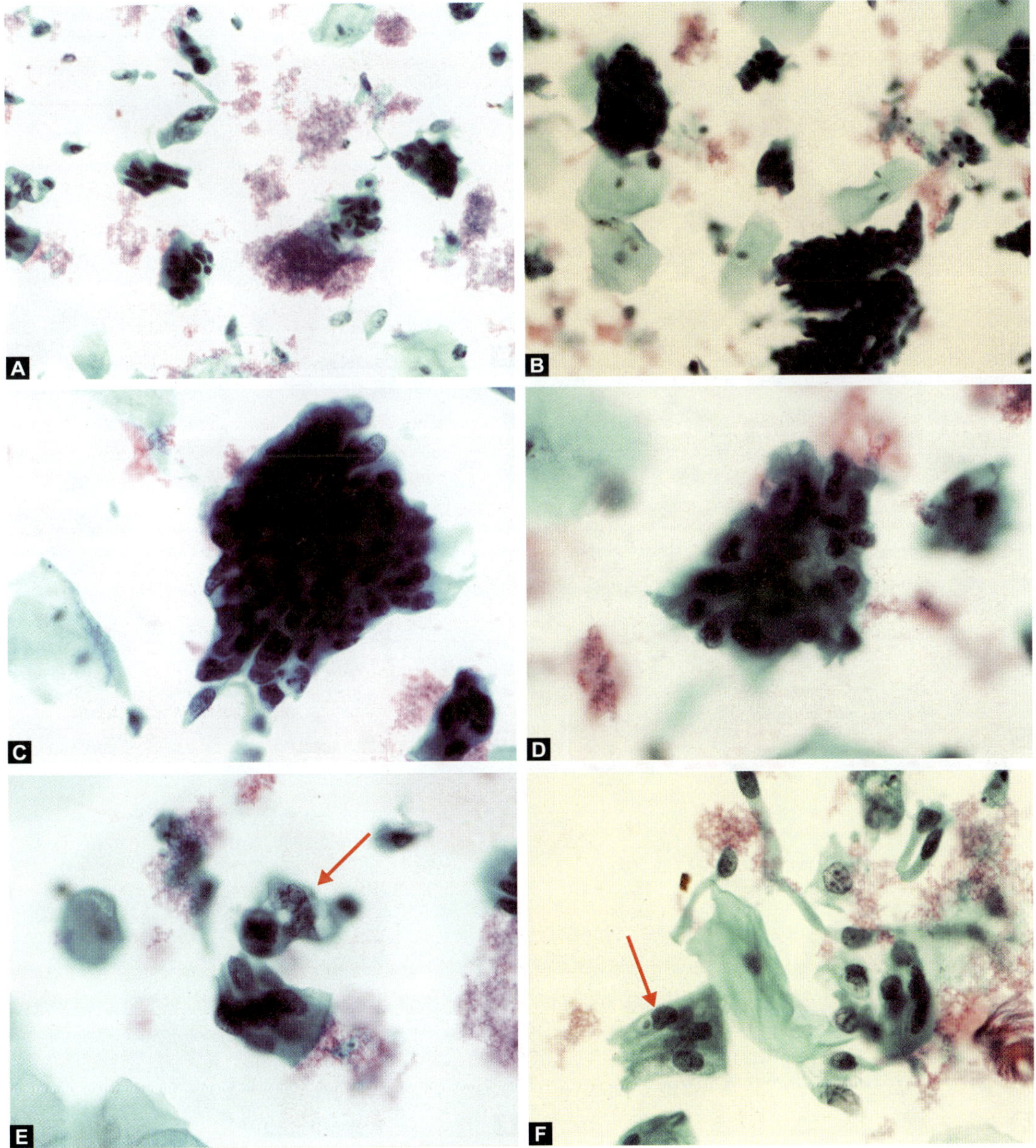

Figures 2.47A to F

Problem 2.48

A 47-year-old female had history of total abdominal hysterectomy 3 year back. Now routine vault smear was done. The most likely diagnosis from Figure 2.48 is:

A. Atypical squamous cells present
B. Squamous cell carcinoma
C. Adenocarcinoma
D. AGCs

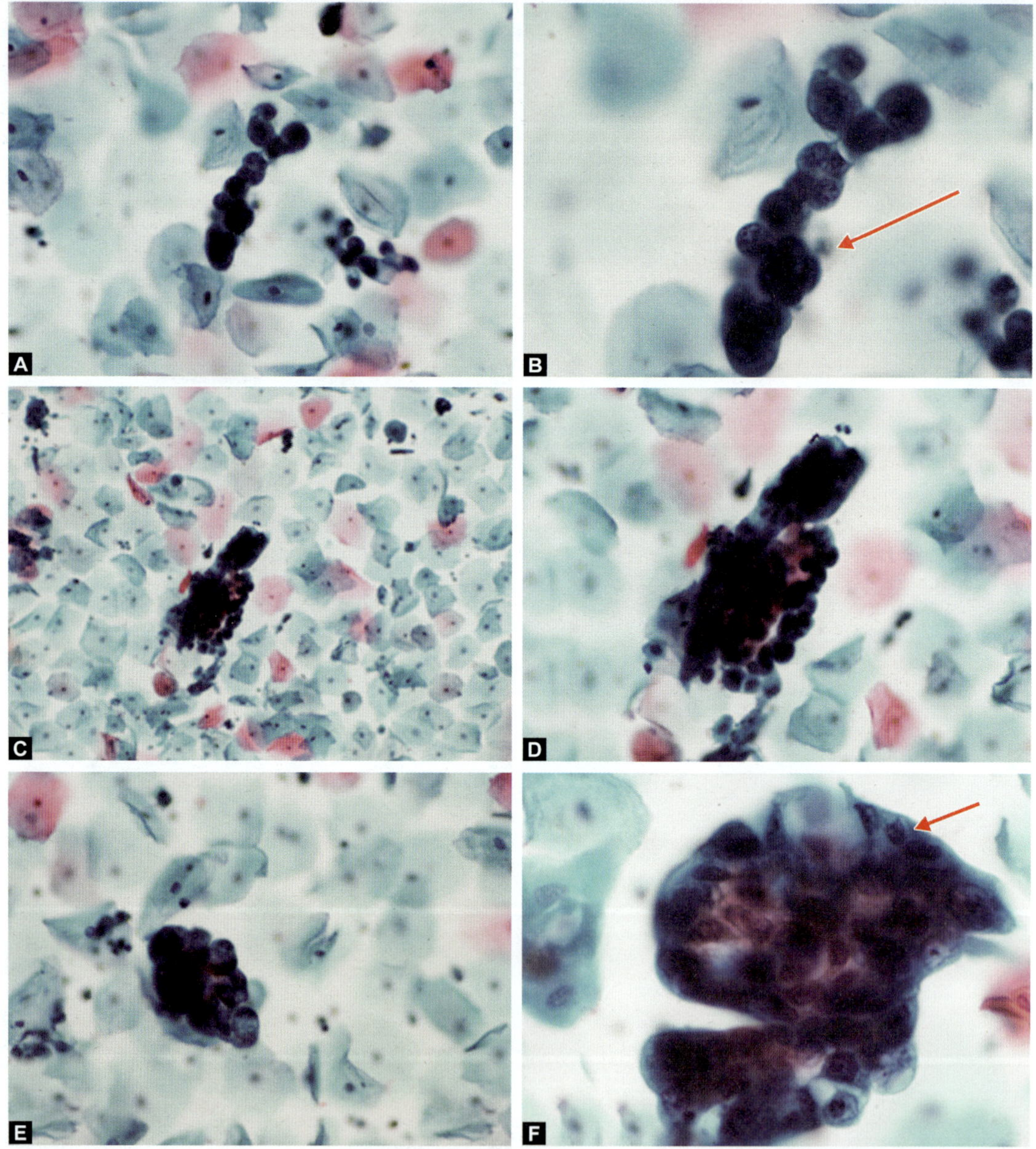

Figures 2.48A to F

Problem 2.49

A 45-year-old female presented with right adnexal mass measuring 13 cm diameter. Solid and cystic on USG examination. Her CA 125 blood level was 3500 IU/mL. Cervical smear examination was done (Fig. 2.49). The most likely diagnosis is:

A. HSIL
B. AGUS
C. Adenocarcinoma cervix
D. Adenocarcinoma extrauterine (ovary)

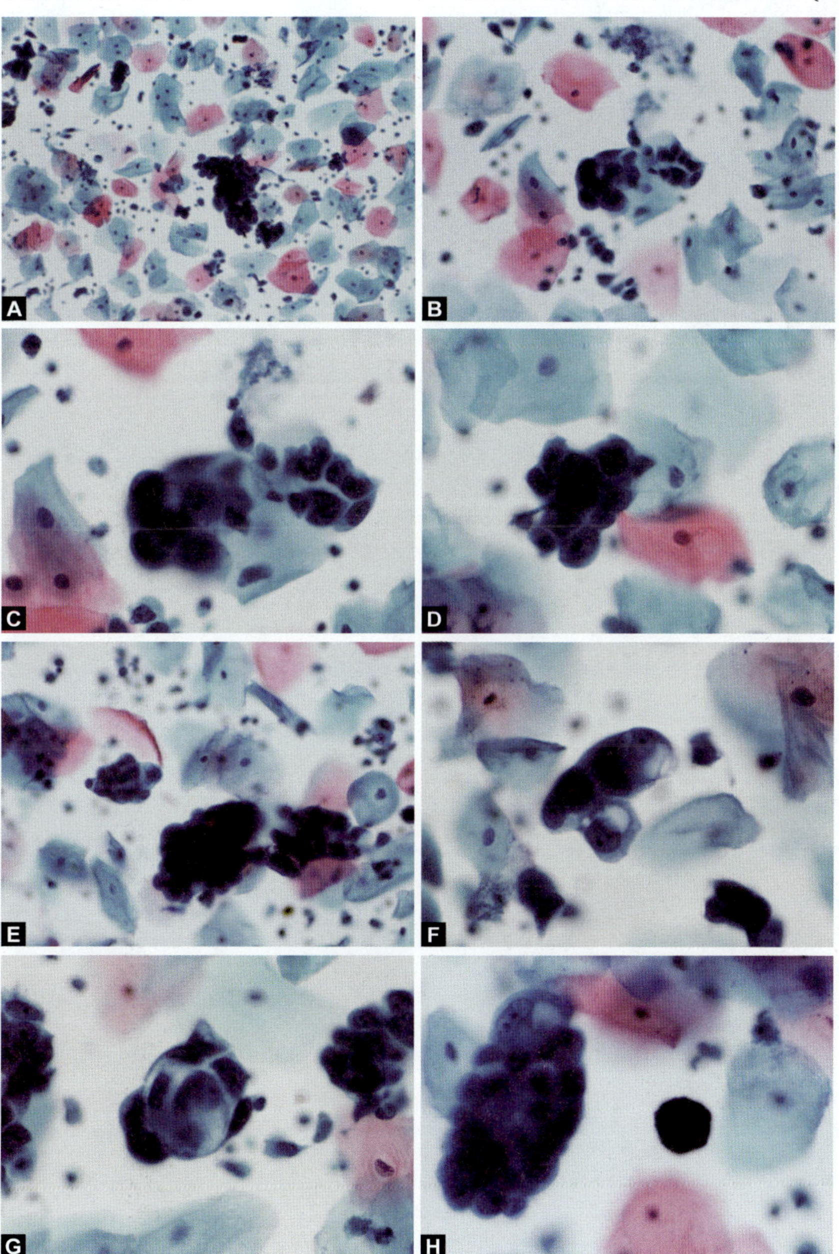

Figures 2.49A to H

Problem 2.50

Routine cervical smear in a 38-year-old female. The most likely diagnosis from the SurePath preparation (Fig. 2.50) is:

A. ASCUS
B. ASC-H
C. LSIL
D. HSIL

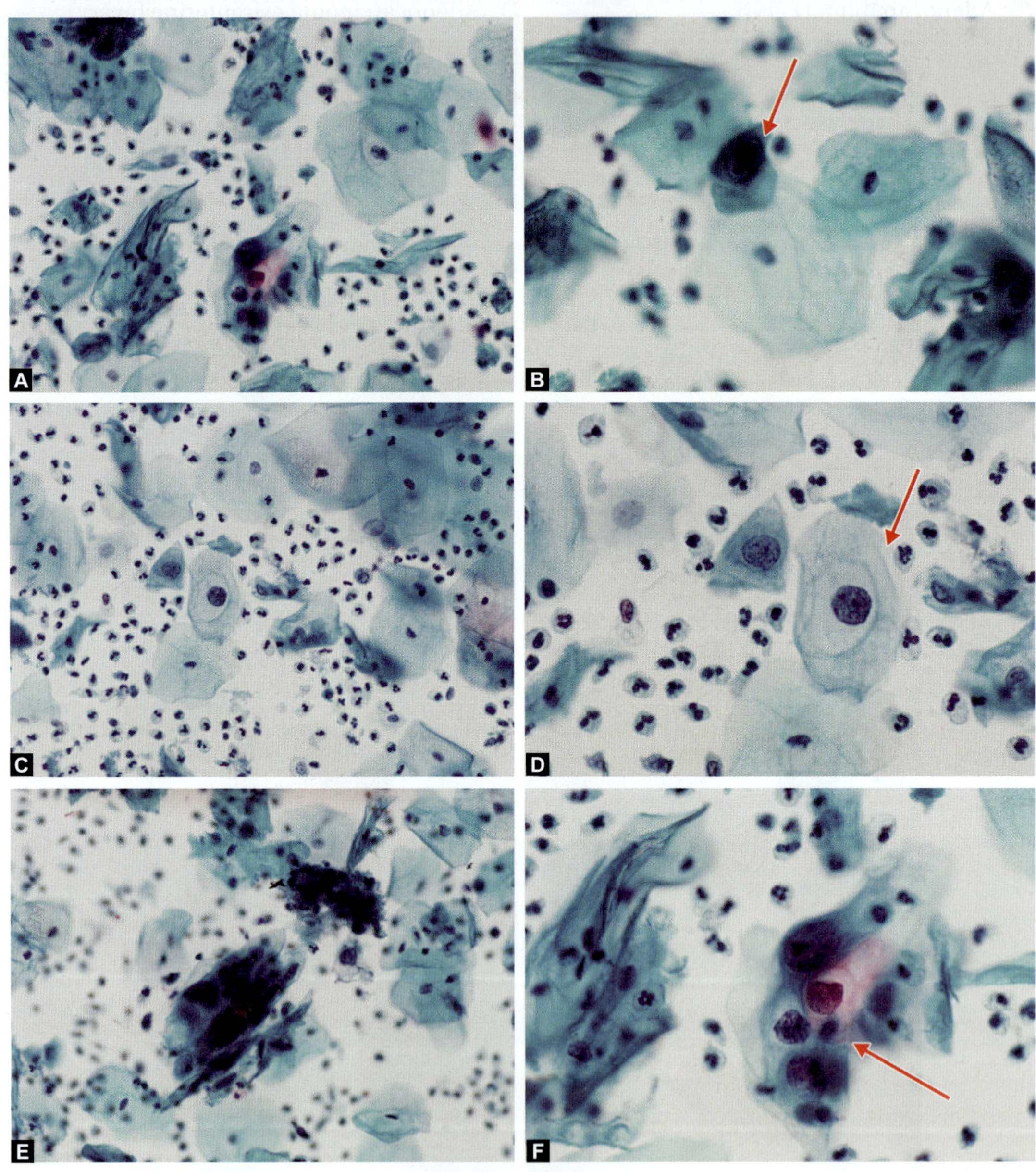

Figures 2.50A to F

Problem 2.51

A 51-year-old female with history of menorrhagia for 3 months had cervical smear. The most likely diagnosis from the SurePath preparation (Fig. 2.51) is:

A. HSIL
B. LSIL
C. AGC (Atypical glandular cell)
D. Adenocarcinoma possibly endometrial

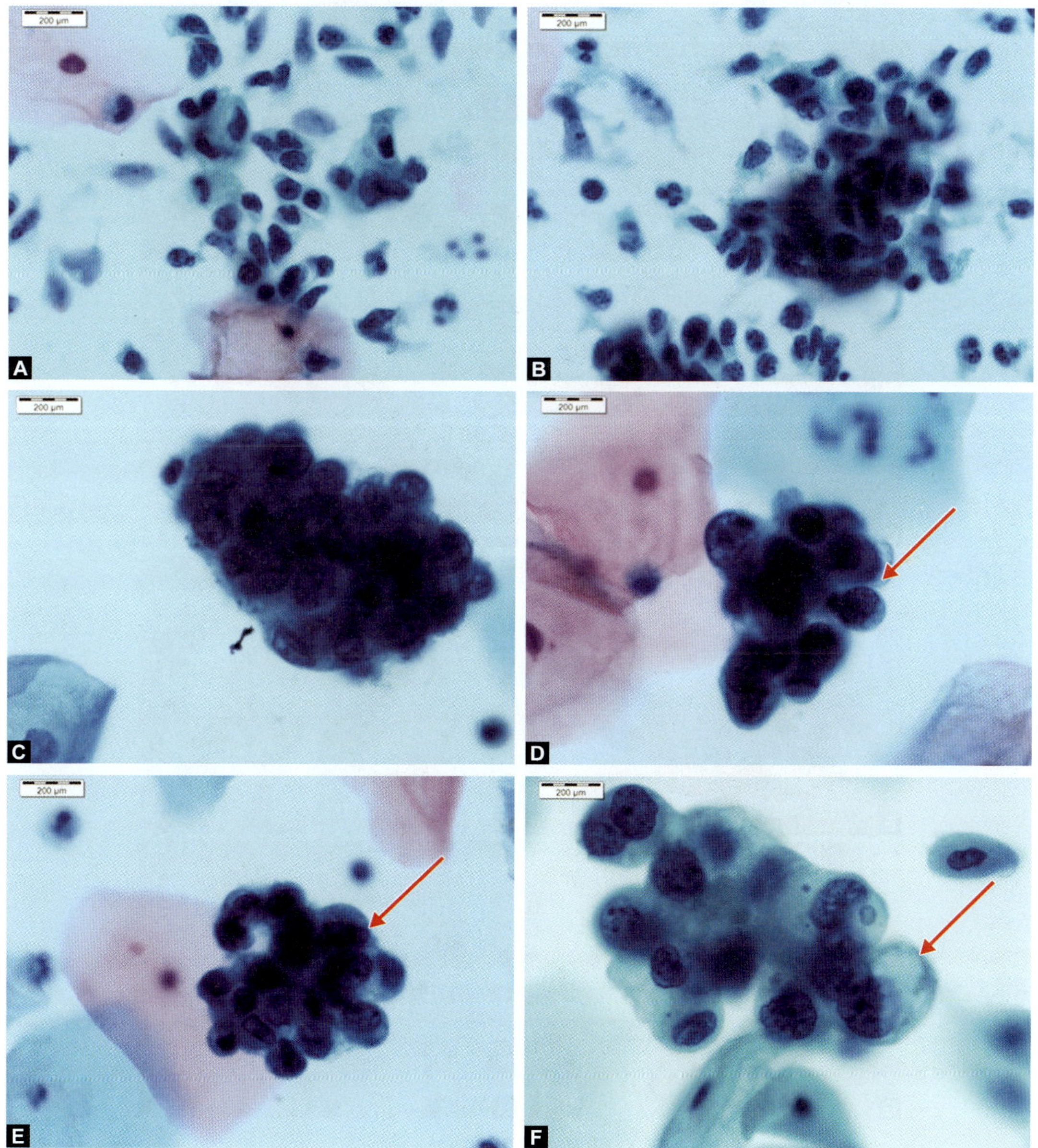

Figures 2.51A to F

Problem 2.52

A 31-year-old female was treated for Hodgkin lymphoma 6 months back. The routine cervical smear was done during her follow up period. What is your likely diagnosis from the Figures 2.52A to H?

A. ASCUS
B. HSIL
C. Squamous cell carcinoma
D. Adenocarcinoma

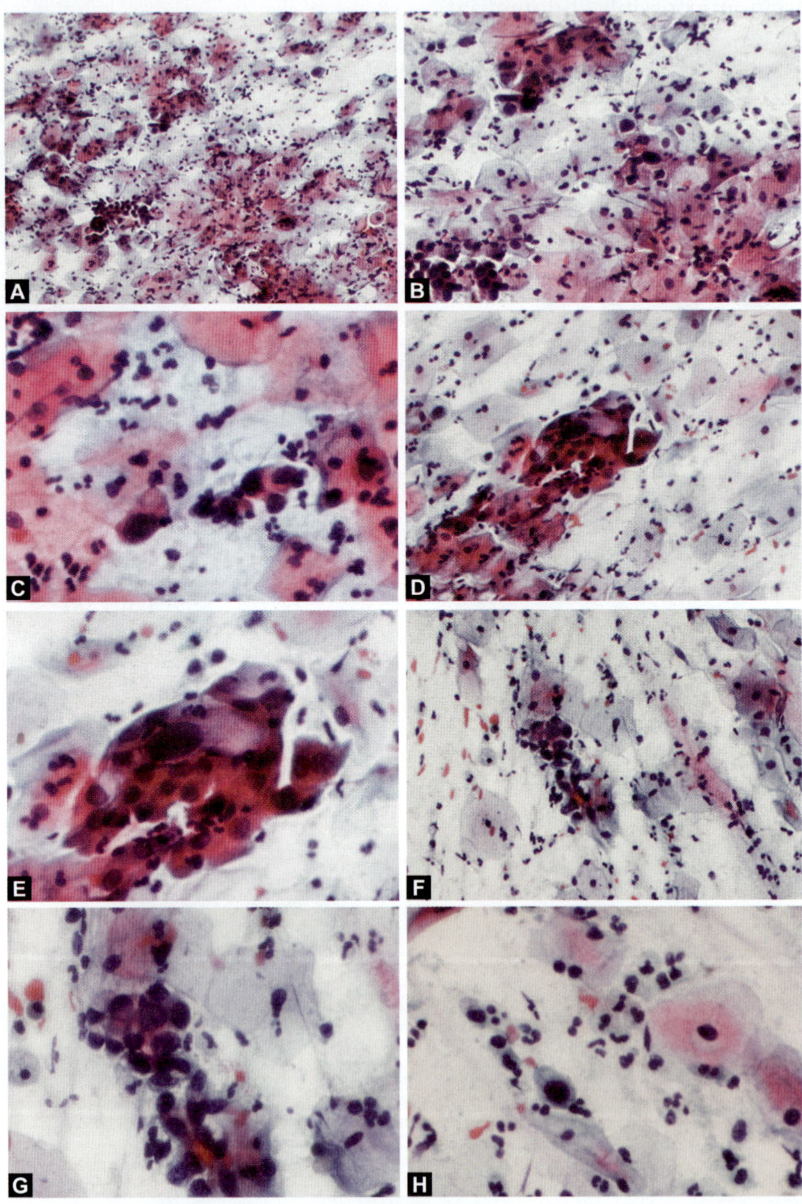

Figures 2.52A to H

Problem 2.53

Routine cervical smear in a 49-year-old female. What is your interpretation from the microphotograph (Fig. 2.53) of SurePath preparation?

A. Within normal limit
B. Endometrial cells present
C. AGCUS
D. Adenocarcinoma

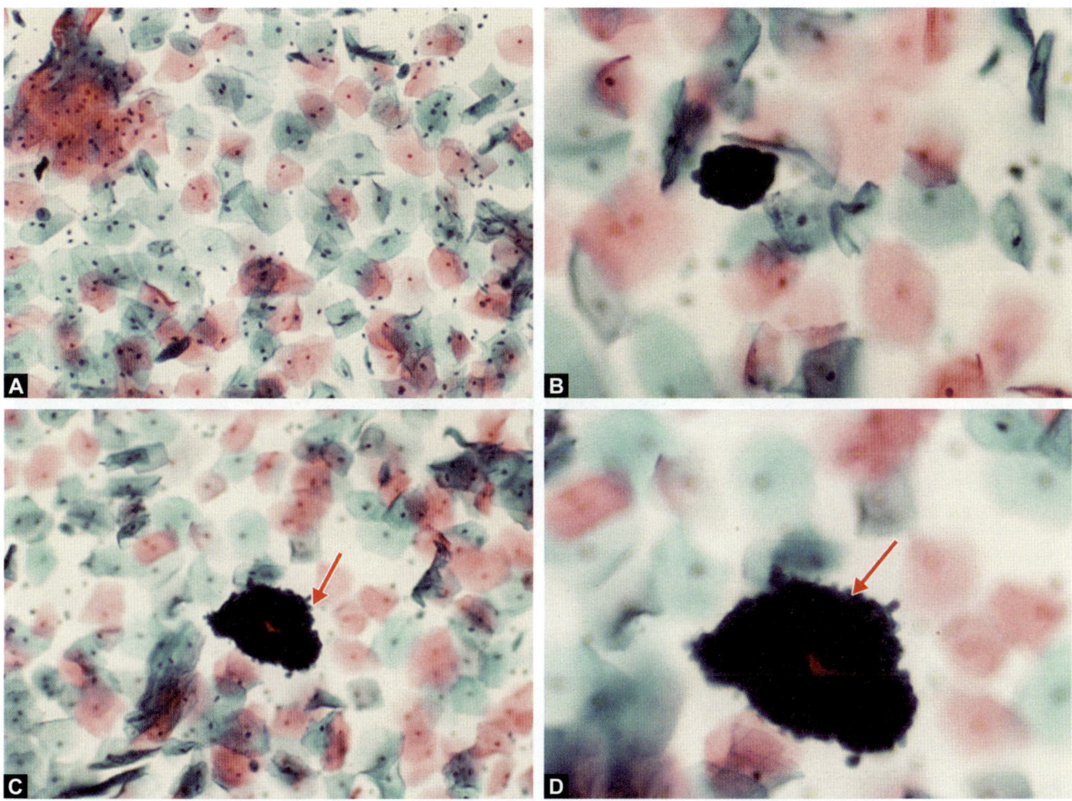

Figures 2.53A to D

Problem 2.54

A 57-year-old female has 25 cm diameter solid-cystic right adnexal swelling. Her CA-125 level is 865 IU/mL. Routine cervical smear was done. The most likely diagnosis from the microphotographs of SurePath preparation (Fig. 2.54) is:

A. Endometrial cells present in the post-menopausal patient
B. AGCUS
C. Metastatic adenocarcinoma from ovary
D. HSIL

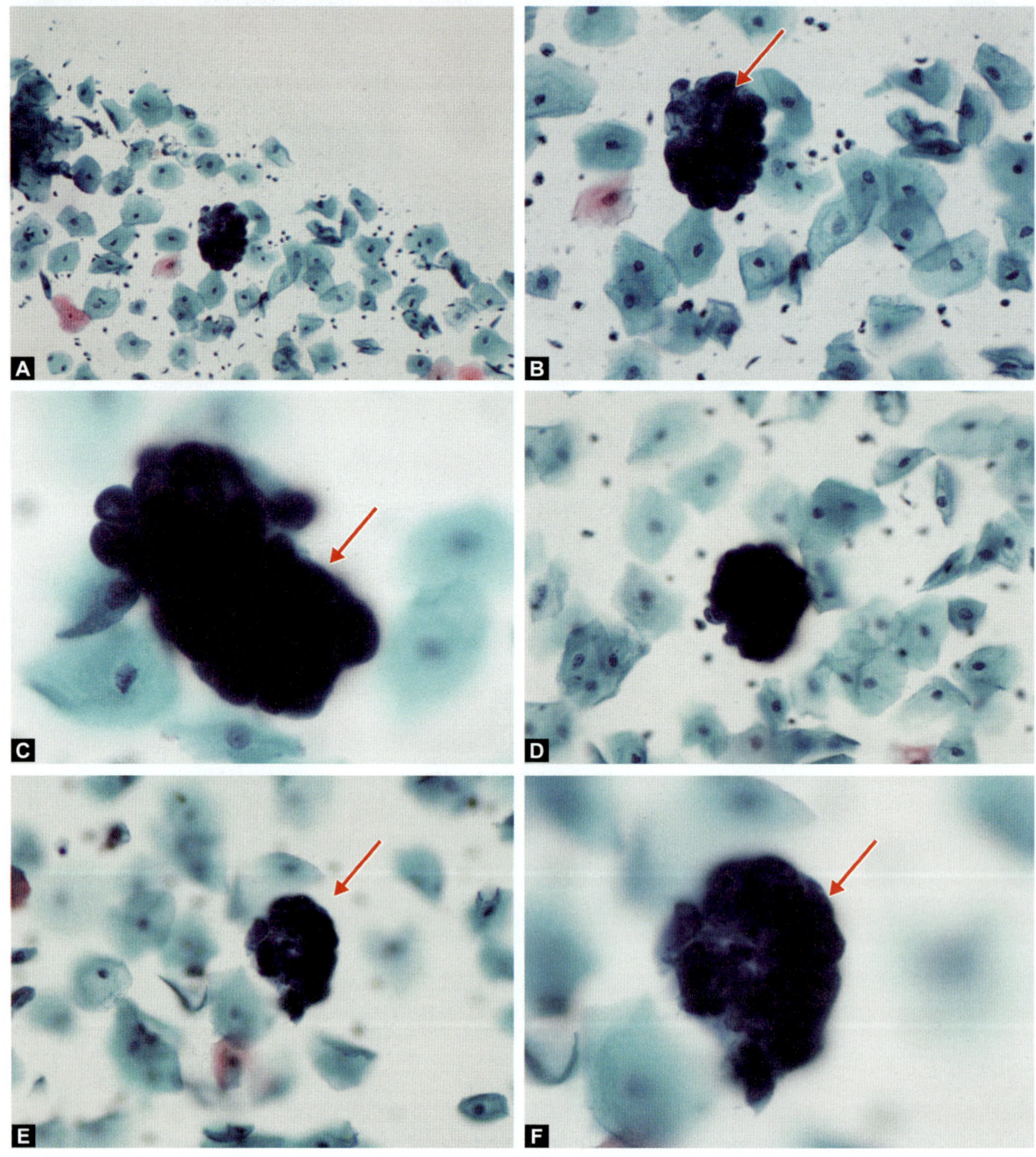

Figures 2.54A to F

Problem 2.55

Routine cervical cytological examination was done in a 51-year-old female. The most likely diagnosis from the SurePath preparation (Figs 2.55A to F) is:

A. LSIL
B. HSIL
C. Atypical glandular cell
D. Endometrial cell present

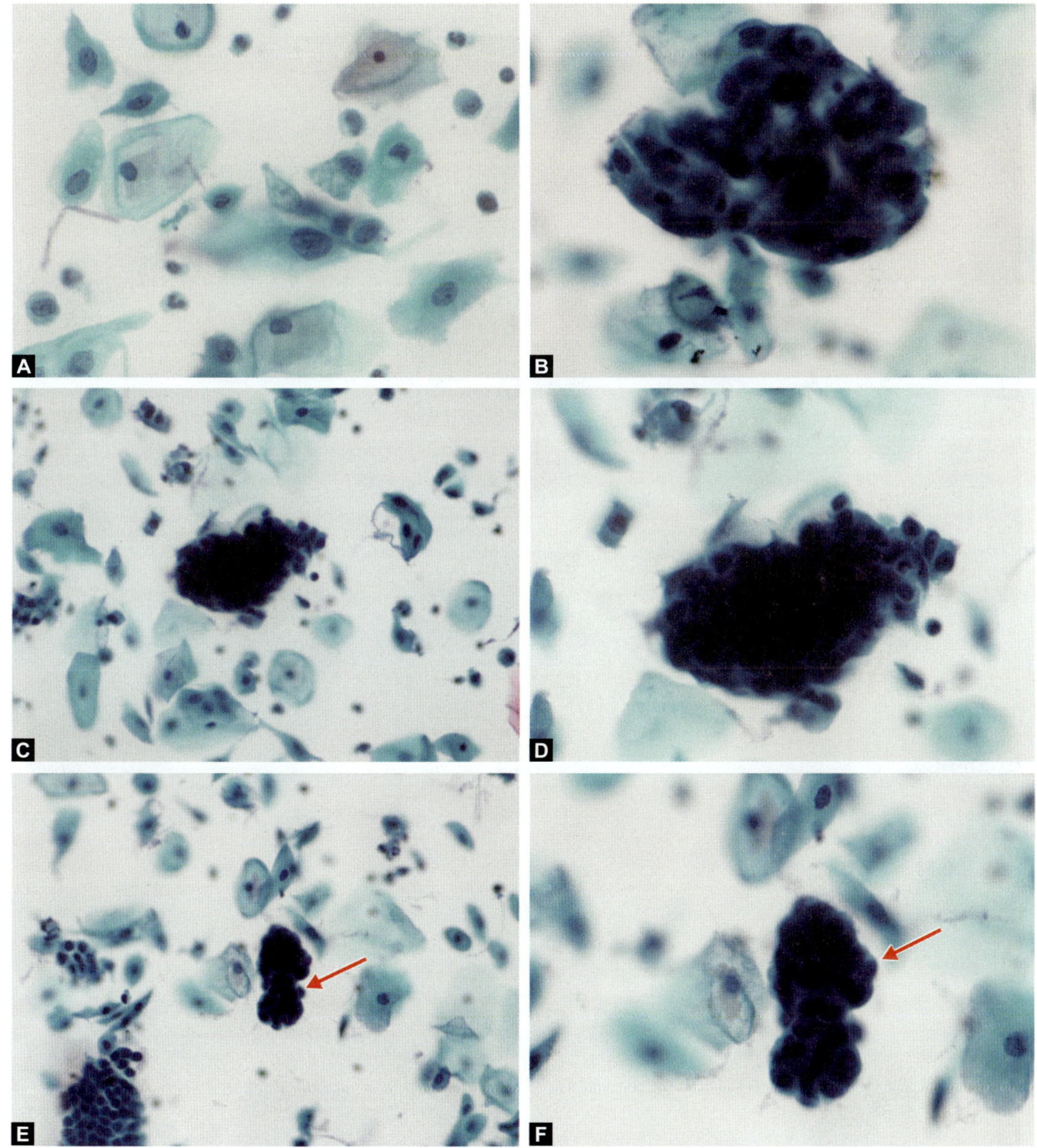

Figures 2.55A to F

Problem 2.56

A 55-year-old female had huge swelling in right ovary. USG showed solid cystic growth in the right ovary. Her CA 125 level is 500 IU/mL. Cervical cytology smear was done. What is the most likely interpretation from this SurePath smear (Fig. 2.56)?

A. AGCUS
B. Adenocarcinoma
C. HSIL
D. Benign looking endometrial cells in postmenopausal patient

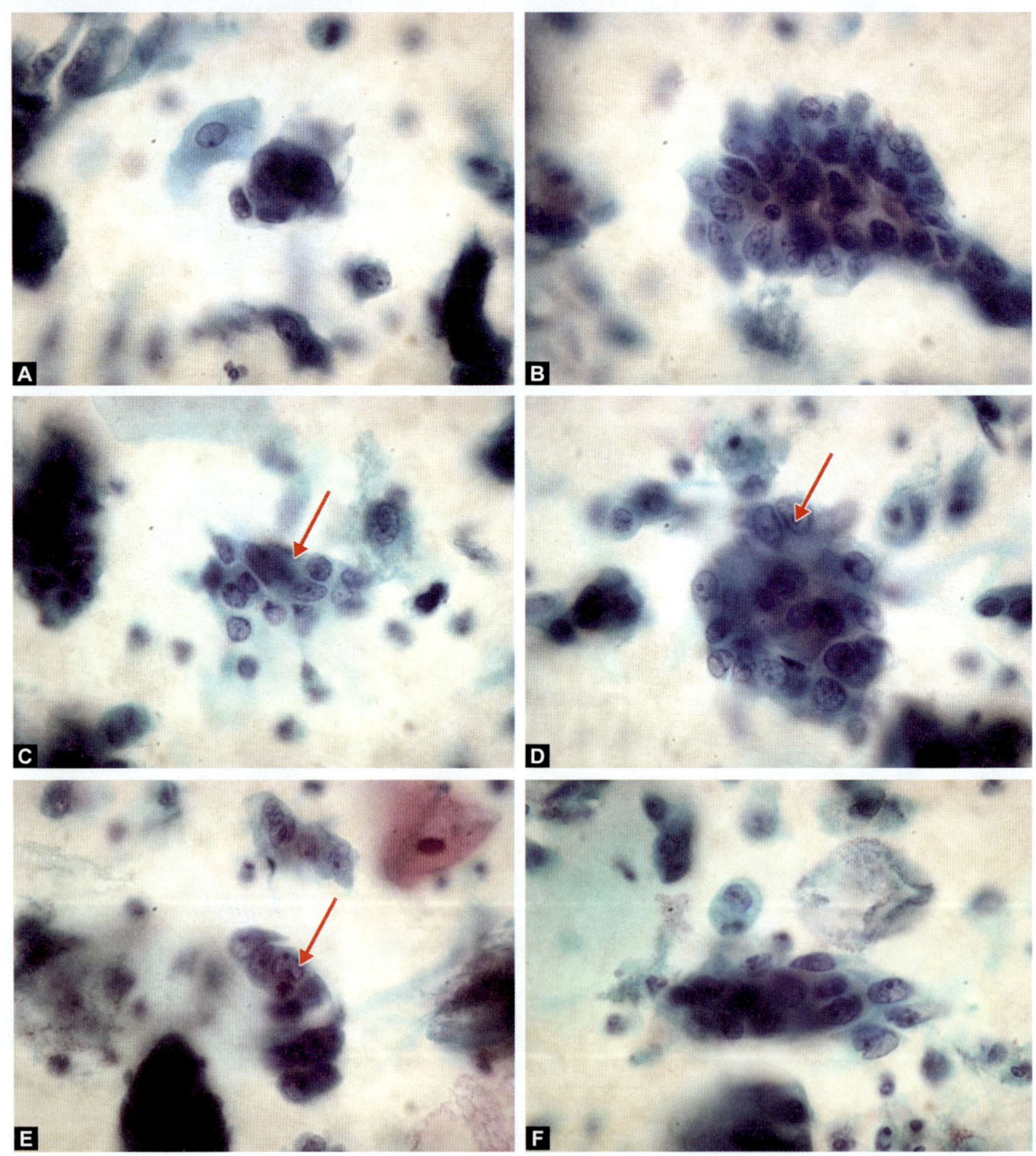

Figures 2.56A to F

Problem 2.57

Routine cervical smear in a 37-year-old female. The most likely diagnosis from Figure 2.57 is:

A. ASCUS
B. ASC-H
C. LSIL
D. HSIL

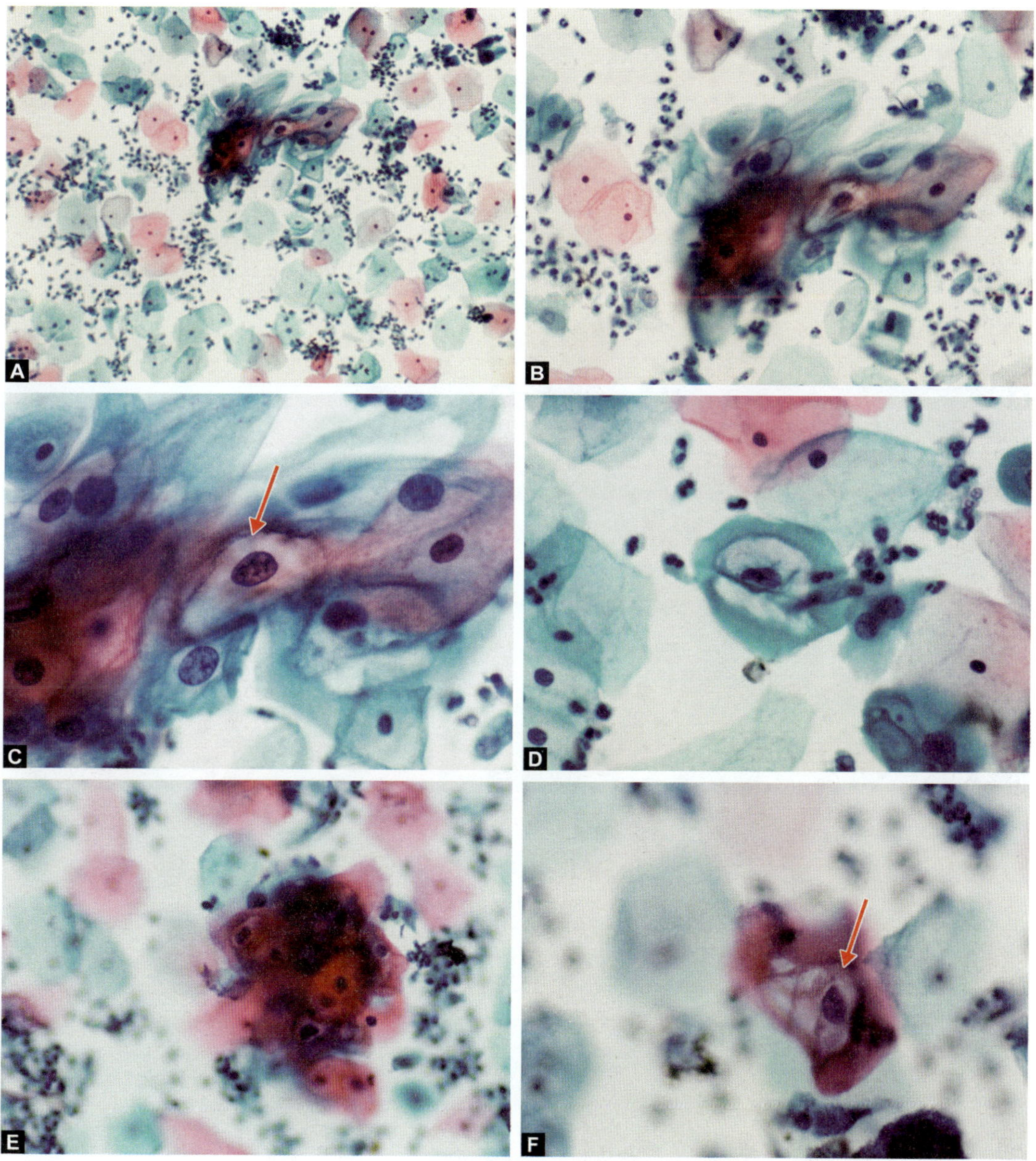

Figures 2.57A to F

Problem 2.58

A 32-year-old female ulcerated cervical lesion. The most likely diagnosis from the SurePath preparation (Fig. 2.58) is:

A. LSIL
B. LSIL (koilocytes) with Herpes simplex viral infection
C. Herpes simplex viral infection
D. ASCUS

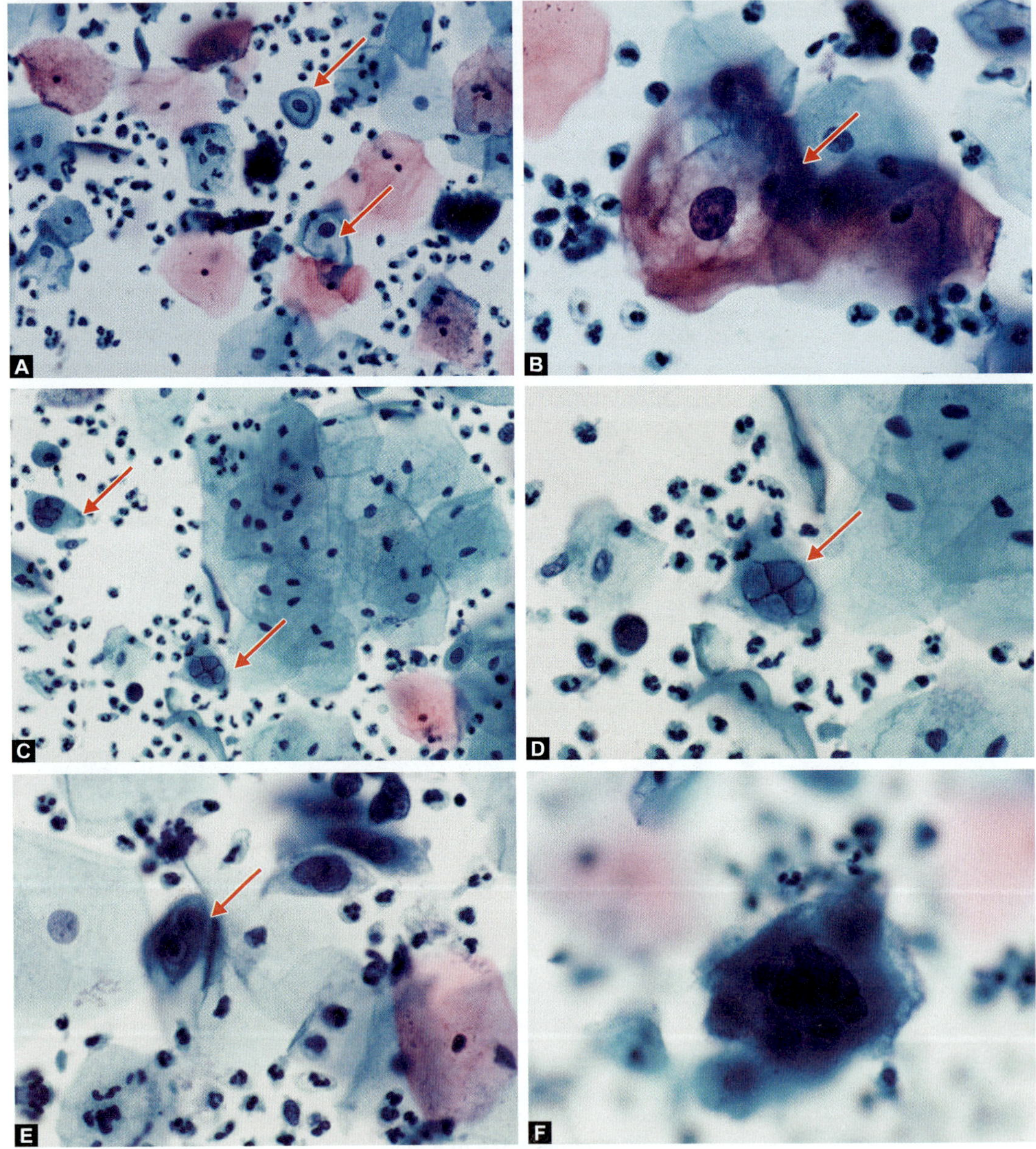

Figures 2.58A to F

ANSWERS

Problem 2.1

A.	6. Parabasal and basal cells
B.	5. Endocervical cells
C.	2. Superficial squamous cells
D.	3. Intermediate cells
E.	1. Metaplastic cells
F.	4. Endometrial cells

Discussion

A. **Parabasal and basal cells:** These are small round to oval cells with dense cytoplasm and relatively large nucleus. Nuclei of the cells are round with condensed chromatin. Basal cells have almost the same appearance as that of parabasal cells except they are small in size. Parabasal cells are frequently seen in atrophic smear of post-menopausal patient.

B. **Endocervical cells:** These cells are small round to columnar shaped with vacuolated cytoplasm and basally placed nuclei. The diameter of endocervical cell is about 18–20 μ and nucleus is 8–10 μ. The nucleus is vesicular with finely granular chromatin having occasional small nucleoli. These cells are usually present in tight honey comb shaped clusters. On side view, the cells may give the appearance of picket fence.

C. **Superficial squamous cells:** These are large polyhedral cells with abundant thin orangeophilic cytoplasm and centrally placed small pyknotic nuclei. The diameter of the superficial cell is usually 30–45 μ and nucleus is about 5–7 μ.

D. **Intermediate cells:** These are large polyhedral cells with abundant pale green cytoplasm having centrally placed vesicular nucleus. Nucleus is the main differentiating point between superficial and intermediate cell. The nucleus of the intermediate cell is vesicular with fine granular chromatin. Nuclear groove may be seen in intermediate cells as deep longitudinal fold. Intermediate cells with peripheral boat like folding are known as navicular cells.

E. **Metaplastic cells:** Metaplasia is the most common change of endocervical cells. The metaplastic cells are composed of immature parabasal cell. The cells are arranged in interlocking manner and one cell fits within another. This gives a mosaic like pattern. The cells of squamous metaplasia are relatively small with scanty cytoplasm. The nuclei are enlarged with slightly irregular contour and mildly hyperchromatic.

F. **Endometrial cells:** Endometrial cells are usually arranged in small tight spherical clusters. Individual cells are small round shaped with round dark nuclei. The nuclei often show molding. Nucleoli are not seen in endometrial cells. Cytoplasm of the cells is scanty and occasionally may show vacuolation.

Further Reading

1. Dey P. Diagnostic Cytology. Jaypee Brothers Medical Publishers (P) Ltd. New Delhi, 2014.

Problem 2.2

Unsatisfactory for evaluation for intraepithelial lesion or malignancy due to obscuration by inflammation.

Discussion

The Bethesda system 2001 classified specimen adequacy in two part:

Satisfactory for Evaluation

- Adequate number of squamous cells:
  - Conventional smear: 8000 to 12000 squamous cells
  - Liquid based cytology: 5000 squamous cells
- Mention the presence or absence of endocervical cells or transformation zone components: at least 10 endocervical or metaplastic cells should be present, no need for cell clusters.
- Any other quality elements such as drying artifact, obscured by blood or inflammation etc. Partially obscured: 50 to 75% epithelial cells cannot be visualized, Unsatisfactory: when more than 75% epithelial cells cannot be visualized.

Unsatisfactory for Evaluation

- Specimen rejected: Specimen rejected due to broken slide or mislabelled specimen
- Fully evaluated unsatisfactory specimen: Specimen processed but unsatisfactory because:
  - Insufficient squamous cells
  - More than 75% of smear is obscured by blood or inflammatory cells.

The unsatisfactory specimens also should be screened properly as these smears may also show the presence of endometrial cells in post-menopausal women or organisms. The simple presence of benign endometrial cells does not fulfil the criteria of adequacy.

In the index case the smear is totally obscured by inflammatory cells and therefore unsatisfactory for evaluation. In this situation the cervical smear should be repeated.

Problem 2.3

Satisfactory for Evaluation

Interpretation: Negative for intraepithelial lesion or malignancy.

Discussion

The Bethesda system (TBS) 2001 classified adequacy criteria as mentioned in problem 2.2:

In case of specimen adequacy, TBS 2001 maintains only "satisfactory for evaluation" and "unsatisfactory" categories and eliminates "satisfactory but limited" category.

TBS 2001 provides the quantitative numerical estimate for the adequacy of squamous cells. An adequate conventional smear should contain approximately 8000 to 12000 well-preserved and well visualized squamous cells. The endocervical cells and inflammatory cells should not be included in the estimation of cellularity. However, the metaplastic squamous cells can be included during the assessment of cellularity for adequacy.

An adequate LBC preparation should contain 5000 well visualized squamous cells. This cellularity can be assessed by observing 10 microscopic filed of 40 X magnification.

Endocervical Component

The smear should contain at least 10 well preserved endocervical or squamous metaplastic cells to represent the transformation zone. These criteria should be applied both in LBC and conventional smear. The presence of mucus or parabasal cells should not be considered in the assessment of transformation zone. In each smear, the presence of transformation zone component should be mentioned. The absence of transformation zone component is not important in case of HSIL or hysterectomy specimen.

Further Reading

1. Solomon D, Davey D, Kurman R, et al. for the Forum Group Members and the Bethesda 2001 Workshop. The 2001Bethesda system—terminology for reporting results of cervical cytology. JAMA. 2002;287:2114-9.

Problem 2.4

B. Inflammatory smear

Discussion

The criteria of the diagnosis of inflammatory smear include:

- Polymorphonuclear leucocytes infiltration
- Necrotic debris
- Inflammatory cells stick to the epithelial cells
- Increased number of parabasal cells
- Eosinophilia of the cell cytoplasm
- Degenerative changes
  - *Cytoplasm*: Vacuoles
  - *Nucleus*: Pyknosis, karyorrhexis, karyolysis.

In the index case the smear shows adherence of polymorphs with the squamous cells and nuclear changes of inflammation such as perinuclear halo, nuclear pyknosis and karyorrhexis.

Atrophic smear shows good number of parabasal/basal cells. There is no nuclear atypia to suggest the diagnosis of ASC-US. The diagnosis of *Trichomonas vaginalis* is excluded in absence of any such organism.

Further Reading

1. Dey P. Diagnostic Cytology. Jaypee Brothers Medical Publishers (P) Ltd., New Delhi 2014.

Problem 2.5

Satisfactory for Evaluation

Interpretation: Negative for intraepithelial lesion or malignancy.
Other non-neoplastic finding: Atrophic smear.

Discussion

The criteria of atrophic cervical smear include:

- Predominant parabasal and basal cell population. The cells are usually present in flat monolayer sheets.

- In liquid based cytology (LBC) preparation the parabasal cells may be discrete
- The parabasal cells show generalized nuclear enlargement
- The nuclei may be 3 to 5 times of the area of the intermediate cells
- Mild hyperchromasia of the nuclei
- Uniformly distributed chromatin
- "Blue blobs": Dark blue round shaped amorphous material
- Drying artifact with cytoplasmic eosinophilia and uniform nuclear enlargement
- Elongated spindle shaped cells may be confused with malignancy
- In addition the smear may show degenerated orangeophilic or eosinophilic parabasal cells and histiocytes.

The index case shows abundant parabasal and basal cells with mild nuclear enlargement and hyperchromasia.

Problem 2.6

Satisfactory for Evaluation

Interpretation: Negative for intraepithelial lesion or malignancy.

Organism: Clue cells present, suggestive of bacterial vaginosis.

Discussion

Bacterial vaginosis is a fairly common infection. The patients may be asymptomatic or may complaint of thin milky vaginal discharge with foul fishy smell.

The characteristic features of bacterial vaginosis include:

- Filmy background
- Absence of lactobacilli
- Numerous small coccobacilli and curved bacilli and
- *Clue cells*: Polygonal squamous cells filled with small coccobacilli.

In LBC preparation the filmy background may be missing. Clue cells are helpful diagnostic feature of BV; however, they are not specific for BV.

In the index case, there are multiple clue cells and scanty polymorphs. The background is relatively clear compared to conventional cervical cytology smear.

Problem 2.7

Satisfactory for evaluation

Interpretation: Negative for intraepithelial lesion or malignancy.

Organism: Fungal organism morphologically consistent with *Candida* species.

Discussion

The diagnostic criteria of *Candida* include:

- Candidal pseudohyphae formed by long structures along with constrictions.
- Budding yeasts: The yeasts are 3–5 μ in diameter
- The organisms may stick with the polymorphs.

Occasionally the squamous cells may show mild nuclear atypia and perinuclear halo. These cells may simulate atypical cells.

The index case shows typical candidal pseudohyphae.

Problem 2.8

B. Leptothrix

Discussion

Leptothrix is a type of *Lactobacillus* and is a gram-negative, non-spore forming anaerobic organism. They are thin, segmented, long filamentous organism. *Leptothrix* are commonly associated with Trichomonas infection.

Problem 2.9

D. *Trichomonas vaginalis*

Discussion

The smears show relatively dirty background with many green colored pear shaped organisms. Each organism has a central nucleus and scanty cytoplasm.

Trichomonas vaginalis (TV) is the most common infection in the female genital tract. The patient usually complains of burning sensation, severe itching, and foul smelling vaginal discharge. Considerable number of patients (up to 50%) may be asymptomatic.

TV is characterized by:

Morphology of TV

- 15–30 μ long
- Round-to-oval "pear-shaped" organisms
- Indistinct vesicular nucleus: important feature
- Red cytoplasmic granules

Background

- Dirty, fuzzy
- "Cannon ball" or "pus ball": Aggregates of neutrophils around the surface of the squamous cells
- Associated with *Leptothrix* (80% cases).

TV can be mistaken as cytoplasmic fragments and vice versa. However the presence of nucleus indicates it as TV. Occasionally parabasal cells may be mistaken as TV. The cytoplasm of the parabasal cells are much more and thick in nature.

Further Reading

1. Dey P. Diagnostic Cytology. Jaypee Brother Medical Publishers (P) Ltd. New Delhi. 2014.

Problem 2.10

C. LSIL

Discussion

Description: The smear shows many discrete superficial and intermediate cells. The nuclei of the intermediate cells are enlarged (more than 3 times of normal intermediate cells), hyperchromatic and mildly pleomorphic.

Disease description: LSIL is caused by both low- and high-risk HPV. However, the majority of LSIL is caused by low-risk HPV 6 and 11. The large majority of the LSIL lesions (60%) regress over the course of time and only 1% of LSIL cases may progress to invasive cancer if left untreated.

The cytological diagnostic criteria of LSIL are:

- Single or sheets of superficial and intermediate cells
- Large cells with abundant cytoplasm. Cytoplasmic differentiation is evident
- Well defined cytoplasmic margin of the atypical cells
- Enlarged nucleus: Nuclear area is more than three times of that a normal intermediate cell nucleus
- Mildly increased nucleocytoplasmic (N/C) ratio
- Mild hyperchromatic nuclei
- Variably irregular nuclear margin
- Chromatin is usually evenly distributed coarsely granular
- Nucleoli are commonly absent or inconspicuous
- Koilocytosis: Distinct perinuclear cytoplasmic vacuoles with sharp condensed thick cytoplasmic margin. Nuclei are enlarged. The perinuclear halos along with the nuclear abnormalities are essential for the qualification of koilocytosis to diagnose LSIL. The presence of koilocytosis is the characteristic feature of LSIL but it is not an essential feature to interpret LSIL.

In LBC preparation the atypical cells may not show significant hyperchromasia. The single discrete atypical cells are easily identifiable in LBC preparation.

Differential diagnosis of LSIL:

- Reactive inflammatory changes in squamous epithelial cells: Nuclear enlargement is often seen in squamous cells due to the effect of inflammation. However other nuclear abnormalities are not noted in these cells.
- Air drying artefact: Air drying may cause nuclear enlargement and may be mistaken as LSIL. Loss of nuclear detail is an indicator of the air drying artefact.
- Nonspecific perinuclear halos: Nonspecific perinuclear halos present in inflammation or trichomonas infection may not have any nuclear atypia.
- Reactive changes in endocervical cells: Reactive endocervical cells may have enlarged nucleus and polygonal shape. These cells may mimic LSIL. However, accompanying more recognizable endocervical cells may indicate their character.
- ASC-US: The distinguishing features of ASC-US and LSIL are both qualitative and quantitative.

Management of LSIL cases:

- Colposcopy is recommended in all LSIL cases.
- If there is no lesion in colposcopy, unsatisfactory colposcopy, or the patient is not pregnant then endocervical sampling is preferred.

- If colposcopy does not show CIN 2 or 3 then HPVDNA testing or repeat cervical cytology should be done.
- If the patient shows high risk HPV DNA or atypical squamous cell in cytology smear, then a repeat colposcopy is advised.
- If colposcopic guided biopsy shows CIN 2 or 3, then surgical ablation could be done.
- In post-menopausal or adolescent patients colposcopy could be deferred and repeat cytology testing at 6 monthly intervals can be done.

Further Reading

1. Wright TC Jr, Cox JT, Massad LS, Twiggs LB, Wilkinson EJ; ASCCP-Sponsored Consensus Conference. 2001 Consensus Guidelines for the management of women with cervical cytological abnormalities. JAMA. 2002;287(16):2120-9.
2. Dey P. Diagnostic Cytology. Jaypee Brothers Medical Publishers (P) Ltd. New Delhi, 2014.

Problem 2.11

Satisfactory for Evaluation

B. LSIL

Discussion

Description: The smear shows many scattered discrete cells with centrally placed enlarged nuclei having perinuclear haloes. The margin of the cytoplasm near the halo is thick. These cells are koilocytotic cells. In addition occasional multinucleated cell is also seen.

Diagnostic feature of LSIL in this index case:

- Many koilocytotic cells with nuclear enlargement and perinuclear halo
- Many bi and multinucleated cells
- The nuclei of the cells are enlarged and more than three times of the area of normal intermediate cells
- Mildly increased N/C ratio
- Mild hyperchromasia
- Evenly distributed coarse granular chromatin.

The presence of koilocytosis along with bi and multinucleated cells indicate the features of HPV infection.

Differential Diagnosis: See problem 2.10.

Problem 2.12

A.	3. Koilocyte like cell due to glycogen
B.	2. Koilocyte like cell in inflammation
C.	1. True koilocytotic cell

Discussion

Figure 2.12A shows false koilocytotic cell. This is a glycogen rich squamous cell. There is no nuclear abnormality in this cell.

Figure 2.12B is also a false koilocytotic cell. The perinuclear halo in this cell is due to inflammatory change. The nuclei are normal in the cell.

Figure 2.12C is a true koilocytotic cell. The cell shows enlarged atypical nuclei with clear halo around nucleus. The cytoplasmic margin of the cell is also thick.

It is important to identify the true koilocytotic cell as the presence of koilocytotic cell indicates the diagnosis of LSIL. The false koilocytotic cell is seen in case of inflammation, trichomonas vaginalis infection, and glycogenated squamous cells. The following features help to identify true koilocytotic cell:

- Enlarged atypical nucleus
- Thick cytoplasmic margin of the halo
- The presence of other abnormal cells.

Problem 2.13

A. ASCUS

Discussion

Description: The cytology smears show discrete intermediate squamous cells with enlarged (2.5 times than normal intermediate cell nuclei) mildly hyperchromatic nuclei. Nuclear margin of the cell is regular.

Disease description: Atypical squamous cells (ASC) are classified as:

- Atypical squamous cells of undetermined significance (ASC-US)
- Atypical squamous cell, cannot exclude high-grade squamous intraepithelial lesion (ASC-H).

The term ASCUS is applied when cytology smear is suggestive of SIL but there are insufficient quantitative or qualitative evidences to have a definitive interpretation of SIL. ASCUS represents an equivocal case and it is a diagnosis of exclusion. The interpretation of ASCUS is subjective and depends on pathologist's diagnostic threshold. ASC-US represents changes that are more likely to have LSIL or SIL, difficult to classify. It has been shown that 10 to 20% of ASC-US cases show either LSIL or HSIL.[1]

The criteria of ASC-US include[2]:

- Nuclear atypia in the squamous cells: The nuclear area of the atypical cells is 2.5 to 3 times more than the nucleus of the normal intermediate cell
- Mildly increased nucleo-cytoplasmic ratio
- Nuclear hyperchromasia is minimal
- Regular margin with fine granular chromatin.

Diagnostic criteria of the index case: The index case shows mild nuclear enlargement, hyperchromasia and regular nuclear margin. The chromatin is fine granular.

The management of ASCUS include:

- Repeat cervical cytology and HPV DNA testing are recommended
- If the patient is positive for high-risk HPVDNA then colposcopy is recommended
- If the woman is negative for high-risk HPVDNA then repeat cervical cytology at 6 monthly intervals should be done

- If two successive Pap test are negative then the patient can return to routine screening schedule
- If any of the two Pap test shows ASCUS or greater, then colposcopy is recommended.

References

1. The ALTS Group: Results of a randomized trial on the management of cytology interpretations of atypical squamous cells of undetermined significance. Am J Obstet Gynecol. 2003;183:1383-92.
2. Schiffman M, Solomon D. Findings to date from the ASCUS-LSIL Triage Study (ALTS). Arch Pathol Lab Med. 2003;127(8):946-9.

Problem 2.14

B. LSIL

Discussion

Description: The smear shows many discrete intermediate cells with more than 3 times enlarged nuclei than normal intermediate cells. The nuclei are hyperchromatic with regular margin and mildly pleomorphic.

Diagnostic points in this index case: The following features suggest the interpretation of LSIL:

- Discrete atypical superficial and intermediate cells
- Abundant cytoplasm
- Mildly enlarged nucleus: Nucleus occupy more than 1/3rd of the cytoplasmic area (3 to 5 times of a normal intermediate cell nucleus)
- Mild hyperchromasia
- Evenly distributed fine granular chromatin
- Mildly increased N/C ratio.

Differential Diagnosis: See problem 2.10

Problem 2.15

A. ASCUS

Discussion

Description: The smears show many discrete intermediate cell with mildly hyperchromatic enlarged nuclei (2.5 times than normal intermediate cells). The nuclear margin is regular and chromatin is homogenous.

Diagnostic Features: The index case shows following diagnostic features of ASCUS:

- Squamous differentiation of atypical cells
- Mildly increased nucleo-cytoplasmic ratio
- Mild nuclear enlargement, hyperchromasia and regular nuclear margin.

In this present case the occasional cells show perinuclear clearing, however there is no true koilocytotic cells. The cases of ASCUS should be distinguished from the inflammatory changes, air drying artefact resulting in nuclear enlargement, and degenerative changes.

Further Reading

1. Dey P. Diagnostic Cytology. Jaypee Brothers Medical Publishers (P) Ltd. New Delhi, 2014.

Problem 2.16

D. HSIL

Discussion

Description: The cytology smear shows many discrete and also loose clusters of intermediate and parabasal cells. The cells show large round nuclei occupying more than 2/3rd of the cytoplasm. The nuclei are moderately pleomorphic with coarsely clumped chromatin.

Description of the disease: HSIL in Bethesda system encompasses CIN 2 (moderate dysplasia) and CIN 3 (severe dysplasia/carcinoma in situ). Approximately 0.5 % of all cervical smears show HSIL. 97% HSIL cases are associated with high-risk HPV infection.[1,2] The diagnostic criteria of HSIL include:

- Single, sheets or syncytial aggregates of cells
- Parabasal or basal types of cells are commonly seen. These cells are smaller and less mature
- Cytoplasm is dense or lacy and orangeophilic in keratinized variant
- Enlarged nuclei with marked increased N/C ratio
- Usually coarsely granular chromatin, but it may be fine granular
- Marked irregular nuclear membrane with frequent indentation
- Nucleoli is commonly absent. However, the dysplastic cells may show prominent nucleoli if the endocervical glands are involved by the dysplastic cells.

Diagnostic criteria in the index case: The following features favor the diagnosis of HSIL:

- Small immature atypical cells
- High N/C ratio
- Moderate nuclear pleomorphism
- Coarse granular chromatin
- Marked nuclear margin irregularity.

Differential Diagnosis:

- **Atrophic smear:** Small parabasal cells with increased N/C ratio in atrophic smear may simulate HSIL at times. However, coarse granular nuclear chromatin and irregular nuclear margin in HSIL helps to distinguish it from atrophic smear.
- **Squamous metaplastic cells:** Squamous metaplastic cells with relatively large nucleus and dense chromatin may simulate HSIL. Metaplastic cells have regular chromatin pattern compared to coarse granular chromatin in HSIL.
- **Histiocytes:** Mononuclear histiocytes with enlarged nucleus may also simulate HSIL. However nuclei of histiocytes have fine chromatin texture.
- **ASC-H:** The presence of atypical squamous cells with nuclear enlargement and pleomorphism simulate HSIL. The qualitative and quantitative changes in ASC-H are less compared to HSIL.
- **Endometrial cells:** At times, clusters of endometrial cells may be difficult to distinguish from HSIL. However endometrial cells are always present in tight cohesive well circumscribed clusters and individual cells are smaller than that of HSIL.

Management of HSIL

- Colposcopy with endocervical assessment
- If colposcopy confirms CIN 2/3, then surgical excision is done
- If colposcopy is negative then either a diagnostic excisional biopsy or repeat Papanicolaou's smear should be done.

References

1. Solomon D, Schiffman M, Tarone R. Comparison of three management strategies for patients with atypical squamous cells of undetermined significance: Baseline results from a randomized trial. J Natl Cancer Inst. 2001;93:293-9.
2. Dey P. Diagnostic Cytology. Jaypee Brothers Medical Publshers (P) Ltd. New Delhi. 2014.

Problem 2.17

C. LSIL

Discussion

Description: The smears show occasional intermediate cells with hyperchromatic enlarged nuclei (4 times more than normal intermediate nucleus). In addition superficial squamous cells also show nuclear enlargement and pleomorphism.

Diagnostic Features: In this index case the following features indicate the interpretation of LSIL:

- Discrete atypical superficial and intermediate cells
- Cells show moderate amount of cytoplasm
- Mildly increased nucleocytoplasmic (N/C) ratio
- Mildly enlarged nucleus: Nucleus occupy more than 1/3rd of the cytoplasmic area (3 to 5 times of a normal intermediate cell nucleus)
- Mild hyperchromasia with evenly distributed fine granular chromatin.

Differential Diagnosis: See problem 2.10.

Problem 2.18

D. HSIL

Discussion

Description: The cytology smear shows many discrete parabasal and intermediate cells with markedly enlarged nuclei occupying more than 2/3rd of the cytoplasmic area. The nuclei show hyperchromasia, irregular margin and moderate pleomorphism.

Diagnostic criteria in the index case: The features suggestive of HSIL are:

- Parabasal or basal type of atypical cells
- High nucleo-cytoplasmic ratio
- Nucleus:
  - Moderately pleomorphic
  - Hyperchromatic
  - Irregular nuclear margin
  - Coarse chromatin.

Differential diagnosis and management: Discussed in problem 2.16.

Problem 2.19

A. ASCUS

Discussion

Description: The smear shows occasional large intermediate and superficial squamous cells with enlarged nuclei (more than 3 times of the normal intermediate cells). The nuclei show mild pleomorphism. Chromatin is homogenous and granular.

Diagnostic Features: The index case shows following features that favor the interpretation of ASCUS:

- Squamous differentiation of the atypical cells
- Increased nucleo-cytoplasmic ratio of the atypical cells
- The area of the nuclei more than 2.5 times of the normal intermediate cells
- Mild hyperchromasia, regular nuclear margin and fine granular chromatin.

Further Reading

1. The ALTS Group: Results of a randomized trial on the management of cytology interpretations of atypical squamous cells of undetermined significance. Am J Obstet Gynecol. 2003;183:1383-92.

Problem 2.20

D. HSIL

Discussion

Description: The cytology smear shows abundant discrete and loose clusters of parabasal and intermediate squamous cells with markedly enlarged nuclei occupying more than 2/3rd of the cytoplasmic area. The individual cells show hyperchromatic nuclei, moderate nuclear pleomorphism and irregular nuclear margin.

Diagnostic features in the index case: The following features favor the diagnosis of HSIL:

- Clusters and discrete abundant atypical cells
- High N/C ratio
- Moderate to marked nuclear enlargement
- Moderate nuclear pleomorphism
- Coarse granular chromatin
- Marked nuclear margin irregularity.

Differential Diagnosis:

- LSIL: Differentiating features between LSIL and HSIL have been highlighted in the Table 2.1.

Table 2.1: Comparison of LSIL and HSIL

Cytological features	*LSIL*	*HSIL*
Cell arrangement	Singly and in clusters	Singly and in clusters
Cell type	Superficial and intermediate	Parabasal, basal and metaplastic
Koilocytes	Present	Usually absent

Contd...

Contd...

Cytological features	*LSIL*	*HSIL*
Cell cytoplasm	More	Less
N/C ratio	Low	High
Nucleus		
Size	Mild enlargement	Moderate enlargement
Pleomorphism	Mild	Moderate
Contour	Mildly irregular	Moderately irregular
Hyperchromasia	Mild	Moderate
Chromatin	Fine	Coarse
Nucleoli	Absent	Present
Mitosis	Rare	Frequent

- **Atrophic smear:** Discussed in problem 2.16
- **Squamous metaplastic cells:** Discussed in problem 2.16
- **ASC-H:** Discussed in problem 2.16

Endometrial cells: Discussed in problem 2.16

Problem 2.21

B. ASC-H

Discussion

Description: The cytology smear shows many discrete and loose clusters of parabasal cells. The cells have enlarged nuclei more than 2/3rd of the cytoplasmic area. The nuclei are hyperchromatic and moderately pleomorphic.

Disease Description: The Bethesda system (2001) classified atypical squamous cells (ASC) as:

- Atypical squamous cells of undetermined significance (ASC-US).
- Atypical squamous cell, cannot exclude high-grade squamous intraepithelial lesion (ASC-H).

  ASC represents cytological changes suggestive of squamous intraepithelial lesion (SIL), however, the changes are qualitatively or quantitatively insufficient to provide a definitive interpretation of SIL.

  ASC-H cases are uncommon and consist of 5–10% cases of ASC [1]. The follow up of ASC-H cases show HSIL in 30 to 40% cases. The characteristic cytological features of ASC-H are[2]:

- Single and loose cohesive group of cells.
- Immature small squamous cells.
- Nucleus: Mild to moderate enlargement, hyperchromasia and pleomorphism. Nucleoli are usually absent.
- Nuclear size is 1.5 to 2.5 times larger than normal metaplastic cells.
- Increased nucleocytoplasmic ratio.

Diagnostic features in the index case: In the index case the diagnostic features for ASC-H are: squamoid differentiation, enlarged nuclei, mild hyperchromasia and pleomorphism.

Management of ASC-H:

- Colposcopy is always recommended in ASC-H patient
- If the follow up histology of the ASC-H case does not reveal CIN 2 or 3 then repeat cervical smear or HPVDNA test for high-risk viruses should be done. ASC-H is associated with higher risk of oncogenic HPV infection than ASC-US
- If the patient shows ASC-H in repeat smear or high-risk HPVDNA positive then repeat colposcopy is advised.

References

1. Davey DD, Neal MH, Wilbur DC, et al. Bethesda 2001 implementation and reporting rates: 2003 practices of participants in the College of American Pathologists Interlaboratory Comparison Program in Cervicovaginal Cytology. Arch Pathol Lab Med. 2004;128(11):1224-9.
2. Chivukula M, Shidham VB. ASC-H in Pap test—definitive categorization of cytomorphological spectrum. Cytojournal. 2006;3:14.

Problem 2.22

D. Atrophic smear.

Discussion

Description: The cytology smears show abundant parabasal, basal cells and intermediate cell. The cells show mild nuclear enlargement and hyperchromasia. The background shows degenerated cells, nuclear thread and polymorphs.

Diagnostic features in the index case: The following features favor the interpretation of atrophic smear in this index case:

- Abundant parabasal and basal cells
- Many naked nuclei due to autolysis
- Mild nuclear enlargement of the parabasal cells
- Hyperchromasia
- Regular nuclear chromatin.

Differential Diagnosis: The atrophic smear may be mistaken as LSIL and even HSIL. This mistake may occur due to nuclear enlargement, and hyperchromasia. However, regular nuclear margin and evenly distributed chromatin in atrophic smear help to distinguish it from HSIL. Moreover, the N/C ratio of atrophic smear is less than 50% and that also helps to distinguish it from HSIL. The cells with high N/C ratio along with hyperchromasia and irregular nuclear margin should be designated as HSIL or ASC-H depending on the qualitative and quantitative changes.

Problem 2.23

D. HSIL

Discussion

Description: The smears show many discrete intermediate and parabasal cells with moderate nuclear enlargement (nuclear area occupying more than half of cytoplasmic area). The nuclei are hyperchromatic and moderately pleomorphic with irregular nuclear margin. Occasional koilocytes (Fig. 2.23E) are also seen.

Diagnostic Features: In the index case the following features favor the diagnosis of HSIL:

- Many discrete atypical cells
- Bi and multinucleation
- High N/C ratio
- Moderate to marked nuclear enlargement
- Moderate nuclear pleomorphism
- Coarse granular chromatin
- Marked nuclear margin irregularity.

Absence of tumor diathesis, fibre cells and tadpole cells exclude the possibility of squamous cell carcinoma.

Problem 2.24

D. Squamous cell carcinoma

Discussion

Description: The figures show abundant discrete and small clusters of cells with moderate to marked nuclear enlargement and pleomorphism. The nuclei are hyperchromatic, pleomorphic with irregular nuclear margin. In addition there are many fibers like cells present (Fig. 2.24G). Tumor diathesis (Fig. 2.24F) is also present.

Disease description: TBS 2001 does not further subclassify squamous cell carcinoma of cervix. The criteria of squamous cell carcinoma of cervix include:

- Scattered and small clusters of tumor cells
- Moderately pleomorphic cell with high N/C ratio
- Moderate to markedly pleomorphic nuclei
- Moderately hyperchromatic nuclei
- Irregular nuclear membrane
- Irregular coarse chromatin
- Occasionally macronucleoli may be seen
- Cytoplasm: Dense orangeophilic
- **Tadpole cells:** Elongated cell with large abnormal nuclei in one corner
- **Fibre cell:** Spindle shaped hyperchromatic nuclei
- **Tumor diathesis**: Dirty background with necrotic tissue fragments and RBCs.

Tumor diathesis is not always obvious in case of LBC preparation.

Diagnostic feature in this index case: The diagnostic features of the index case include markedly pleomorphic, hyperchromatic nuclei with high N/C ratio, fibre cells along with tumor diathesis.

Differential Diagnosis: The marked nuclear atypia and the presence of fibre cells and tumor diathesis exclude any possibility of HSIL or ASC-H. The nuclear abnormality is too severe and there is no possibility of atrophic smear pattern.

Problem 2.25

B. ASC-H

Description: The Figures 2.25A to F show many discrete and loose clusters of large cells with scanty cytoplasm and enlarged nuclei. The nuclei show high N/C ratio along with moderate hyperchromasia.

Diagnostic features in the index case: In the index case the diagnostic features for ASC-H are moderately enlarged nuclei, mild hyperchromasia and nuclear pleomorphism of the squamous cells.

Disease description and management: Please see problem 2.21.

Problem 2.26

D. Squamous cell carcinoma

Discussion

Description: The microphotographs (Figs 2.26A to H) show many discrete and loose clusters of large cells with moderate nuclear enlargement and pleomorphism. Discrete dyskeratotic squamous cells (Fig. 2.26C) are also seen. Figure 2.26H suggests the evidence of tumor diathesis.

Diagnostic criteria: The following features suggest the possibility of squamous cell carcinoma:
- Many moderately pleomorphic cells with nuclear atypia
- Dyskeratotic cells
- Suggestion of tumor diathesis.

The histopathology of the colposcopic guided biopsy showed squamous cell carcinoma (Fig. 2.59).

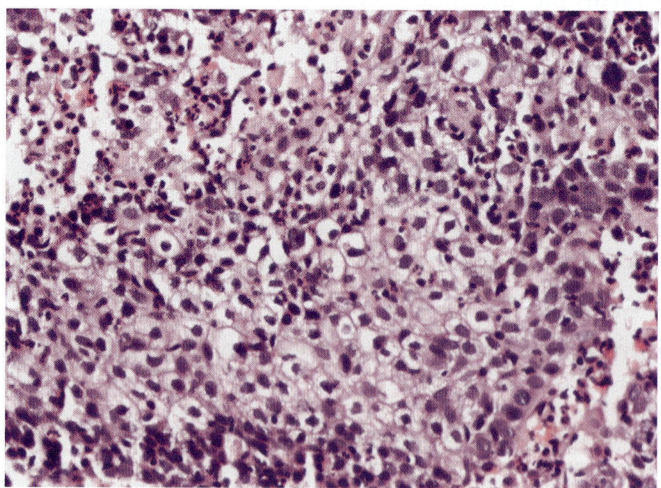

Figure 2.59: Histopathology of cervical lesion

Problem 2.27

D. Tubal metaplasia

Discussion

Description: The Figures 2.27A to F show benign squamous cells and many rows and clusters of columnar cells with tubal metaplasia.

Diagnostic features: The columnar cell with cilia in the cervical smear indicates tubal metaplasia. This is commonly noted in the cervix after conization.

Problem 2.28

A. Squamous cell carcinoma

Discussion

Description: The microphotographs (Fig. 2.28) show many discrete and loose clusters of cells with moderately enlarged hyperchromatic nuclei. Many fiber cells are also seen.

Diagnostic features: The diagnostic features of the index case include:
- Abundant atypical cells with markedly pleomorphic nuclei
- Hyperchromatic nuclei
- high N/C ratio
- Many fiber cells and tadpole cells.

Differential diagnosis: The marked nuclear pleomorphism, presence of fiber cells excludes any possibility of HSIL or ASC-H. The cases of atrophic smear occasionally show nuclear enlargement and hyperchromasia of the parabasal cells and may simulate carcinoma. The index case shows significant nuclear enlargement and pleomorphism and the possibility of atrophic smear is excluded.

Problem 2.29

A. Radiation effect

Discussion

Description: The microphotographs show many multinucleated giant cells along with many scattered squamous cells with both nucleomegaly and cytomegaly.

Radiation changes are characterized by:
- Markedly enlarged cell
- Low nucleo-cytoplasmic ratio as nucleus is mildly enlarged
- Cytoplasm of the cell is abundant and vacuolated
- Multinucleation is frequent
- Mildly enlarged and pleomorphic nuclei
- Single to multiple prominent nucleoli.

Diagnostic features: In the index case the features suggestive of radiation change include many multinucleated giant cells, and enlarged cells with prominent nucleoli having low N/C ratio.

Differential diagnosis: Herpes simplex virus (HSV) infection shows many multinucleated cells along with viral inclusion in the nuclei. The absence of high N/C ratio, nuclear pleomorphism and hyperchromasia eliminate the possibility of squamous cell carcinoma or HSIL.

Problem 2.30

D. High-grade squamous intraepithelial lesion.

Discussion

Description: The microphotograph shows many discrete large intermediate squamous cells with moderate nuclear enlargement (nuclei are occupying more than 50% of the cytoplasmic area) and pleomorphism.

Diagnostic points: In the index case the following features favour the diagnosis of HSIL:

- Abundant discrete atypical cells
- High N/C ratio
- Moderate to marked nuclear enlargement
- Moderate nuclear pleomorphism
- Coarse granular chromatin.

Problem 2.31

D. Atrophic smear

Discussion

Description: The microphotographs show many parabasal and basal cells with mildly enlarged nuclei. Occasional cells show perinuclear halo that simulate koilocytosis. These cells are actually glycogenated parabasal cells.

Diagnostic features: In this index case the following features suggest the interpretation of atrophic smear:

- Abundant parabasal and basal cells
- Mild nuclear enlargement of the parabasal cells
- Mild hyperchromasia
- No abnormality of the nuclear chromatin.

The perinuclear halo in the squamous cells may be mistaken as koilocytosis. However, these cells do not show any nuclear atypia and these cells are only glycogenated squamous cells. Absence of significant change in the nuclear chromatin and regular nuclear margin exclude the possibility of any squamous intraepithelial lesion.

Problem 2.32

D. Squamous cell carcinoma

Discussion

Description: The microphotographs show many intermediate and parabasal cells with markedly enlarged and hyperchromatic nuclei. In addition many dyskeratotic cells and fiber cells are also seen. Figure 2.32D shows tumor diathesis.

Diagnostic points in this index case:

- Marked nuclear atypia
- Fibre cells
- Tumor diathesis.

Differential diagnosis: See 2.28

Problem 2.33

C. Squamous cell carcinoma recurrence with radiation changes.

Discussion

Description: The microphotographs show many discrete and loose clusters of cells. These cells have markedly enlarged nuclei with increased N/C ratio. The nuclei are hyperchromatic and pleomorphic. In addition tumor diathesis is also seen in 2.33B (indicated by red arrow).

Diagnostic points: In the index case the following features suggest the diagnosis of squamous cell carcinoma:

- Many large atypical cells
- Cells with increased N/C ratio
- Markedly enlarged nuclei
- Marked hyperchromatic nuclei
- Many bi- and multinucleated cells indicate associated radiation changes.

Differential Diagnosis: The nuclear abnormality in the present case is significant and thus eliminates the possibility of only radiation effect or other possibilities.

Problem 2.34

A. Radiation induced changes

Discussion

Description: The microphotographs show many discrete large squamous cells with both nuclear and cytoplasmic enlargement. The N/C ratio of these cells is not altered. The nuclei are normochromatic and show mild irregular margin.

Diagnostic points: The following diagnostic points indicate radiation changes only:

- N/C ratio of the cells is not altered
- Normochromatic.

Problem 2.35

A. a. Colposcopy followed by biopsy.
B. g. CIN3

Discussion

Description: The microphotographs (Figs 2.35A to F) show many small clusters and discrete cells with moderate to marked nuclear enlargement and pleomorphism. The chromatin of the cells is coarsely clumped. No definite tumor diathesis is noted.

Diagnostic points in the index case:

- Abundant basal and parabasal cells showing nuclear atypia
- Marked nuclear enlargement
- Coarse clumped chromatin.

All case of HSIL should have colposcopy followed by biopsy from the abnormal area. Follow up is advised only in cases of ASCUS.

The histopathology outcome of this case was CIN3 (Fig. 2.35G). The other figures are squamous cell carcinoma (Fig. 2.35H), adenocarcinoma (Fig. 2.35I) and CIN 1 (Fig. 2.35J).

Problem 2.36

A.	3. Foreign body giant cells
B.	2. Fiber cells
C.	1. Tubal metaplasia
D.	4. Corn flakes cell

Discussion

A. **Foreign body giant cells**: The presence of foreign body giant cells is not uncommon in the cervical smear. This may occur due to stimulation by dust or any other foreign body in the cervical smear.

B. **Fiber cells:** The presence of fiber cell in the cervical smear indicates the possibility of squamous cell carcinoma. These are elongated cells with hyperchromatic elongated nuclei. Occasionally the stromal cells may simulate fiber cell. However, the nuclei of the stromal cells are normochromatic.

C. **Tubal metaplasia:** Cervical lining epithelium may undergo tubal metaplasia due to chronic irritation. The tubal metaplastic cells are characterized by:
   - Groups of columnar cells with terminal plate and cilia
   - The cells are columnar in appearance with round nuclei
   - The nuclei may be enlarged and pleomorphic
   - N/C ratio of the cells may be high
   - Evenly distributed chromatin of the nuclei
   - Nucleoli are absent
   - Vacuolated cytoplasm.

D. **Corn flakes cell:** These cells are the common artefact due to air bubbles under the cover slip. The cells show brownish bubbles in the cytoplasm. The brown bubbles should not be confused with organism.

Problem 2.37

A.	4. Hyperchromatic crowded cells of benign endocervical cells
B.	3. Hyperchromatic crowded cells of endometrial cells
C.	1. Hyperchromatic crowded group of endocervical adenocarcinoma
D.	2. Hyperchromatic crowded cells in HSIL

Discussion

Hyperchromatic crowded group (HCG) is defined as tight ball like three dimensional cluster of cells with hyperchromatic nuclei in the cervical smear. HCG may be seen in both benign and malignant conditions and is the common source of misinterpretation in the cervical smear. The common benign lesions that contain HCG include:

- Endometrial cells
- Endocervical cells
- Tubal metaplastic cells
- Atrophic smear.

The common source of HCG in malignant conditions include:

- Endocervical adenocarcinoma
- Endometrial adenocarcinoma
- Squamous cell carcinoma
- Metastatic carcinoma.

The distinguishing features between the benign and malignant HCG are:

- Irregular frayed margin in malignant HCG compared to regular margin in benign HCG
- Loss of polarity of the cells in malignant HCG
- Frequent mitosis in malignant HCG.

In the present problem:

- Figures 2.37A and B show hyperchromatic crowded cells of benign endocervical cells and benign endometrial cells respectively. In both the groups the clusters have regular well defined border and the cell polarity is maintained.
- In C and D the HCGs are of malignant in origin and indicate endocervical adenocarcinoma and HSIL respectively. The margin of the HCG is irregular and the cells also show considerable nuclear pleomorphism.

Further Reading

1. Demay RM. Hyperchromatic crowded groups: pitfalls in pap smear diagnosis. Am J Clin Pathol. 2000;114 Suppl:S36-43.

Problem 2.38

Figure A	2. Squamous pearl
Figure B	3. Tumor diathesis
Figure C	1. Koilocytosis

Discussion

Squamous pearl is often seen in cervical smear. They may be seen in both benign and malignant condition.

Tumor diathesis indicates invasive squamous cell carcinoma. Tumor diathesis is composed of necrotic tissue fragment entangled with malignant cells.

Koilocytosis represents HPV infection. The presence of koilocytotic cell in a smear qualifies it as LSIL. The koilocytotic cells show perinuclear vacuoles with thick cytoplasmic margin along with nuclear enlargement. The various other cells also mimic koilocytotic cells such as glycogenated cells, inflammatory cells, etc.

Problem 2.39

Question A: Answer is 3
Question B: Answer is 4
Question C. Answer is 1 Figure 2.39G

Discussion

Description: The microphotographs (Figs 2.39A to F) show abundant discrete squamous cells with moderate nuclear enlargement and pleomorphism. Many fiber cells are also seen.

Diagnostic points: The index case shows abundant discrete squamous cells with marked nuclear atypia along with many fiber cells. This indicates the diagnosis of squamous cell carcinoma.

Figure G shows the histopathology of squamous cell carcinoma.

Figures H and I show CIN 3 and adenocarcinoma, respectively.

Cervical squamous cell carcinoma is caused by high risk HPV infection. It has been noted that HPV 16, 18, 45, 31, 33, 52, 58, and 35 are responsible for major bulk (95%) of cervical squamous-cell carcinomas. Smith JS et al[1] in a meta analytic study have shown that HPV 16 is the most common type (52–56 %) followed by HPV 18 (13–22 %) in invasive squamous cell carcinoma.[1]

In this index case the immediate investigation should be biopsy from the suspected area either by visually or by colposcopic guidance.

Reference

1. Smith JS, Lindsay L, Hoots B, et al. Human papillomavirus type distribution in invasive cervical cancer and high-grade cervical lesions: a meta-analysis update. Int J Cancer. 2007;121(3):621-32.

Problem 2.40

Figure A	Figure 4
Figure B	Figure 1
Figure C	Figure 3
Figure D	Figure 2

Discussion

Figure A. Squamous cell carcinoma	Figure 4. Squamous cell carcinoma
Figure B. Adenocarcinoma	Figure 1. Adenocarcinoma
Figure C. LSIL	Figure 3. CIN1
Figure D. HSIL	Figure 2. CIN 3

Problem 2.41

C. Actinomycosis

Discussion

Description: The microphotographs show tight clusters of filamentous bacteria radiating as sun ray appearance.

Disease description: Actinomyces is often associated with the use of intrauterine contraceptive device. The patients are usually asymptomatic. However, the presence of *Actinomyces* in cervical smear along with clinical features of pelvic infection such as fever, back pain or fowl smelling vaginal discharge may raise clinical suspicion of *Actinomyces* abscess.

The organisms are typical clumped of grayish blue filamentous bacteria radiating from a dense center. On Papanicolaou's stain the organisms are long, thick and beaded in appearance. The smear may also show leukocytes, macrophages and giant cells (absent in the index case).

Problem 2.42

A. AGCUS

Discussion

Description: The cytology smear shows many endocervical glandular cells with mild nuclear enlargement and pleomorphism. The cytoplasm of the cells is moderate and vacuolated.

Disease Description: The cases of atypical endocervical cells, show abnormal cells with nuclear atypia more than reactive or reparative changes , however, the morphological abnormality fall short of unequivocal diagnosis of adenocarcinoma in situ or invasive carcinoma.

The terminology of atypical glandular cell should be more precisely defined as:

- Endocervical
- Endometrial
- Glandular cell, not otherwise specified.

The criteria of diagnosis of atypical endocervical cells are:

- Cells are in sheets or small overcrowded groups
- Cytoplasm: abundant, vacuolated with distinct cell border
- Nucleus
  - Mild enlargement: Three to five times the area of normal endocervical nuclei
  - Mildly pleomorphic
  - Hyperchromatic
  - Small nucleoli.
- Rare mitotic figures.

Diagnostic points in the index case: In this index case the features that favor the diagnosis of atypical endocervical cells are: endocervical appearance of the cells with mild nuclear enlargement, pleomorphism and vacuolated cytoplasm with distinct cell border.

Problem 2.43

B. AGC, favors neoplastic

Discussion

Description: The cytology smear shows many loose clusters and discrete round to oval glandular cells. The cells show scanty vacuolated cytoplasm and enlarged nuclei with high N/C ratio.

Disease description: Atypical glandular cell (AGC) abnormality is classified by TBS 2001 as:

- Atypical
  - Endocervical in origin
  - Endometrial in origin
  - Glandular cell, not otherwise specified.
- Atypical
  - Endocervical, favor neoplastic
  - Glandular, not otherwise specified.

The cases of atypical endocervical cells, favor neoplastic show cellular morphology which is abnormal but fall short of unequivocal diagnosis of adenocarcinoma in situ or invasive carcinoma.

The criteria of AGC, endocervical favor neoplastic are:

- Cells are in sheets or small overcrowded groups
- Rarely rosetting or feathering appearance of the cell cluster
- Cytoplasm: abundant and vacuolated
- Nucleus: enlarged, hyperchromatic, pleomorphic with irregular margin
- N/C ratio increased
- Rare mitotic figures.

Diagnostic features in this index case: The diagnostic features are small loose clusters of cell, round moderately enlarged pleomorphic nuclei having prominent nucleoli and vacuolated cytoplasm.

Problem 2.44

A. AGC: Possibly endometrial

Discussion

Description: The smear shows many clusters of endometrial cells. The cells are small in size with scanty cytoplasm and mildly pleomorphic hyperchromatic nuclei.

Disease description: The atypical glandular cells, if possible, should be classified as:

- Endocervical in origin
- Endometrial in origin
- Glandular cell, not otherwise specified.

The criteria of the atypical endometrial cells are:

- Cohesive clusters of small groups of endometrial cells
- Mildly enlarged nuclei compared to normal endometrial cells
- Mild hyperchromasia
- Small nucleoli, more prominent in LBC (liquid based cytology) preparation
- Scanty cytoplasm with ill-defined margin.

Diagnostic criteria of the index case: The diagnostic features in the index case are small groups of endometrial cells with mildly enlarged pleomorphic nuclei.

Problem 2.45

A. Atypical endocervical cells favors neoplasia

Discussion

Description: The smear shows loose clusters and discrete cells with moderate amount of cytoplasm. The nuclei are moderately enlarged with prominent nucleoli.

Diagnostic features in the index case: In this index case the features that favor the diagnosis of atypical endocervical cells, favor neoplasias are: small groups and single cells, endocervical appearance of the cells with mild to moderate nuclear enlargement, nuclear pleomorphism, prominent nucleoli and vacuolated cytoplasm with distinct cell border.

Problem 2.46

C. Adenocarcinoma possibly endometrial origin

Discussion

Description: The cytology smear shows multiple cohesive clusters of cells with round nuclei and high N/C ratio. The nuclei are hyperchromatic and moderately pleomorphic with frequent overlapping.

Disease description: In endometrial adenocarcinoma the following criteria should be present:

- Tight clusters or small groups of cells
- Pleomorphic nuclei
- Nuclear crowding indicating loss of polarity
- Nuclear hyperchromasia, and irregular clumped chromatin
- Multiple prominent nucleoli depending on grade of the tumor
- Scanty vacuolated cytoplasm
- Intracytoplasmic polymorphs
- Tumor diathesis may be present.

Diagnostic features in the index case:

- Cohesive clusters of small round cells
- Hyperchromatic pleomorphic nuclei
- High N/C ratio
- Nuclear crowding
- Scanty cytoplasm.

Problem 2.47

A. Adenocarcinoma, favors endocervical origin

Discussion

Description: The cytology smear shows multiple cohesive clusters and occasional dissociated cells. The cells are occasionally arranged in small row (Fig. 2.47F). The individual cells show cuboidal to columnar appearance with moderate amount of cytoplasm. Nuclei are enlarged and hyperchromatic with prominent nucleoli.

Disease description: Endocervical adenocarcinoma consists of 15% of all cervical carcinomas. Unlike to squamous cell carcinoma, endocervical adenocarcinoma is more frequently associated with HPV 18 than HPV 16. The patients are usually asymptomatic. However, they may have vaginal bleeding.

The diagnostic criteria of endocervical adenocarcinoma are:

- High cellularity: Abundant atypical cells
- Columnar to cuboidal in appearance
- Single cells, loose clusters of cells, three dimensional tight clusters and syncytial aggregates
- Well defined scalloped borders of the cluster
- Nucleus
  - Large and round
  - Pleomorphic
  - Irregular coarse chromatin
  - Parachromatin clearing
  - Irregular nuclear membrane
  - Prominent nucleoli, macronucleoli may be present
- Cytoplasm: abundant and finely vacuolated
- Tumor diathesis present.

Diagnostic features in this index case: The index case shows crowded three dimensional clusters, abnormal columnar cells, finely vacuolated cytoplasm, pleomorphic nuclei and prominent nucleoli.

Further Reading

1. Dey P. Diagnostic Cytology. Jaypee Brothers Medical Publishers (P) Ltd., NewDelhi. 2014.

Problem 2.48

C. Adenocarcinoma

Discussion

Description: The cytology smear shows multiple tight clusters of round to oval cells. The cells show scanty cytoplasm and enlarged mildly pleomorphic nuclei having prominent nucleoli.

Diagnostic feature in this index case:

- Cohesive clusters of cells
- Round cell with high N/C ratio

- Hyperchromatic nuclei
- History of operation.

Problem 2.49

D. Adenocarcinoma extrauterine (ovary)

Discussion

Description: The cytology smear shows many loose clusters and discrete round cells with mild to moderate amount of vacuolated cytoplasm. The nuclei are moderately enlarged with high nucleocytoplasmic ratio.

Diagnostic features in this index case:

- Marked nuclear enlargement and pleomorphism
- Vacuolated cytoplasm
- Clean background indicates extracervical origin
- History and investigations are suggestive of ovarian tumor.

Problem 2.50

B. ASC-H

Description: The cytology smears show many discrete large cells with enlarged nuclei occupying more than 2/3rd of the cytoplasmic area. The nuclei show mild hyperchromasia and mildly irregular margin.

Diagnostic Features:

- Enlarged nuclei
- Mild hyperchromasia
- Mildly irregular nuclear margin.

The cone biopsy was done in this index case and the histopathology section showed CIN3 with involvement of endocervical gland (Fig. 2.60).

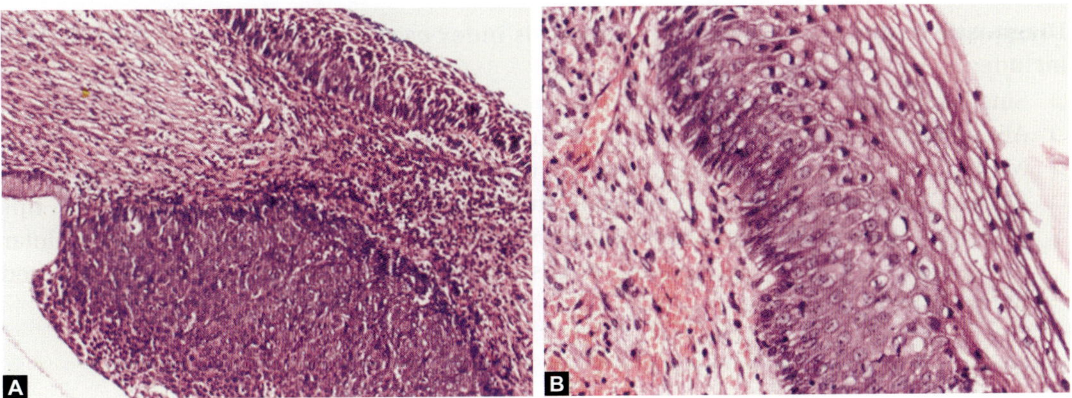

Figures 2.60A and B: Histopathology section of cervical lesion

Problem 2.51

D. Adenocarcinoma possibly endometrial

Discussion

Description: The cytology smear shows many discrete and loose clusters of round cells. The cells have scanty to moderate amount of vacuolated cytoplasm. The nuclei are round, moderately pleomorphic with prominent nucleoli.

Diagnostic features in the index case: The following features favor the diagnosis of endometrial adenocarcinoma in this index case:

- Tight clusters and discrete cells
- The round cells with moderately pleomorphic nuclei
- Nuclear crowding of the cell cluster indicating loss of polarity
- Nuclear hyperchromasia
- Irregular clumped chromatin and multiple prominent nucleoli
- Vacuolated cytoplasm
- Relatively clear background.

Differential Diagnosis: The round cells with vacuolated cytoplasm eliminate the possibilities of LSIL and HSIL. The nuclear characters of these cells are too bad to label the case as AGC.

Problem 2.52

B. HSIL

Discussion

Description: The cytology smear shows many small clusters and discrete cells with moderate nuclear enlargement and pleomorphism. The nuclei are hyperchromatic and also show prominent nucleoli (Fig. 2.52G). The cells show orangeophilic cytoplasm and at places cytoplasmic vacuolation.

Diagnostic features in the index case: In this index case the features suggestive of HSIL include:

- Single and clusters of discrete atypical cells
- Abundant intermediate cells with squamous differentiation
- Moderately pleomorphic nuclei with hyperchromasia
- Occasional clusters of cells show vacuolated cytoplasm and prominent nucleoli. In this index case the prominence of nucleoli and cytoplasmic vacuolation are due to the glandular involvement by the dysplastic epithelial cells. The histopathology of the lesion is highlighted in Figure 2.61.

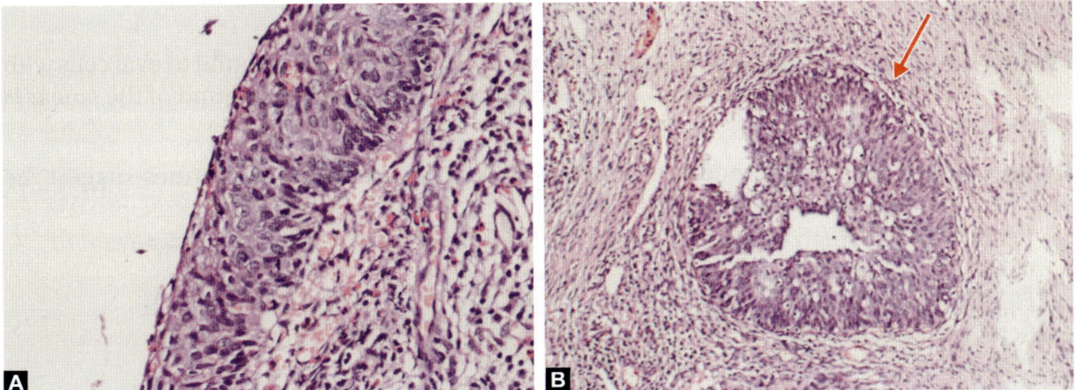

Figures 2.61A and B: Histopathology of the cervical lesion

Problem 2.53

B. Endometrial cells present

Discussion

Description: The cytology smear shows tight cohesive clusters of endometrial cells in a clean background.

Disease description: Endometrial cells are commonly seen during the first half of menstrual cycle in the cervical smear. Both glandular and superficial stromal cells may be present. Spontaneously exfoliated normal looking endometrial cells are also present after birth of baby, abortion, endometrial polyp, endometritis and endometriosis. In 2001 Bethesda system, the presence of endometrial cells in the woman 40 year or older is categorized in the "other". This finding should be mentioned in the report. However, the atypical endometrial cells should be mentioned in the category of "epithelial cell abnormality".

The characteristic features of endometrial cells include:

- Commonly present in ball like tight spherical clusters of cells. Single discrete endometrial cells are rare
- Scanty basophilic cytoplasm
- Cytoplasmic vacuolation may be present
- Nuclei are small and round
- Nuclear molding or angulated contour
- No nucleoli or inconspicuous nucleoli.

In LBC preparation, the single cells are frequently seen along with single cell necrosis. Chromatin detail is better seen in LBC.

Problem 2.54

C. Metastatic adenocarcinoma from ovary

Discussion

Description: The cytology smear shows many tight cohesive clusters of round to oval cells with scanty cytoplasm, high N/C ratio and hyperchromatic nuclei. The background of the smear is clean.

Diagnostic features in the index case: In this index case the following features suggest the diagnosis of adenocarcinoma of ovarian origin:

- Pelvic mass and high CA-125 level indicate the possibility of ovarian malignancy.
- Cohesive three dimensional clusters of tumor cells
- High N/C ratio
- Scanty cytoplasm
- Pleomorphic nuclei
- Large prominent nucleoli
- In addition, the absence of tumor diathesis also indicates the non-cervical origin of the tumor.

Problem 2.55

D. Endometrial cell present

Discussion

Description: The cytology smear shows many tight cohesive clusters of endometrial cells with predominantly scanty cytoplasm and central round monomorphic hyperchromatic nuclei.

Diagnostic points in the index case: The cytology smear shows predominantly benign looking endometrial cells in a postmenopausal patient.

The patient showed well differentiated adenocarcinoma of endometrium in follow up study. Endometrial cells in the post-menopausal patient should always be followed up by endometrial curettings to exclude any neoplastic lesion in the endometrium.

Problem 2.56

B. Adenocarcinoma

Discussion

Description: The cytology smear shows multiple clusters and discrete round cells with moderate amount of vacuolated cytoplasm. The nuclei are enlarged with mildly pleomorphic having prominent nucleoli.

Diagnostic features in the index case: In this index case the following features suggest the diagnosis of adenocarcinoma of ovarian origin:

- History suggesting of ovarian malignancy (pelvic mass and high CA-125 level)
- Clusters and discrete abnormal cells
- Columnar appearance of the cells
- Moderate to abundant vacuolated cytoplasm
- High nucleo-cytoplasmic ratio
- Pleomorphic nuclei

- Large prominent nucleoli
- In addition, the smear shows relatively clean background with no sign of tumor diathesis. This absence of tumor diathesis is also considered as a symbol of non-cervical origin of the tumor.

Problem 2.57

C. LSIL

Discussion

Description: The cytology smear shows many discrete cells with enlarged nuclei along with perinuclear halo with thickened cytoplasmic margin (koilocytosis).

Diagnostic feature in the index case: The presence of many koilocytotic cells qualifies the smear as LSIL in this index case. Koilocytosis indicates HPV infection.

Disease description: See problem 2.10

Problem 2.58

B. LSIL (koilocytes) with herpes simplex viral infection

Discussion

Description: The cytology smears show discrete koilocytotic cells characterized by enlarged nuclei and perinuclear halo. In addition the smear shows many multinucleated and bi-nucleated cells. The nuclei of the cells often show moulding with ground glass appearance. Intranuclear inclusion is also seen (Fig. 2.58E).

Diagnostic Features:

- Koilocytes
- Multinucleated giant cells
- Ground glass nuclei
- Intranuclear inclusion.

CHAPTER 3

Effusion Cytology

Introduction

Effusion may occur due to non-neoplastic or neoplastic causes.

The common non-neoplastic causes of effusion are: Tuberculous effusion, congestive heart failure, cirrhosis, rheumatoid arthritis, systemic lupus erythematosus, etc. There are no specific cytological features to distinguish various non-neoplastic lesions in the effusion cytology.

Neoplastic causes of effusion are: (a) Metastatic malignant tumors and (b) primary neoplastic lesions of the pleura.

Metastatic Malignant Tumor

The diagnosis of malignant cells in effusion depends on the constellation of features. The three main challenges of the cytologist are:

To differentiate:

- Malignant cells from the mesothelial cells
- Adenocarcinoma from mesothelioma
- Mesothelial cells from the mesothelioma.

The cytological features of malignant cells in effusion are:

Cell arrangement

- Cannon ball like tight three-dimensional cell clusters
- Papillary cluster
- Single dissociated cells in small cell carcinoma of lung
- Small rows of cells in lobular carcinoma of breast.

Cytoplasm

- Single to multiple vacuoles in cytoplasm
- Keratin: In squamous cell carcinoma
- Mucus: Adenocarcinoma
- Melanin: Melanoma.

Nucleus
- High N/C ratio
- Nuclear enlargement
- Pleomorphism: Mild to moderate
- Irregular nuclear margin
- Single to multiple prominent nucleoli
- Coarse chromatin.

Mitosis: Atypical tripolar mitosis

Other features:
- Cell cannibalism
- Psammoma bodies.

The following immunocytochemistry markers suggest epithelial malignancy:

CEA: Positive

Ber-Ep4: Positive

Leu M1: Positive

B72.3: Positive.

Metastatic Tumor

Metastatic adenocarcinoma: This is the most common malignancy in the effusion fluid. The most common primary sites are ovary, gastrointestinal tract, lung and breast. The common cytological features of metastatic carcinomas are: ball like three-dimensional clusters, cells with moderate to abundant vacuolated cytoplasm and large pleomorphic nuclei with prominent nucleoli. There may be predominant discrete cells in gastric and breast carcinoma. The metastatic gastric carcinoma often shows round to oval cells with eccentric nuclei containing large vacuoles giving a signet ring like appearance. The cytomorphology of ovarian and lung carcinoma depends on the tumor type.

Squamous Cell Carcinoma

Squamous cell carcinoma uncommonly metastasizes in the serous cavities. The common primary sites are lung, larynx and cervix. The following features suggest metastatic squamous cell carcinoma:
- Small clusters or discrete cell population with hyperchromatic nucleus with inconspicuous nucleoli
- Oval cells in non-keratinizing squamous cell carcinoma
- Polyhedral and occasional spindle cells in keratinizing squamous cell carcinoma
- Dense and orangeophilic cytoplasm.

Small Cell Carcinoma

The common primary site is lung. The predominant cytological features are:
- Discretely or in small chain of cells
- Small cells twice the diameter of the lymphocyte

- The cells show scanty cytoplasm and small hyperchromatic nuclei with granular chromatin and inconspicuous nucleoli
- Cell to cell compression giving vertebra like appearance.

Pseudomyxoma Peritonei (PMP)

This is a low-grade mucinous neoplasm and shows diffuse mucinous implants in the peritoneal surfaces. The most common sources of the primary tumor are mucinous adenocarcinoma of the appendix, mucinous adenocarcinoma of ovary, and gastric carcinoma. The characteristic cytological features of PMP are:

- Thick and gelatinous mucoid materials
- Round to columnar shaped cells with usually bland nuclei
- Clear cytoplasm
- Occasionally the cells show moderate nuclear pleomorphism.

The common immunocytochemistry to point out the primary malignancy are (Table 3.1):

Table 3.1: Markers of malignancy to find out the primary

Site	*Markers*
Lung adenocarcinoma	TTF1, CK7, napsin
Lung squamous cell carcinoma	CK5/6, P63, P40
Ovary	CK7, CEA
Colon	CK20, CDX
Prostate	PSA
Breast	ER/PR receptors, EMA, Gross cystic disease fluid protein 15 (GCDFP-15)
Thyroid	TTF-1, thyroglobulin
Bladder	Positive for both CK 7 and CK20

Hematological Malignancies

Near about 10 to 20% of malignant effusion is caused by non-Hodgkin lymphoma (NHL) and Hodgkin lymphoma (HL). Cytomorphological feature of NHL depends on the specific subtypes of lymphoma. Hodgkin's lymphoma usually shows a polymorphic population of cells consisting of lymphocytes, plasma cells, and eosinophils. The typical Reed-Sternberg cell may be difficult to find out. Large mononuclear type of Hodgkin's cell and multinuclear cells may be helpful in a known case of HL.

In case of lymphocyte rich effusion flow cytometry is helpful to distinguish mature reactive lymphocytes and NHL. The panel of flow cytometry includes:

- Kappa and lambda light chain: Only one type of light chain expression (light chain restriction) is helpful to diagnose a B-NHL. In FCM the kappa/lambda ratio more than 4:1 or more than 1:2 is usually considered as an evidence of monoclonality
- CD3, CD 2, CD5 and CD 7: The aberrant expression of CD2, CD5 and CD7 expression may suggest T-NHL
- In addition, for the subtyping of NHL: CD 20, CD 10, CD 23 are also needed.

Mesothelioma

Pleural surface is the most common site of development of malignant mesothelioma (MM). It usually occurs in 6th or 7th decade of life. Radiograph of MM shows large plaque like nodule on the pleural or peritoneal surface. These patients often have history of asbestos exposure.

The characteristic cytological features of mesothelioma include:

- Spherical clusters with knobby edges and papillae like clusters along with many dissociated cells
- The tumor cells look like mesothelial cell and window like gaps in between the cells may be seen
- The cells are enlarged with low nucleocytoplasmic ratio
- Centrally placed nucleus with large prominent nucleoli (macronucleoli)
- Condensed cytoplasm with ring like peripheral clear zone.

The immunocytochemical markers of mesothelioma are:

Calretinin: Positive

WT1: Positive

Vimentin: Positive

The following immunocytochemical markers may help to distinguish a reactive versus malignant mesothelial cells (favors malignancy):

- Strong EMA positivity and negative desmin
- Strong membranous positivity for GLUT-1
- Strong nuclear staining for p53.

QUESTIONS

Problem 3.1

A 55-year-old man with massive ascites, weakness, mild hepato-splenomegaly. The microphotograph of the cytology smear of the effusion fluid is shown in Figures 3.1. The most likely diagnosis is:

A. No malignant cell in effusion
B. Lymphocyte rich effusion
C. Metastatic adenocarcinoma
D. Mesothelioma

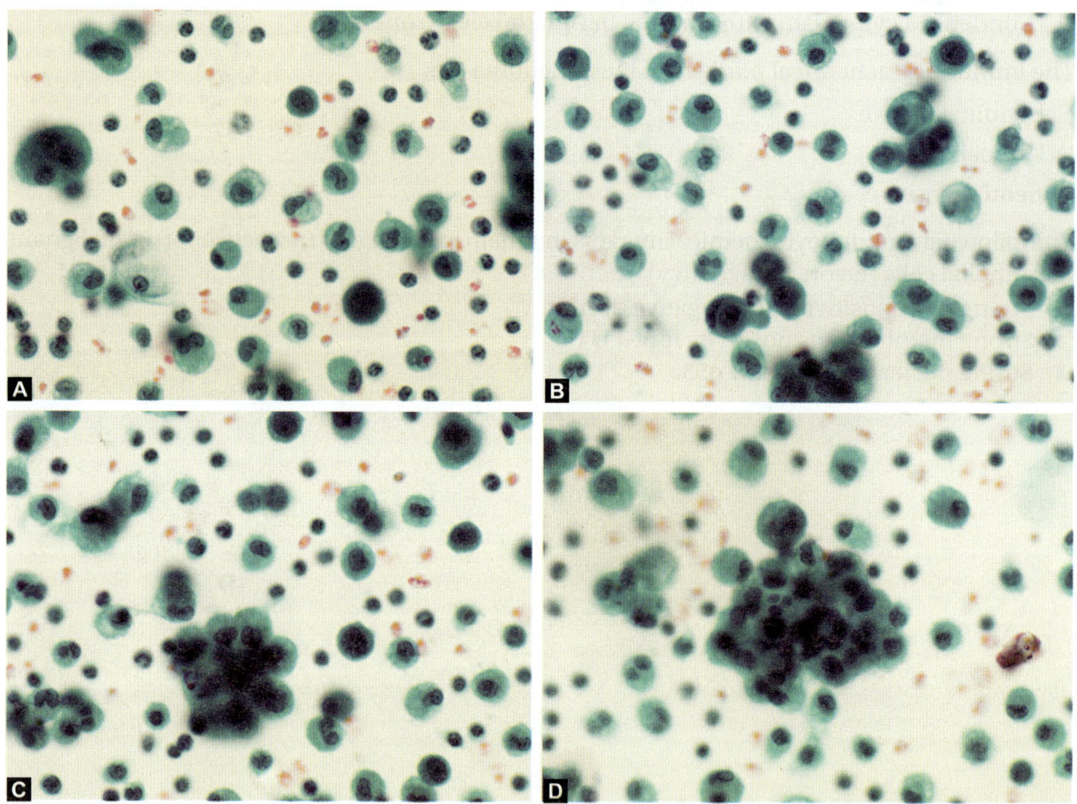

Figures 3.1A to D

Problem 3.2

A 50-year-old female had high fever and chest pain. The microphotograph of the cytology smear of pleural effusion is shown (Fig. 3.2). The most likely diagnosis is:

A. Acute inflammation (empyema)
B. Hyper-eosinophilia
C. Metastatic carcinoma
D. Lymphocyte rich effusion

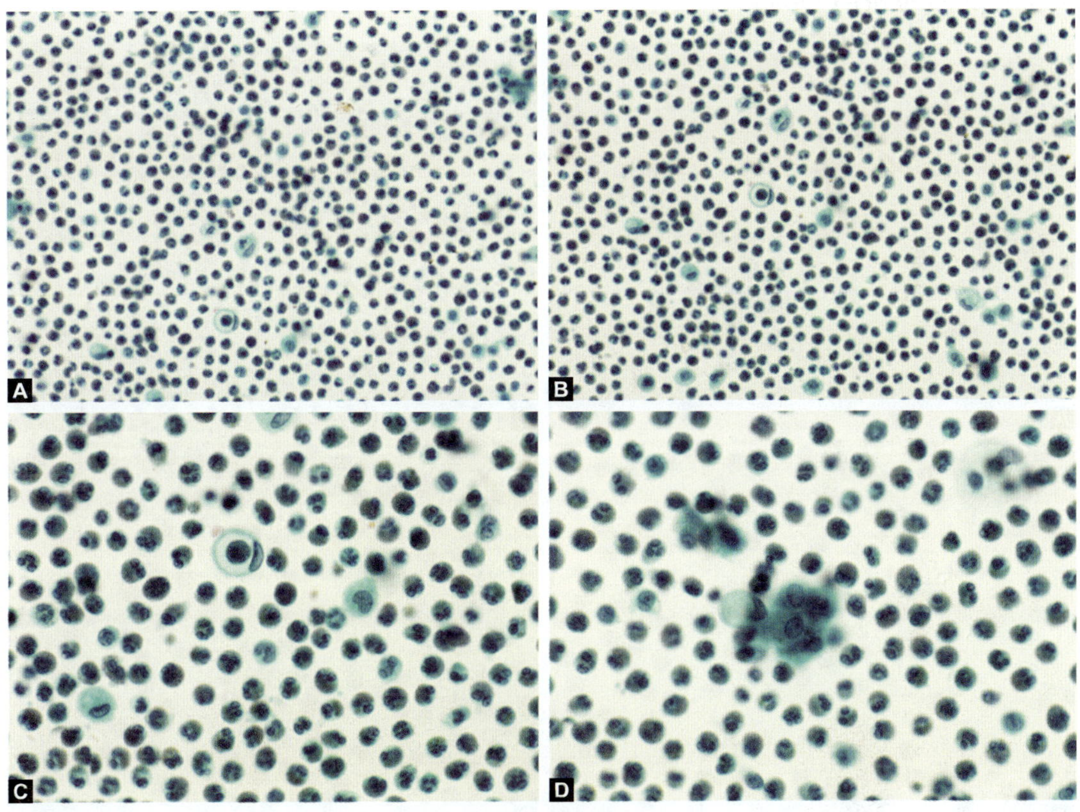

Figures 3.2A to D

Problem 3.3

A 45-year-old female had dyspnea and heaviness in chest along with right sided pleural effusion. The microphotographs of cytology smear of the effusion fluid are shown in Figure 3.3. The most likely diagnosis is:

A. Lymphocyte rich effusion
B. Metastatic carcinoma
C. Empyema
D. Eosinophilic effusion

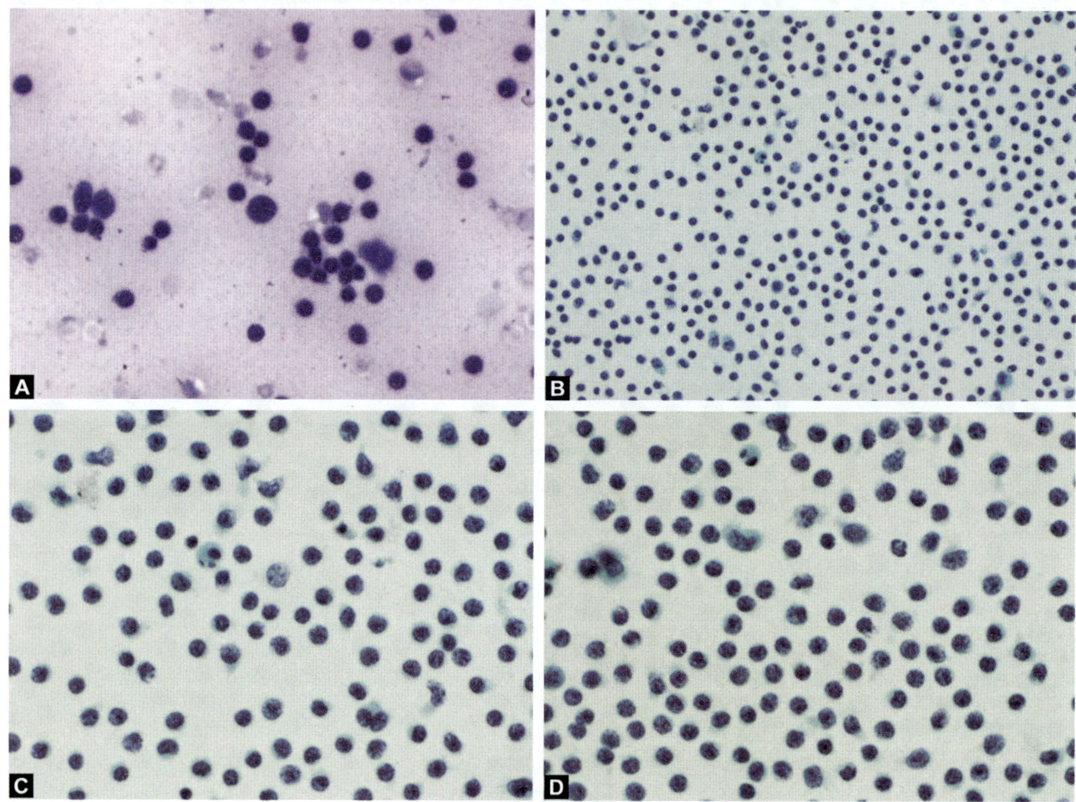

Figures 3.3A to D

Problem 3.4

A 35-year-old male patient had cough and breathlessness for 6 months. He had right sided pleural effusion for last 15 days. The microphotographs of the cytology smear are shown in the Figure 3.4. The most likely diagnosis is:

A. Positive for malignancy
B. Tuberculosis
C. Acute inflammation
D. Eosinophilic effusion

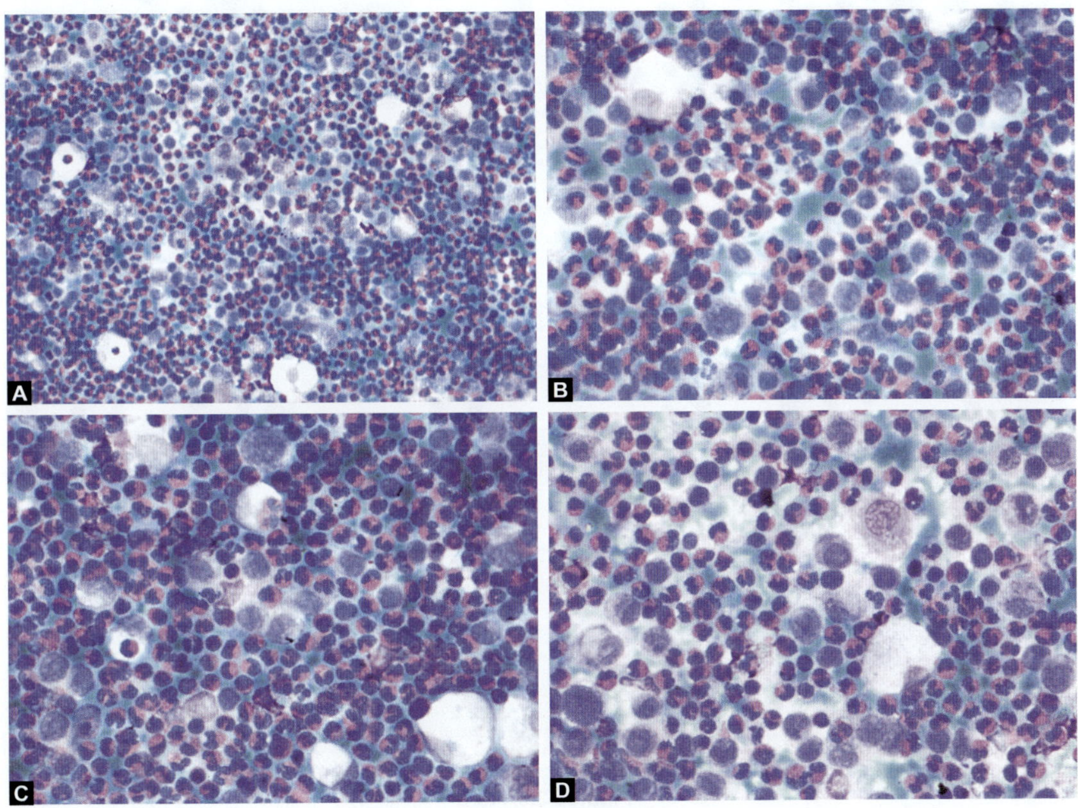

Figures 3.4A to D

Problem 3.5

A 46-year-old male presented with ascites along with chronic liver disease for 6 months. The most likely diagnosis from Figure 3.5 is:

A. Metastatic adenocarcinoma
B. Hepatocellular carcinoma
C. Reactive mesothelial cells
D. Non-Hodgkin lymphoma

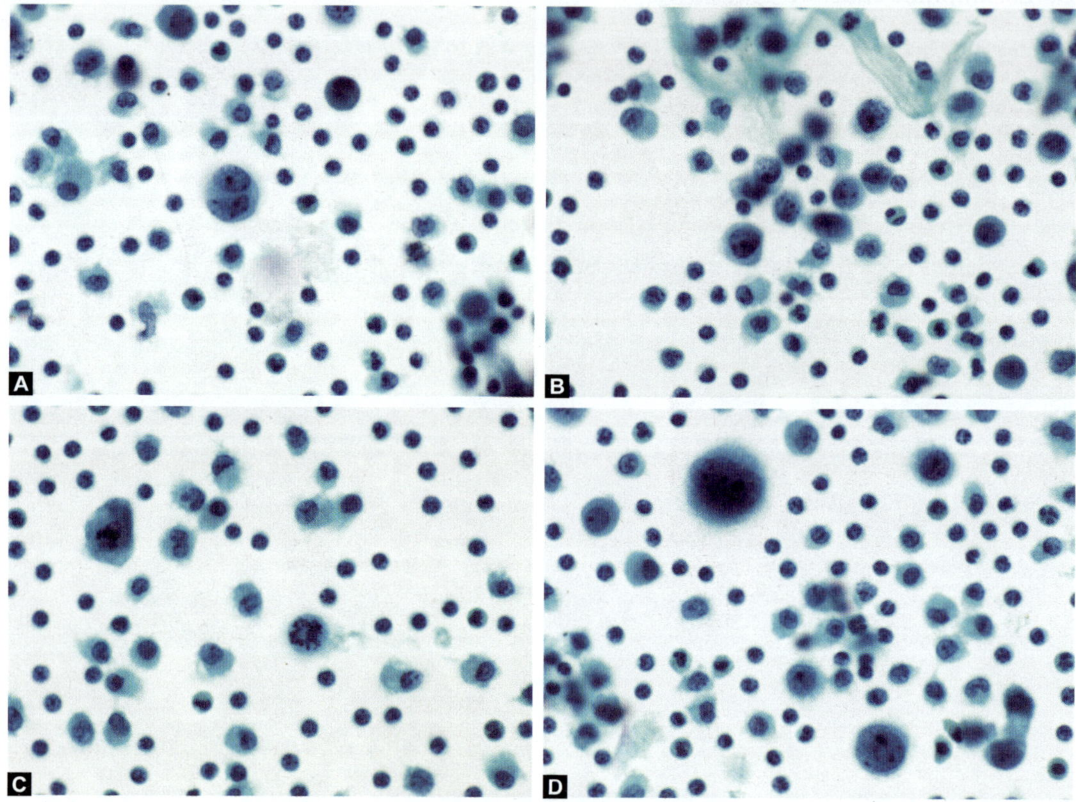

Figures 3.5A to D

Problem 3.6

A 46-year-old man presented with chronic liver disease and ascites. His peritoneal fluid examination was done. The most likely diagnosis from Figure 3.6 is:

A. Histoplasma infection
B. Candida infection
C. Reactive mesothelial cells only
D. *Leishmania donovani* bodies

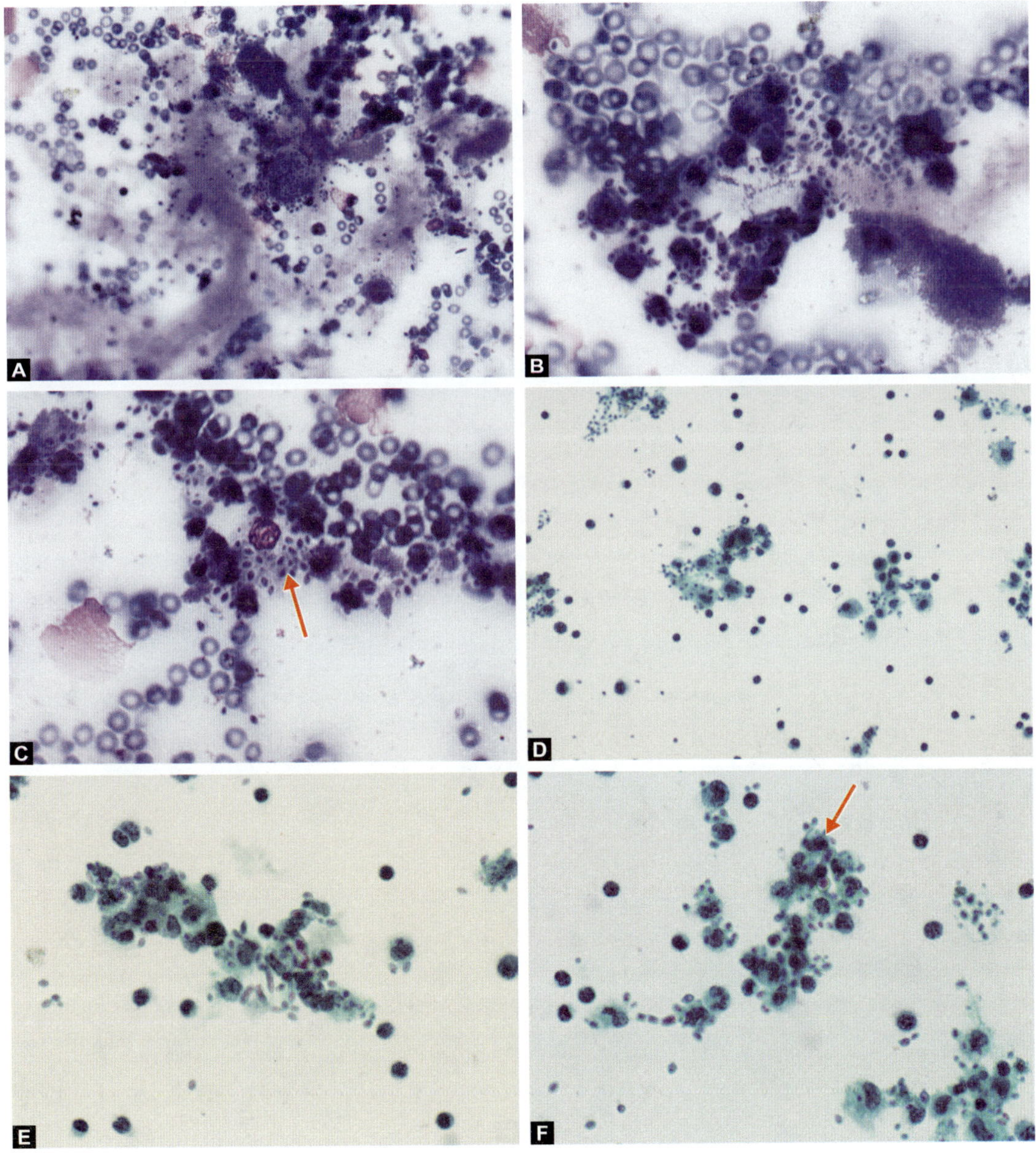

Figures 3.6A to F

Problem 3.7

A 28-year-old male of renal transplant recipient presented with right sided pleural effusion. What is your diagnosis from the microphotographs in Figure 3.7 of the effusion cytology?

A. Candidal infection
B. Mucormycosis infection
C. Aspergillus infection
D. Acute inflammation only

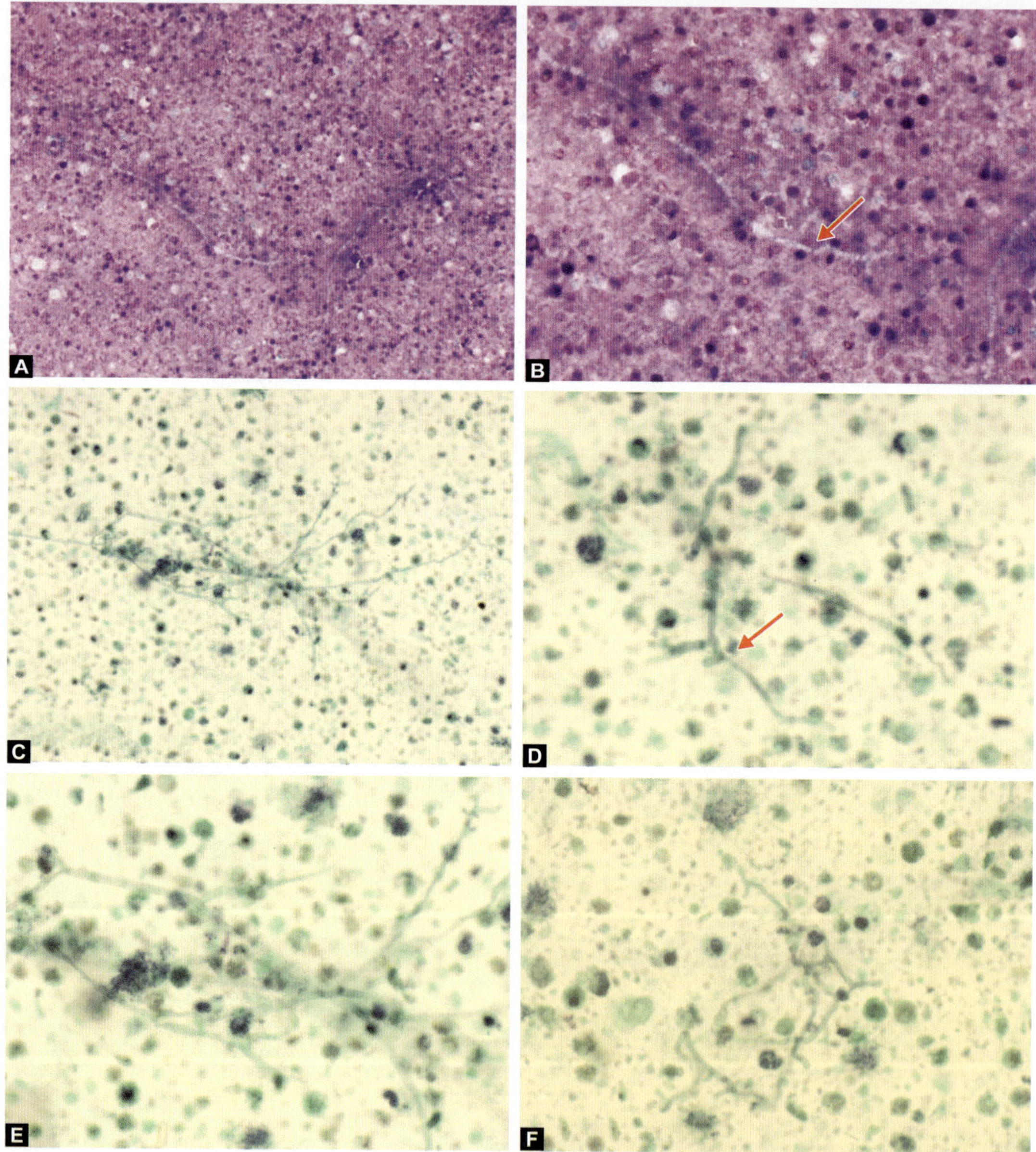

Figures 3.7A to F

Problem 3.8

A 51-year-old female presented with massive ascites and ill-defined mass in the lower abdomen. The most likely diagnosis from the microphotographs in Figure 3.8 of the effusion cytology is:

A. Metastatic adenocarcinoma
B. Lymphocyte rich effusion
C. Reactive mesothelial cells
D. Mesothelioma

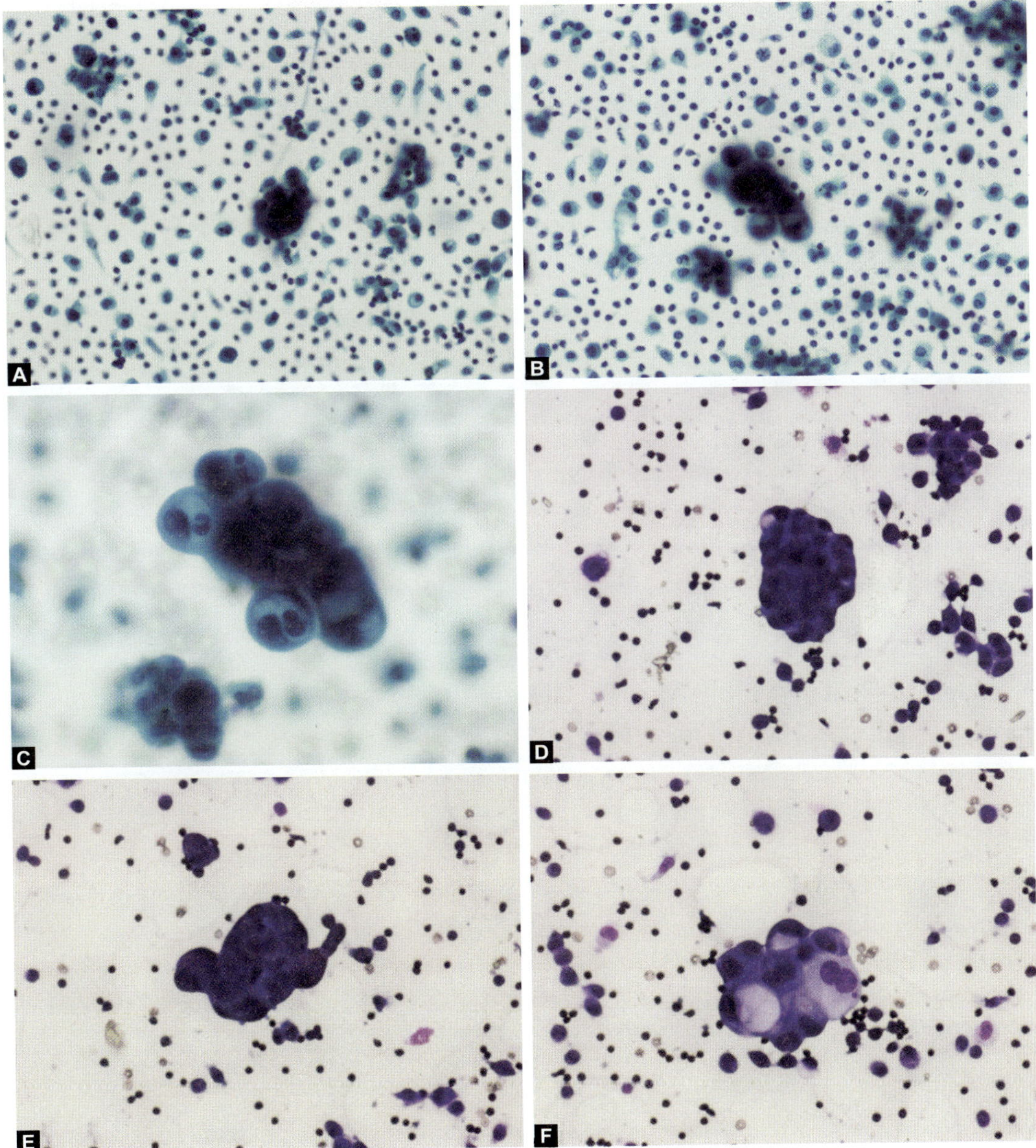

Figures 3.8A to F

Problem 3.9

A 55-year-old female presented with back pain and right sided pleural effusion. The most likely diagnosis from the microphotographs in Figure 3.9 of the effusion cytology is:

A. Lymphocyte rich effusion possibly tuberculosis
B. Metastatic small cell carcinoma
C. Plasma cell neoplasm
D. Mesothelial cells only

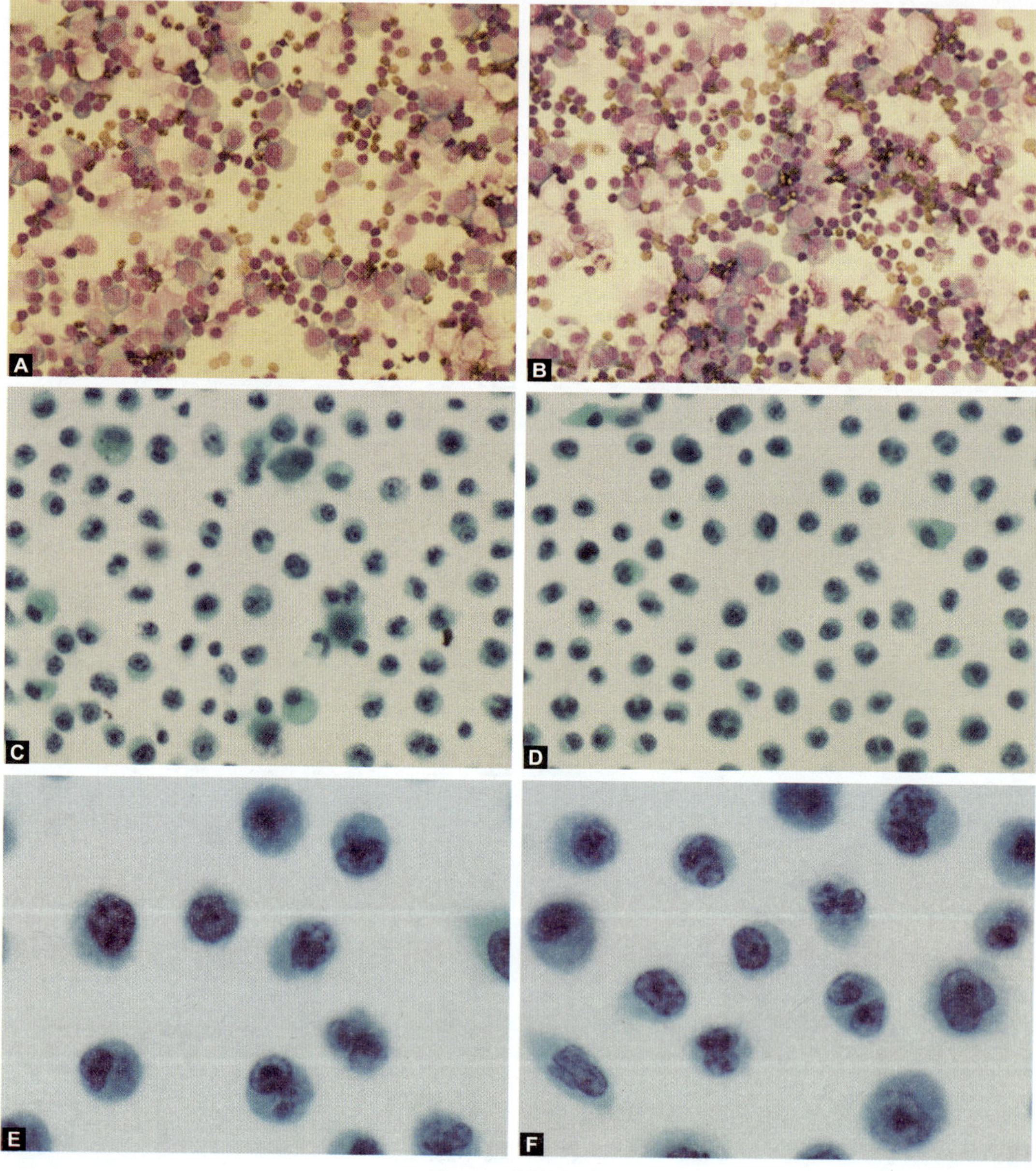

Figures 3.9A to F

Problem 3.10

A 23-year-old female presented with huge adnexal mass in the lower abdomen. Her serum α-fetoprotein level was 2300 IU/mL. The patent also had ascites. Peritoneal fluid cytology was done preoperatively (Fige 3.10A to F). The adnexal tumor was removed. Gross and histopathology section of the ovarian mass were also shown in Figures 3.10G to I. The most likely diagnosis is:

A. Dysgerminoma
B. Mucinous adenocarcinoma
C. Reactive mesothelial cells
D. Yolk sac tumor

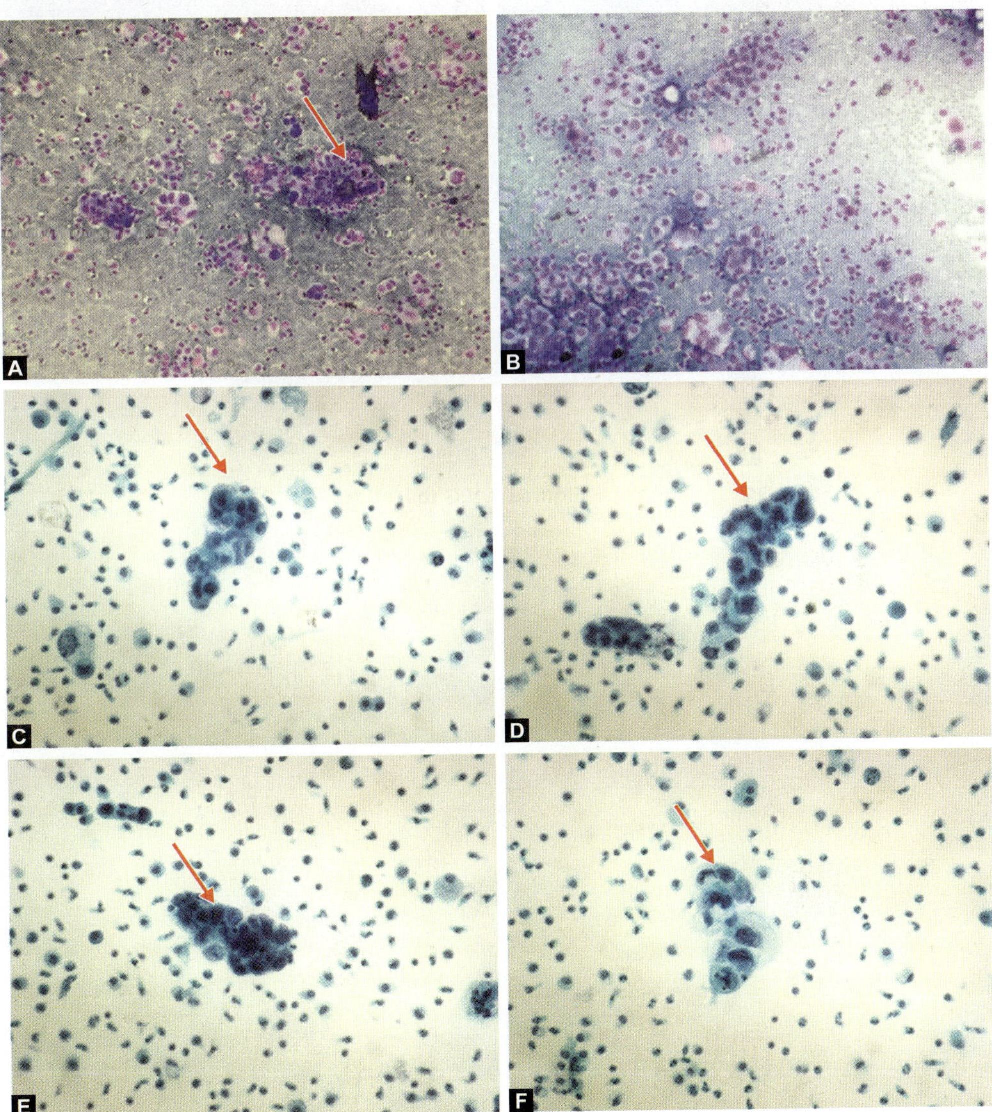

Figures 3.10A to F

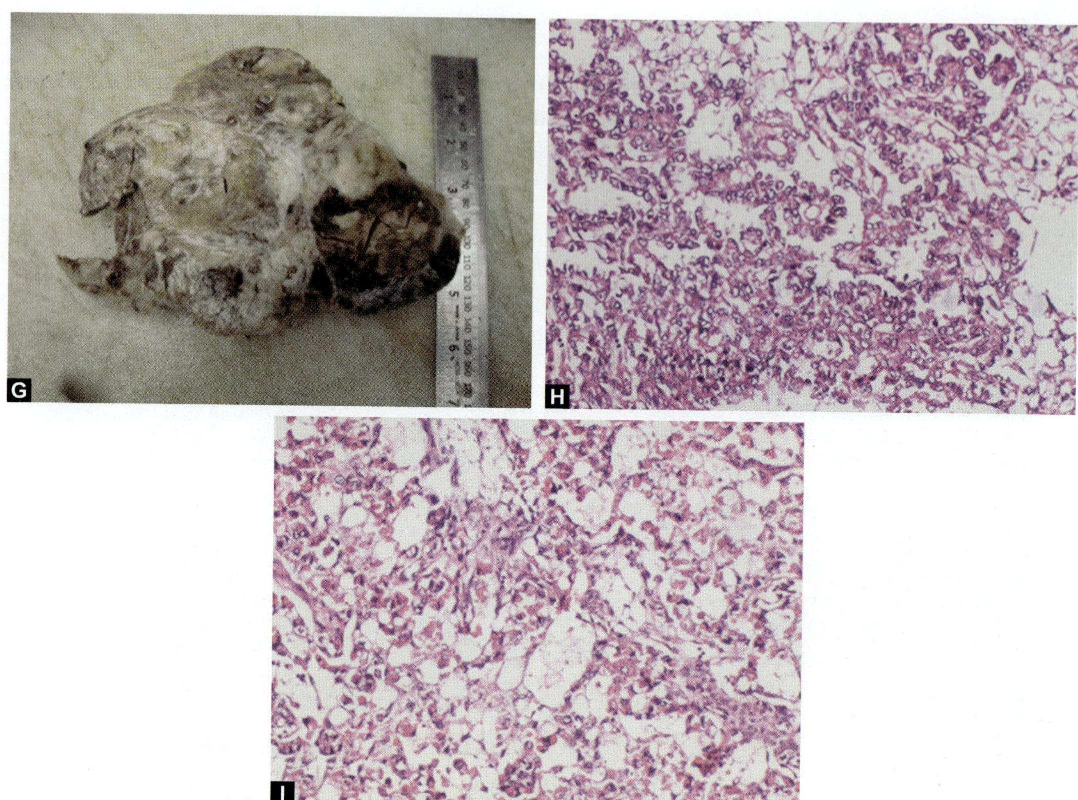

Figures 3.10G to I

Problem 3.11

A 16-year-old male had multiple bilateral cervical lymph nodes 2 to 3 cm diameter for 18 days. He also developed heaviness of right chest due to right sided pleural effusion. FNAC of the lymph node was done (Figs 3.11A and B) followed by cytological examination of the effusion fluid (Figs 3.11C to F). What is the most likely diagnosis?

A. Non-Hodgkin lymphoma
B. Reactive mesothelial cell reaction
C. Tuberculosis
D. Dysgerminoma

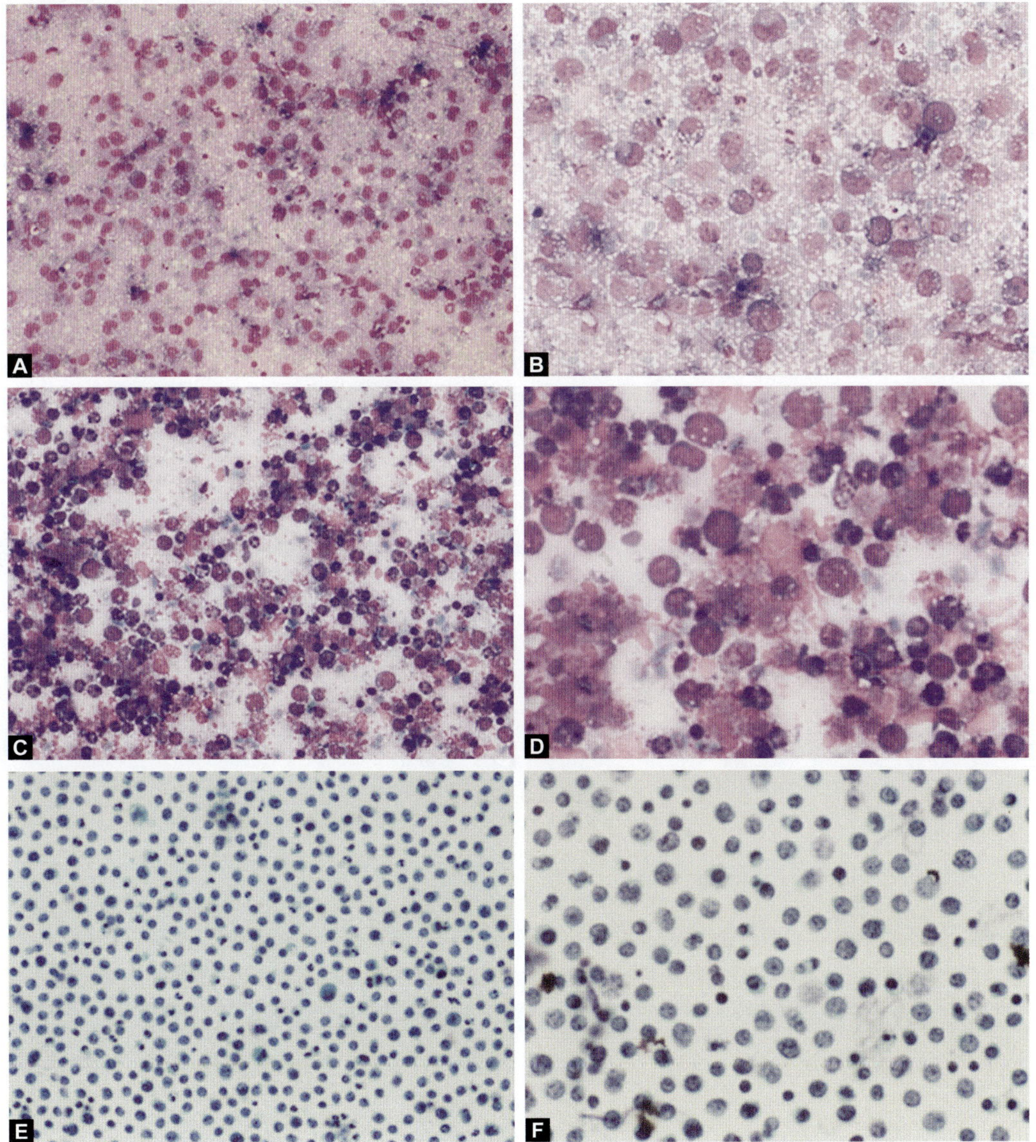

Figures 3.11A to F

Problem 3.12

A 61-year-old male patient had breathlessness along with right sided pleural effusion. Chest X-ray of the patient shows diffuse thickening of the surface of right lung. The pleural cavity is rapidly filling even after repeated paracentesis. The most likely diagnosis from Figure 3.12 is:

A. Adenocarcinoma
B. Reactive mesothelial cells only
C. Mesothelioma
D. Metastatic renal cell carcinoma

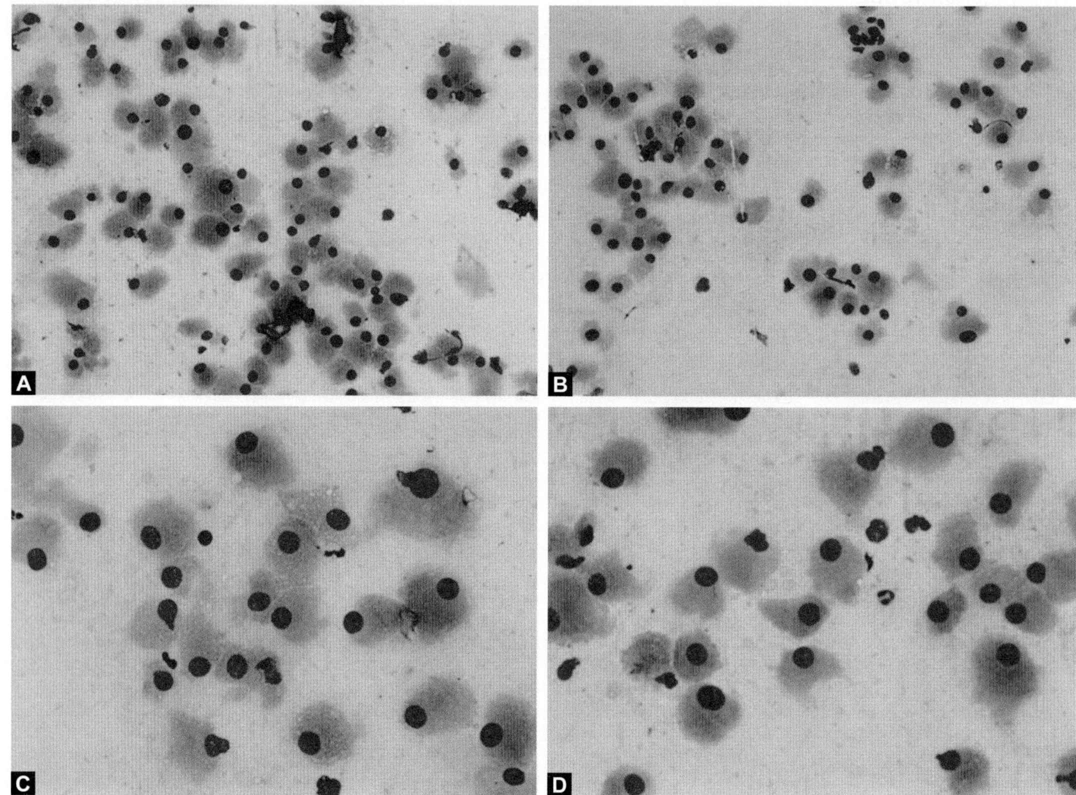

Figures 3.12A to D

Problem 3. 13

A 6-year-old boy presented with dyspnea and mediastinal lymphadenopathy along with bilateral pleural effusion. The most likely diagnosis from the microphotographs of the effusion cytology (Figs 3.13A to D) and flow cytometry (Figs 3.13E to J) is:

A. Lymphocyte rich effusion possibly benign
B. B cell non-Hodgkin lymphoma possibly Burkitt's lymphoma
C. Metastatic round cell tumor possibly embryonal rhabdomyosarcoma
D. T-Lymphoblastic lymphoma

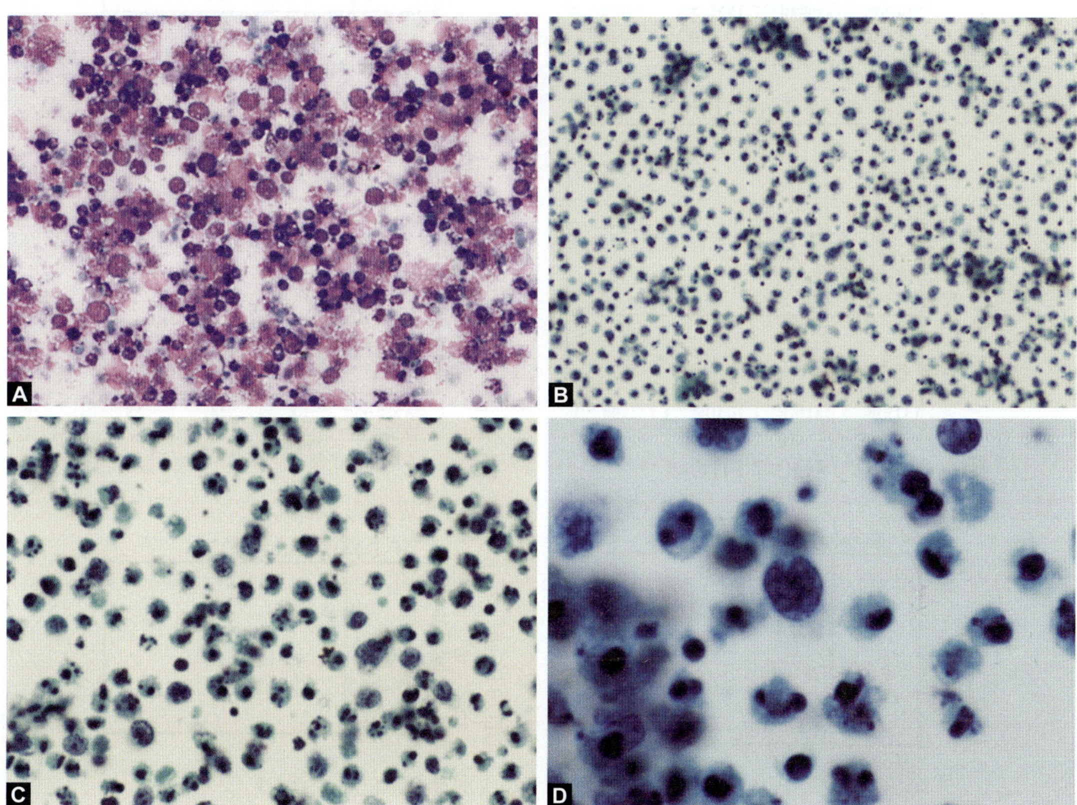

Figures 3.13A to D

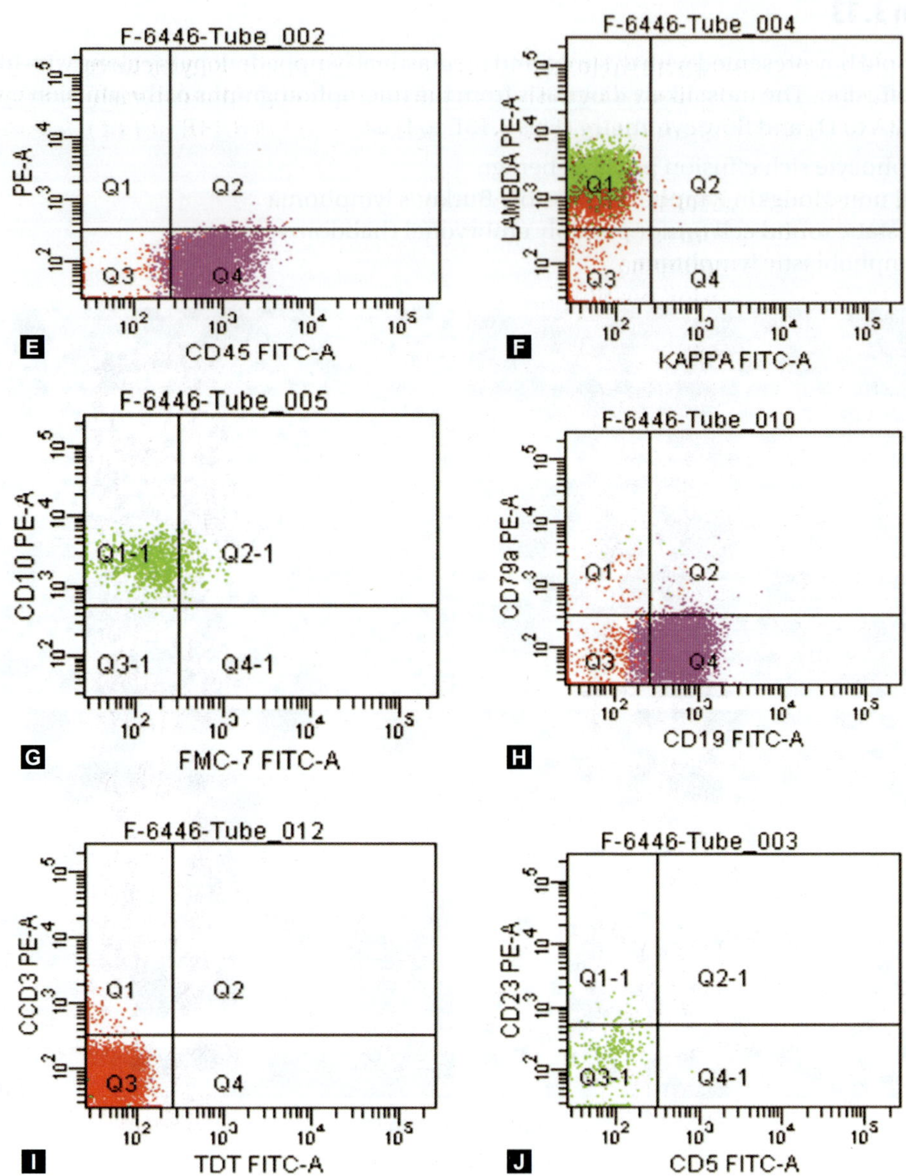

Figures 3.13E to J

Problem 3.14

A 17-year-old female was operated for mass in the ovary. Histopathology pictures of the ovarian mass are shown in the Figures 3.14A to D. The patient also had peritoneal wash cytology fluid. What is your diagnosis from the microphotographs in Figures 3.14E to J of the peritoneal cytology?

A. Reactive mesothelial cells
B. Infiltration by lymphoma
C. Infiltration by dysgerminoma
D. Lymphocyte rich effusion possibly benign

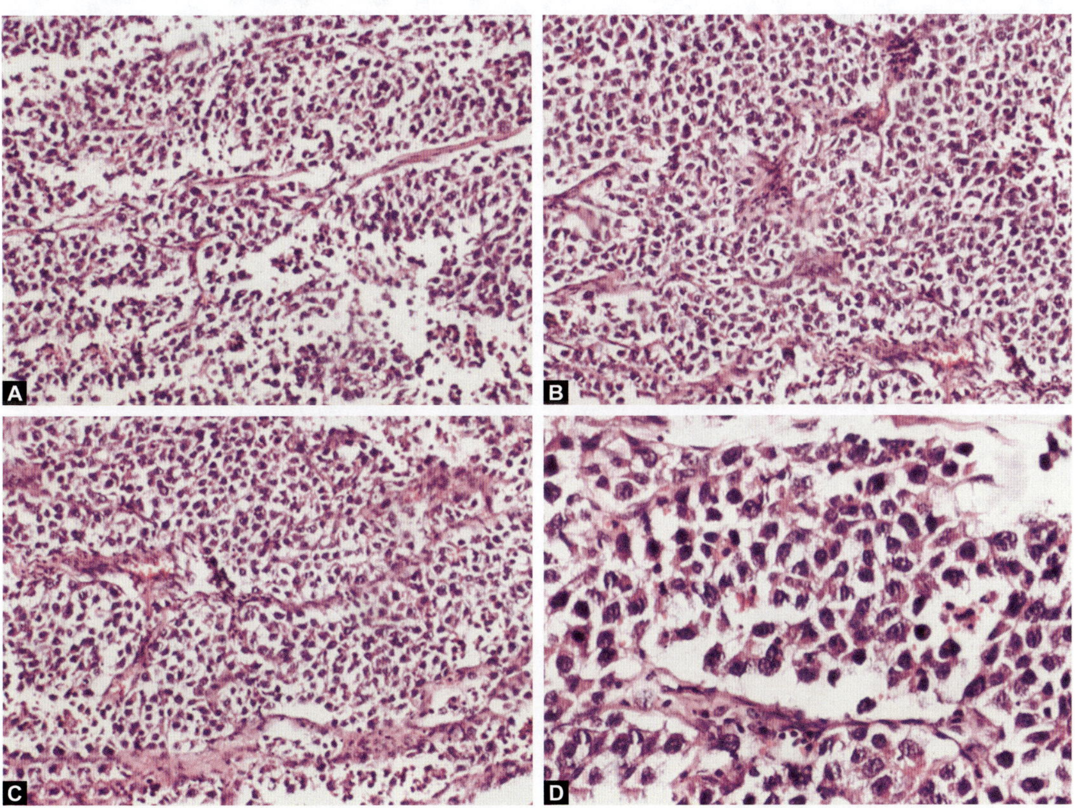

Figures 3.14A to D

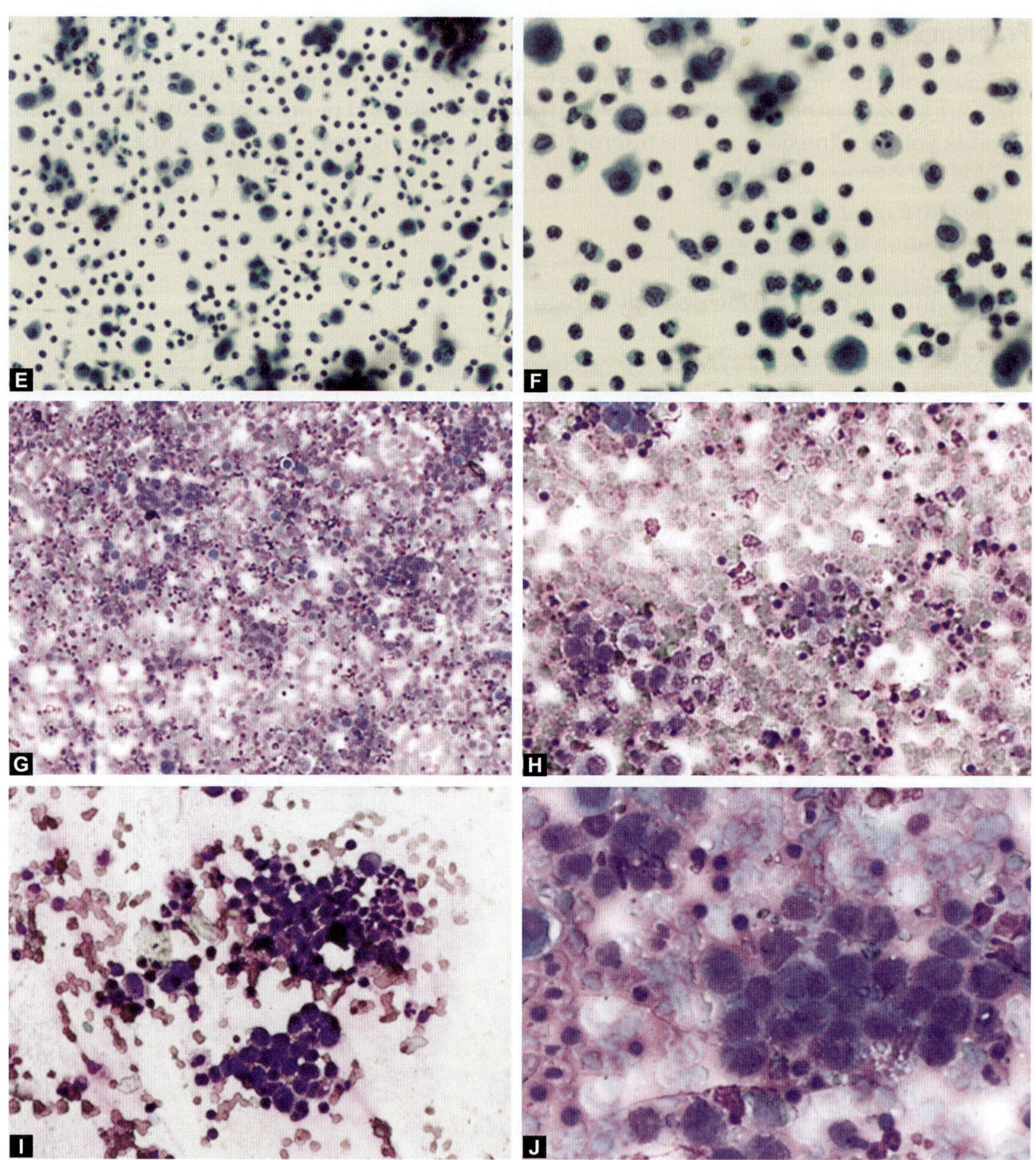

Figures 3.14E to J

Problem 3.15

A 6-year-old boy had 5 cm diameter soft to firm right chest wall swelling for 15 days. On physical examination, he has only pallor. There was no peripheral lymphadenopathy. His liver and spleen were not enlarged. Fine needle aspiration cytology of the chest wall swelling was done and microphotographs are given (Figs 3.15A to H). The boy developed right sided pleural effusion. The cytological examination of the pleural fluid was done (Figs 3.15I to N). The most likely diagnosis is:

A. Non-Hodgkin lymphoma
B. Reactive mesothelial cells
C. Rhabdomyosarcoma
D. Infiltration by soft tissue Ewings' tumor/peripheral neuroectodermal tumor

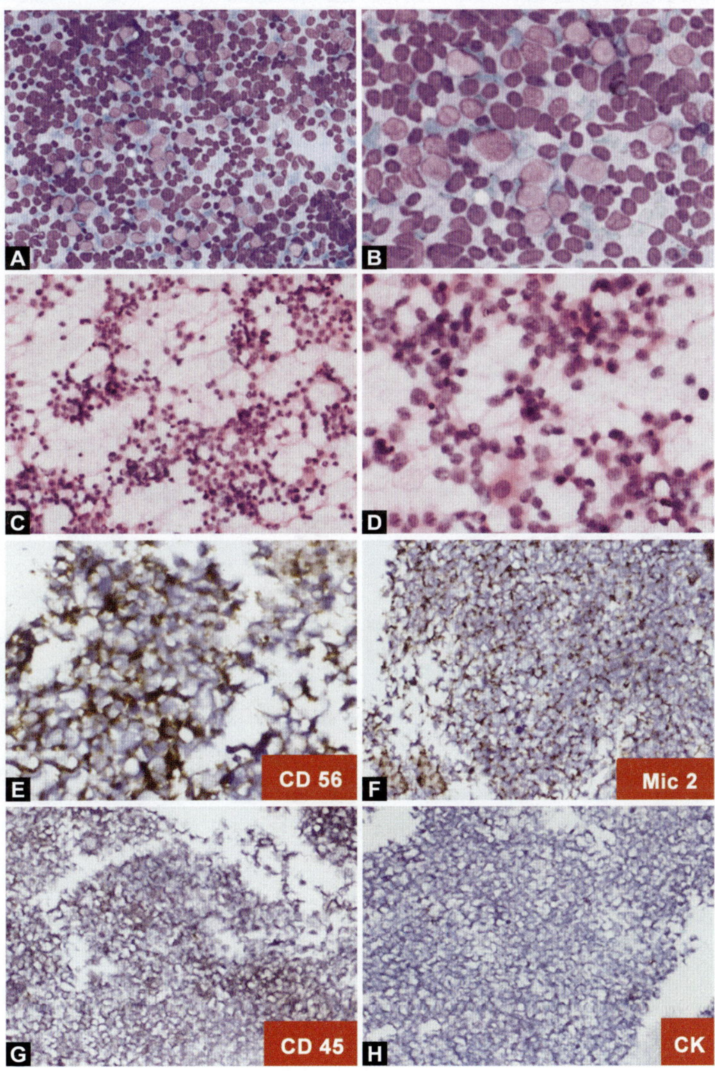

Figures 3.15A to H

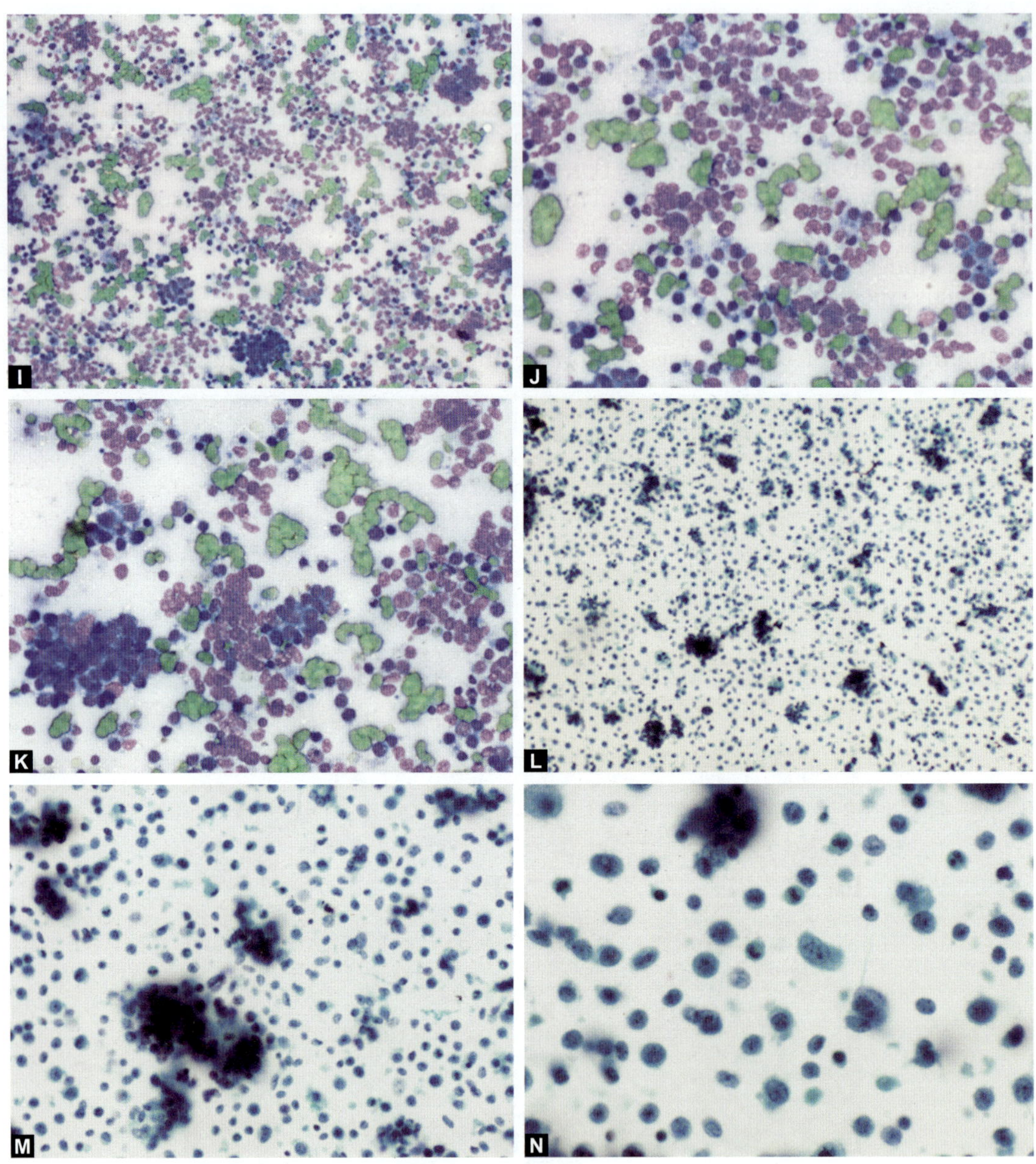

Figures 3.15I to N

Problem 3.16

A 45-year-old female presented with heaviness in abdomen, ascites and pelvic mass for 2 months. Abdominal paracentesis yielded thick gelatinous material. The most likely diagnosis from the microphotograph in Figure 3.16 of the effusion cytology is:

A. No malignant cells seen
B. Adenocarcinoma
C. Pseudomyxoma peritonei
D. Mesothelioma

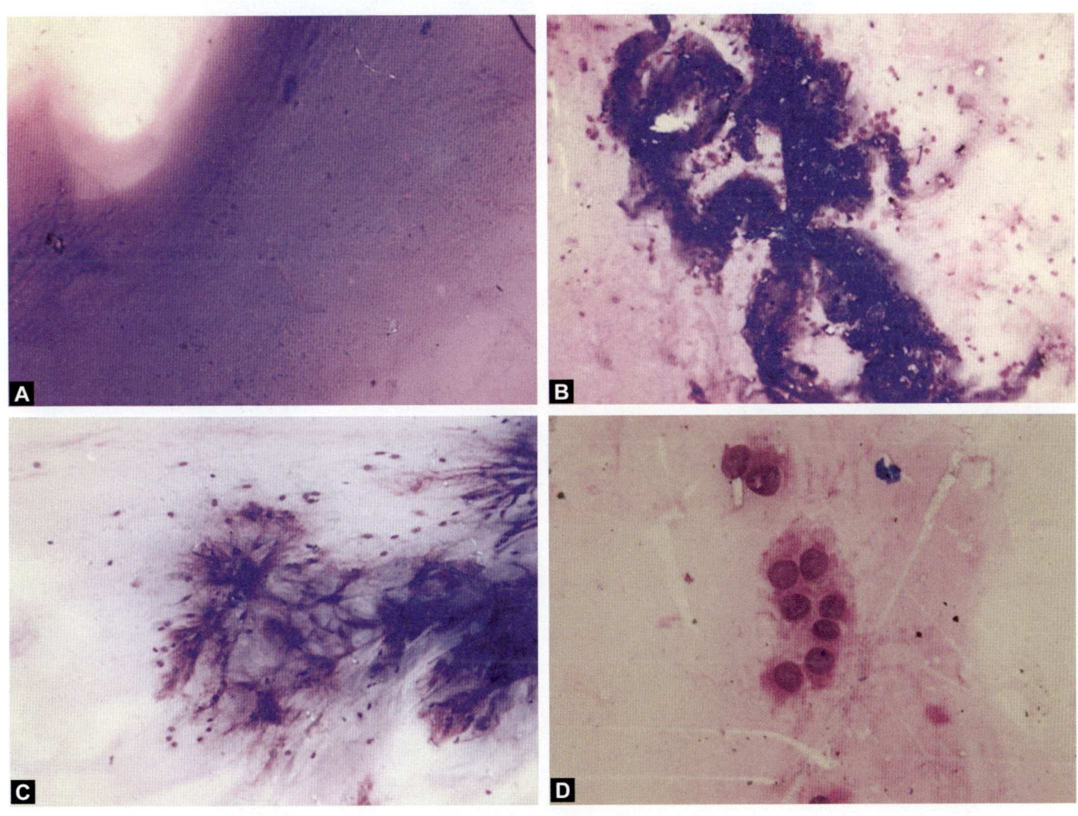

Figures 3.16A to D

Problem 3.17

A 53-year-old female had right upper abdominal mass, ascites and haematuria for 3 months. The most likely diagnosis from the microphotographs in Figure 3.17 of the effusion cytology is:

A. Metastatic renal cell carcinoma
B. Reactive mesothelial cells
C. Mesothelioma
D. Metastatic carcinoma from gastrointestinal tract

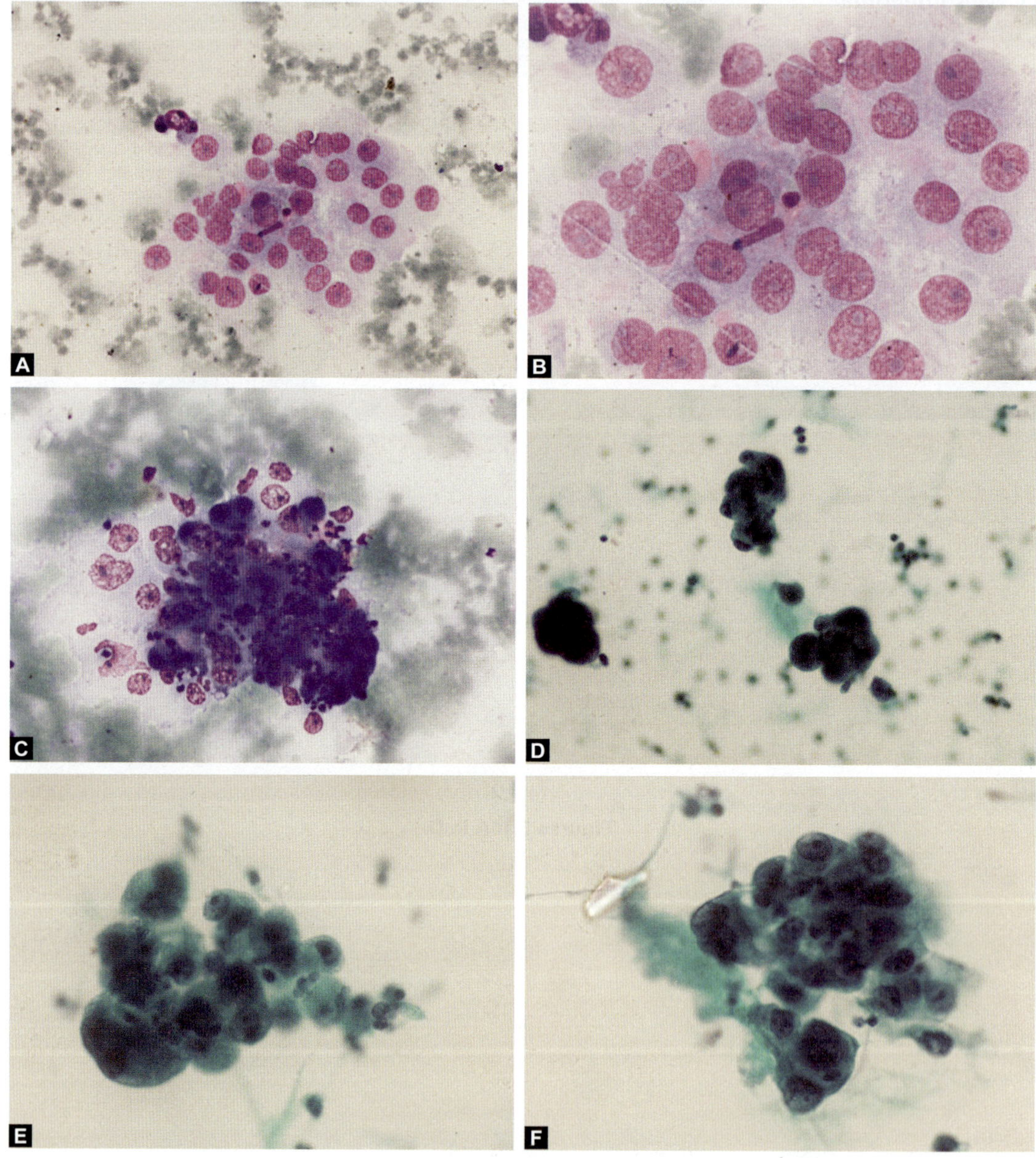

Figures 3.17A to F

Problem 3.18

A 38-year-old male presented with projectile vomiting, anorexia, weight loss and ascites. The most likely diagnosis from the microphotograph in Figure 3.18 of the effusion cytology is:

A. No malignant cell seen
B. Lymphocyte rich effusion
C. Metastatic signet ring carcinoma
D. Mesothelioma

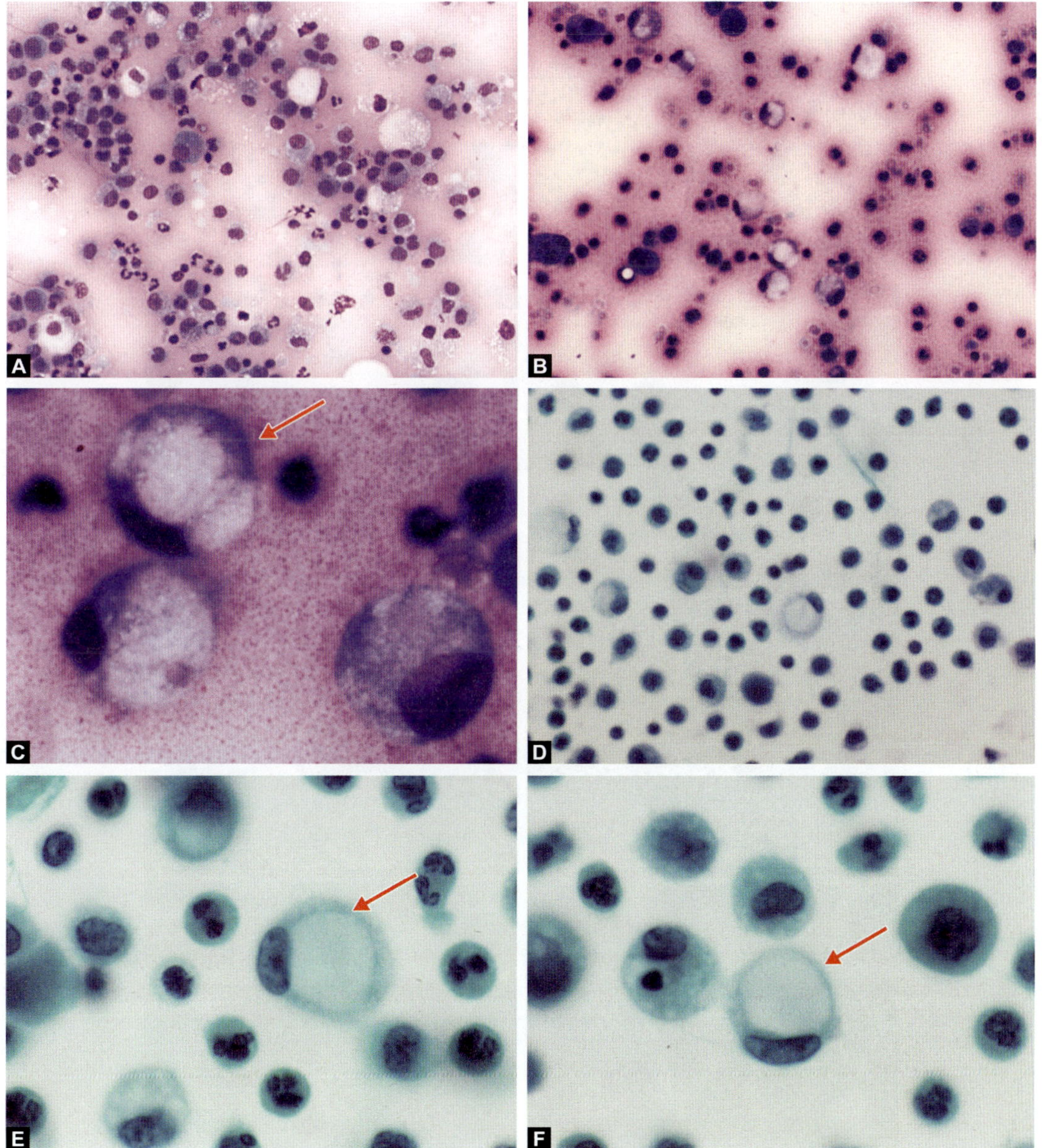

Figures 3.18A to F

Problem 3.19

A 49-year-old female presented with hemoptysis, dry cough and chest pain and left sided pleural effusion. The most likely diagnosis from the microphotograph (Figure 3.19) of the effusion cytology is:

A. Metastatic adenocarcinoma possibly from lung
B. Metastatic small cell carcinoma possibly from lung
C. Reactive mesothelial cells
D. Mesothelioma

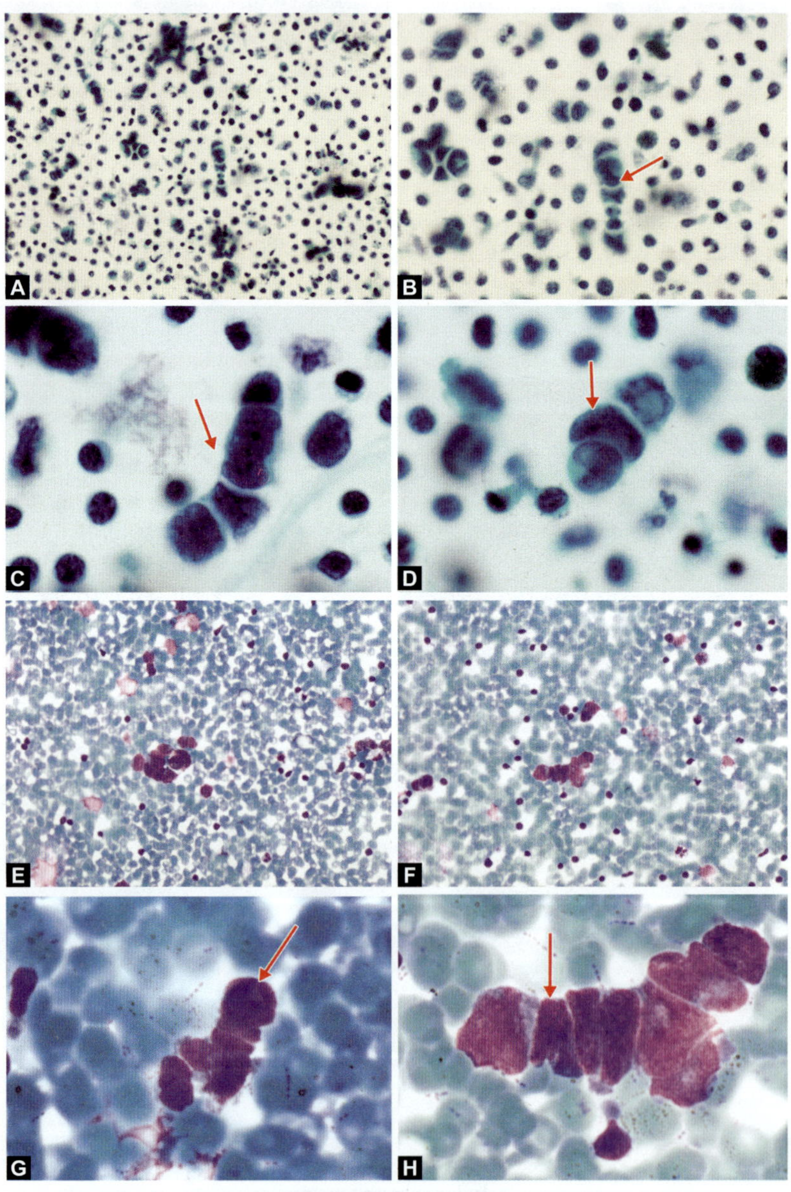

Figures 3.19A to H

Problem 3.20

A 55-year-old male had peritoneal effusion and mass in the upper abdomen. The most likely diagnosis from the microphotographs in Figure 3.20 of the effusion cytology is:

A. Metastatic small cell carcinoma possibly from lung
B. Reactive mesothelial cells: no malignant cells
C. Mesothelioma
D. Metastatic renal cell carcinoma

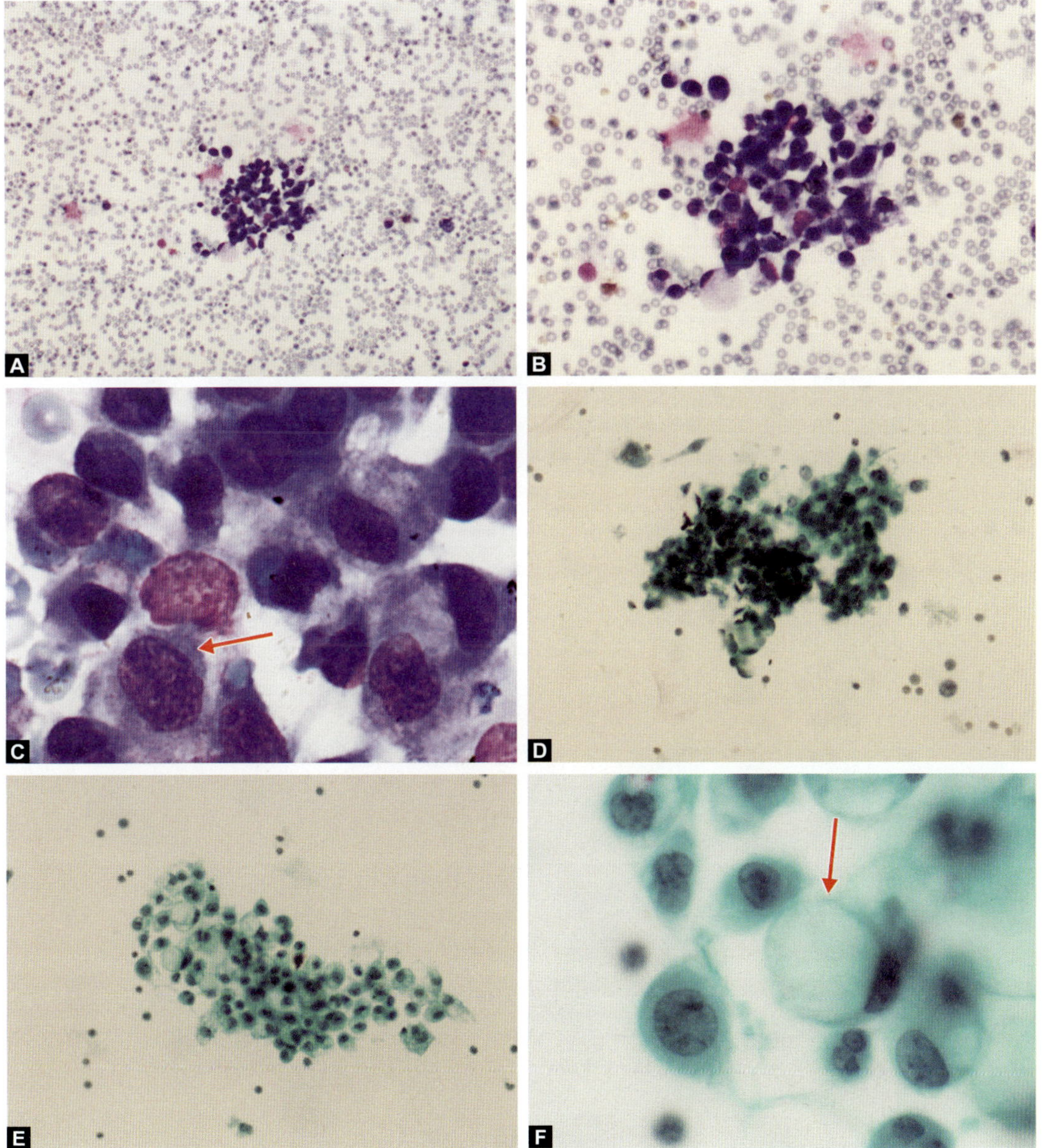

Figures 3.20A to F

Problem 3.21

A 59-year-old male presented with hemoptysis, loss of weight and right sided pleural effusion. The most likely diagnosis from the microphotograph (Figure 3.21) of the effusion cytology is:

A. Metastatic squamous cell carcinoma
B. Reactive mesothelial cells: no malignant cell seen
C. Mesothelioma
D. Metastatic adenocarcinoma

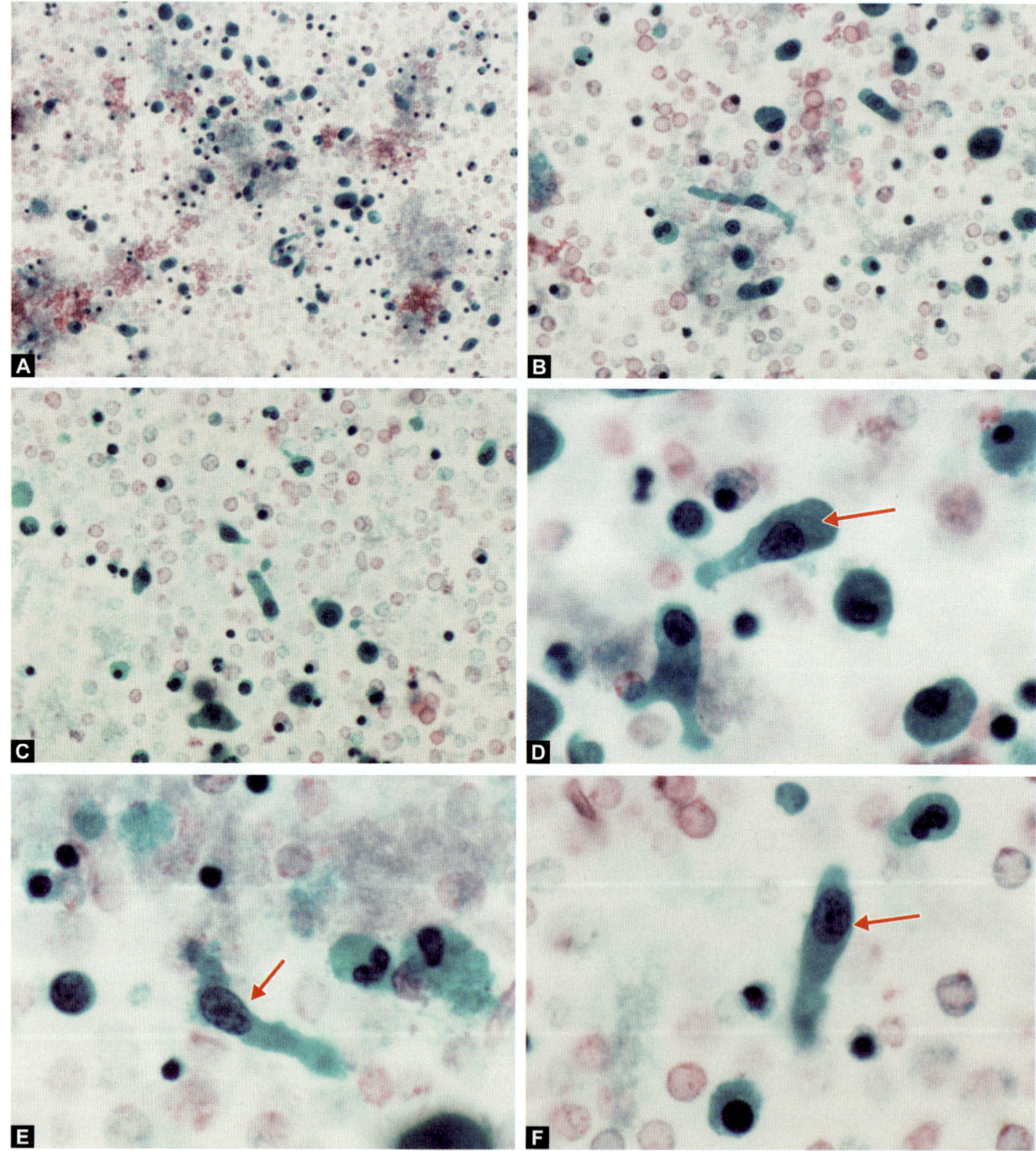

Figures 3.21A to F

Problem 3.22

A 59-year-old male presented with cough and hemoptysis for 25 days. His upper right lobe of lung showed 6 cm diameter space occupying lesion on radiological examination. The patient also developed right sided pleural effusion. The microphotographs of the lung FNAC are shown in Figures 3.22A to C. Figures 3.22D to I show microphotographs of the SurePath preparation of the right pleural effusion. The most likely diagnosis of effusion cytology is:

A. Reactive mesothelial cells
B. Mesothelioma
C. Metastatic small cell carcinoma
D. Metastatic adenocarcinoma

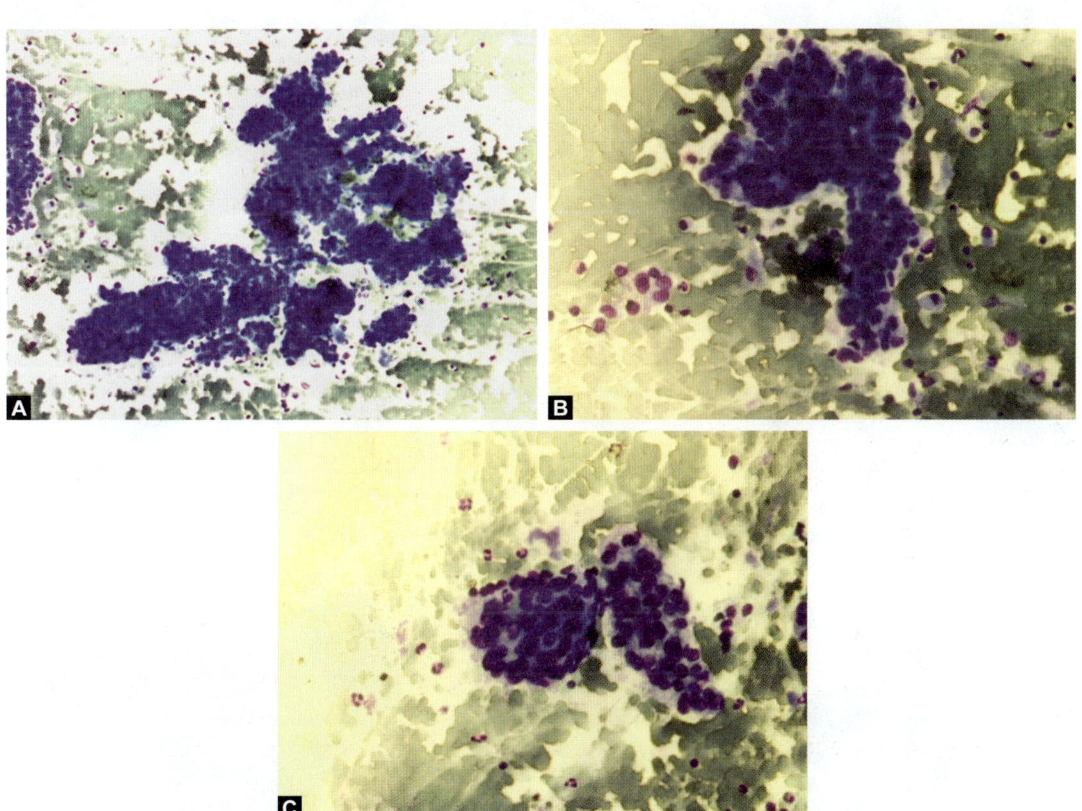

Figures 3.22A to C

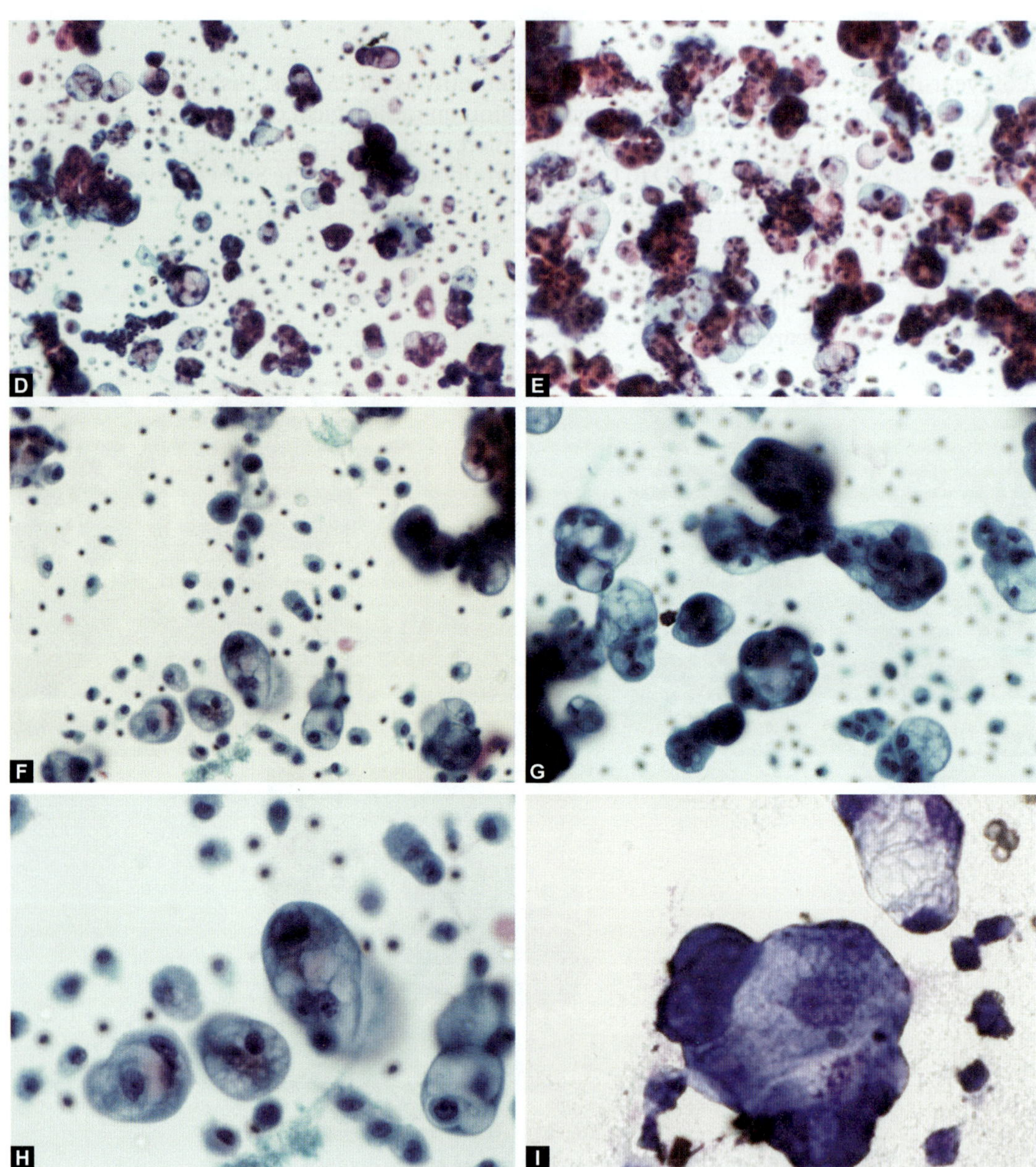

Figures 3.22D to I

Problem 3.23

A 38-year-old female was operated for carcinoma of left breast 2 year back. Histopathology of the tumor is shown in Figures 3.23A to C. Now she presented with left sided pleural effusion. The most likely diagnosis from the microphotographs in Figures 3.23D to I of the effusion cytology is:

A. Metastatic lobular carcinoma of breast
B. Mesothelial cells only: no malignant cells
C. Non-Hodgkin lymphoma
D. Mature lymphocytes only possibly tuberculosis

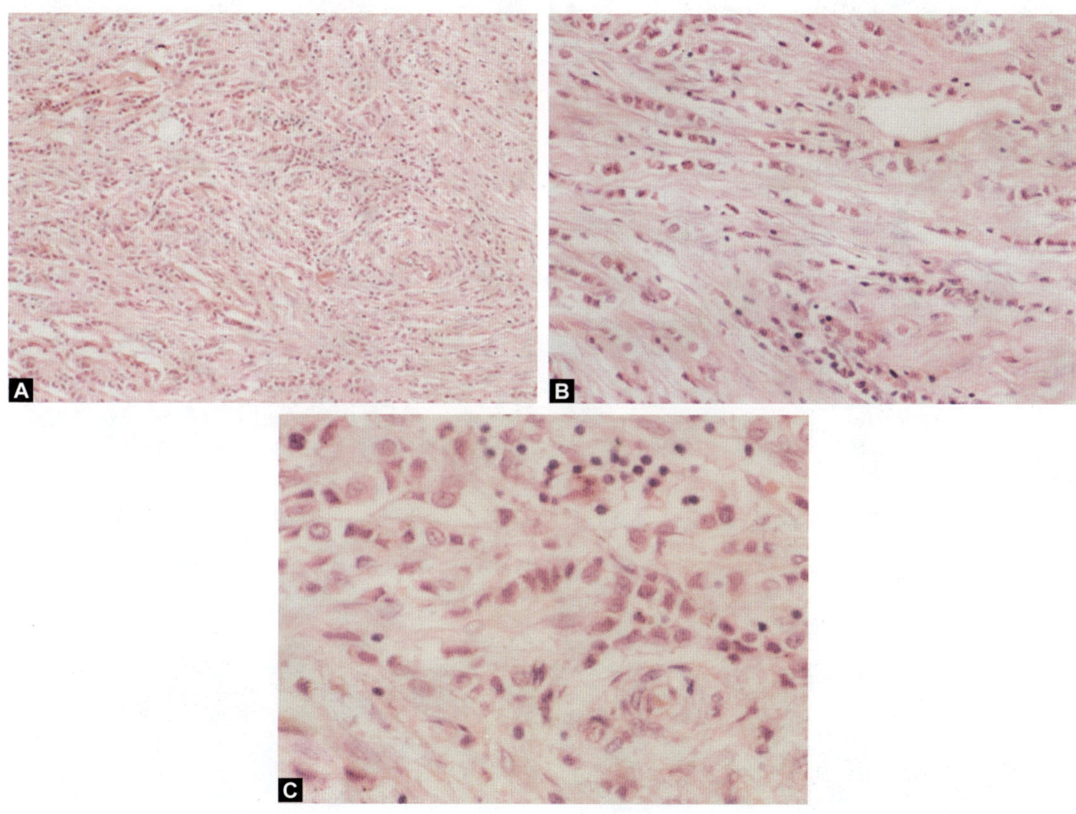

Figures 3.23A to C

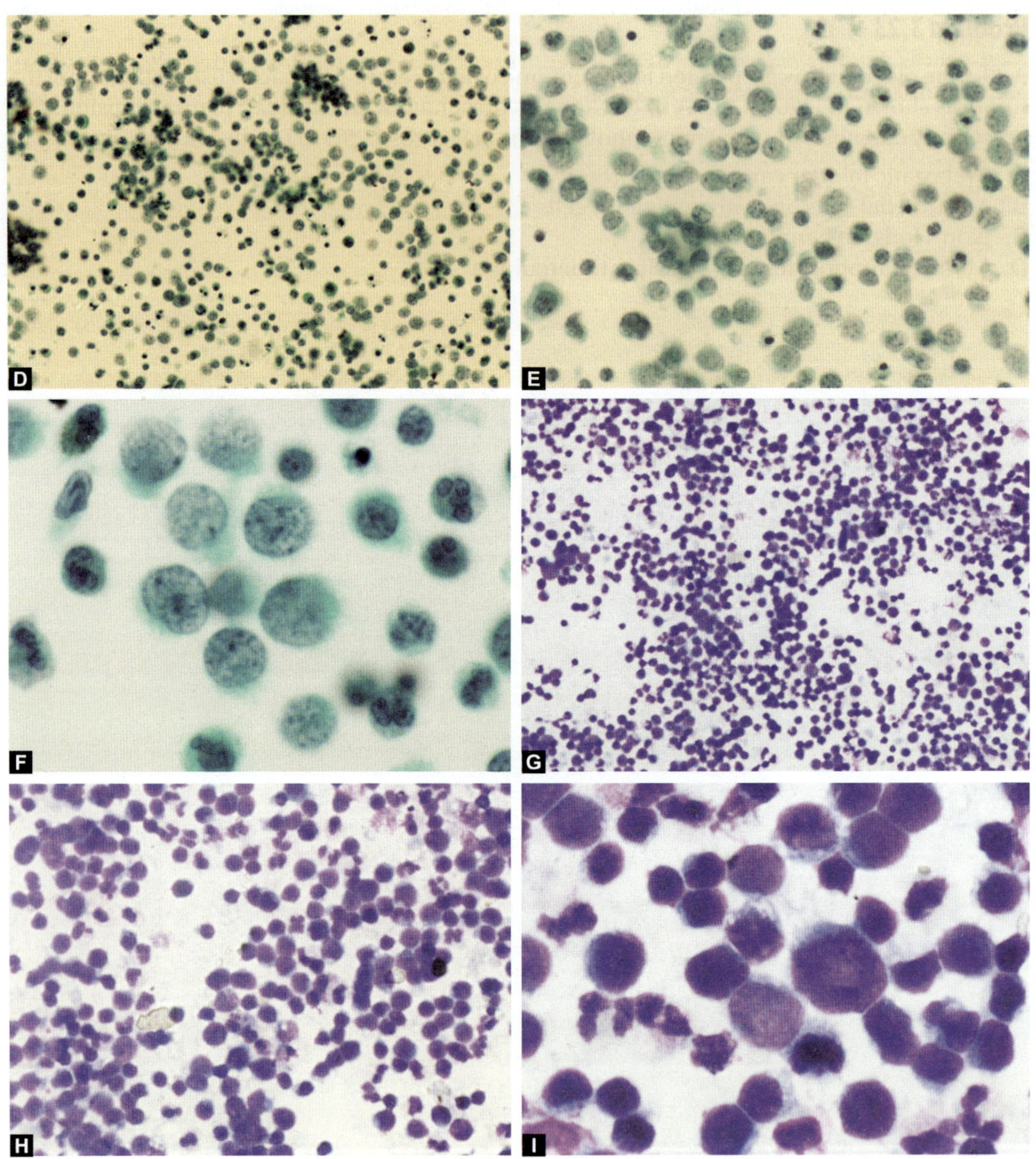

Figures 3.23D to I

Problem 3.24

A 39-year-old male presented with dysphagia, breathlessness and right sided pleural effusion for 2 months. The most likely diagnosis from the microphotographs in Figure 3.24 of the effusion cytology is:

A. Empyema (acute inflammation)
B. Reactive mesothelial cells only, no malignancy
C. Metastatic adenocarcinoma
D. Metastatic squamous cell carcinoma

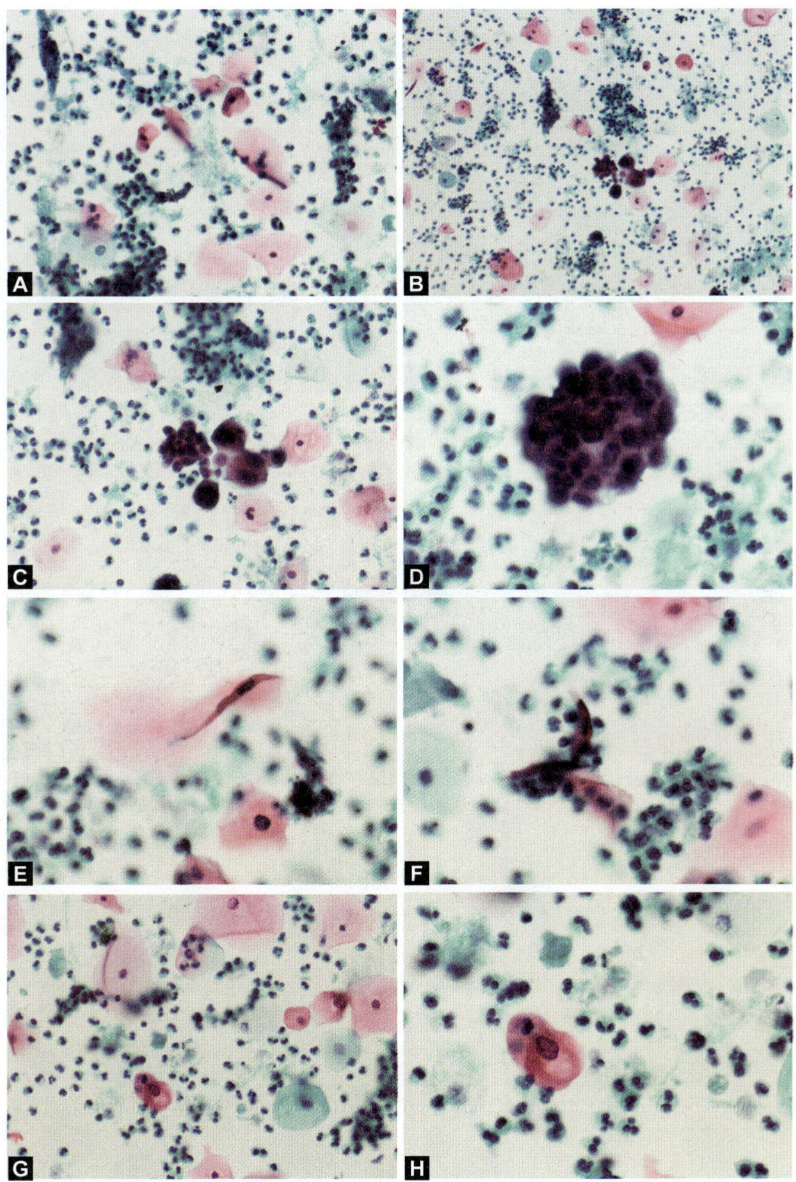

Figures 3.24A to H

Problem 3.25

A 26-year-old male was a diagnosed case of diffuse large B cell lymphoma 2 year back. He was treated by chemotherapy. Now he had history of anemia and weakness. He also developed right sided pleural effusion. The most likely diagnosis from Figure 3.25 is:

A. Lymphoma (recurrence)
B. Lymphocyte rich effusion
C. Metastatic carcinoma
D. Mesothelioma

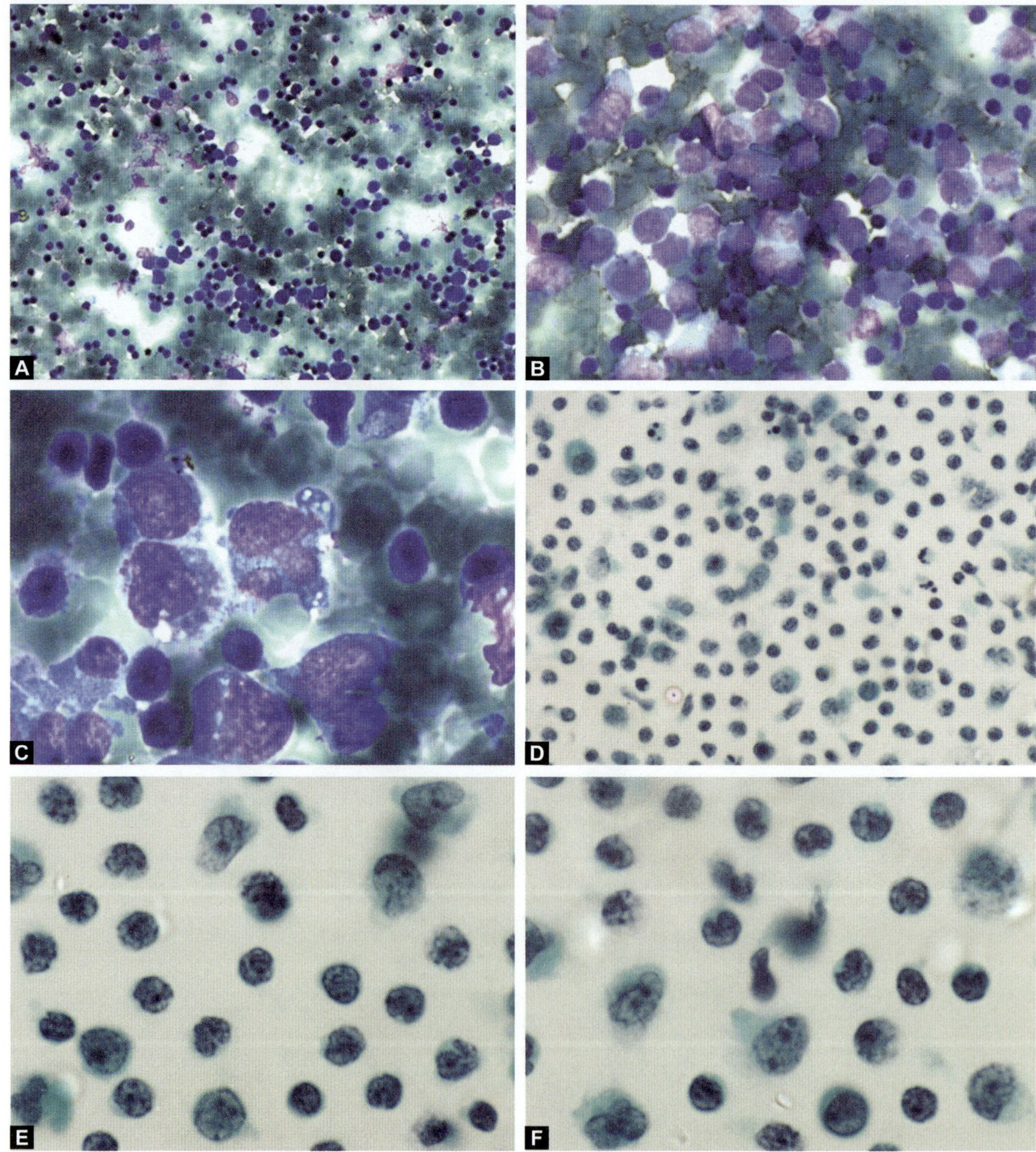

Figures 3.25A to F

Problem 3.26

A 39-year-old female had 3.5 cm diameter hard swelling over the left breast. Histopathology of the swelling are shown in Figures 3.26A to D. The patient was treated accordingly. After 2 years she developed ascites. The microphotograph of the SurePath preparation (Figs 3.26E to J) of the effusion fluid are shown.

The most likely diagnosis is:

A. Reactive mesothelial cells
B. Metastatic ductal carcinoma of breast
C. Mesothelioma
D. Metastatic lobular carcinoma of breast

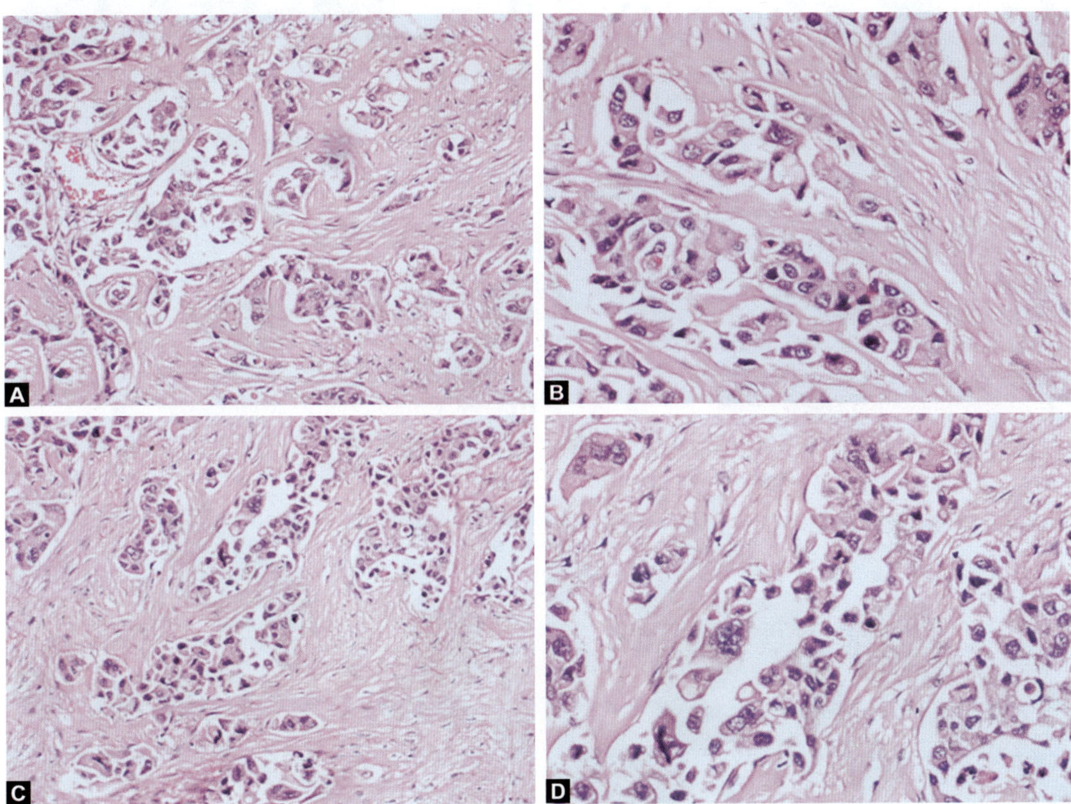

Figures 3.26A to D

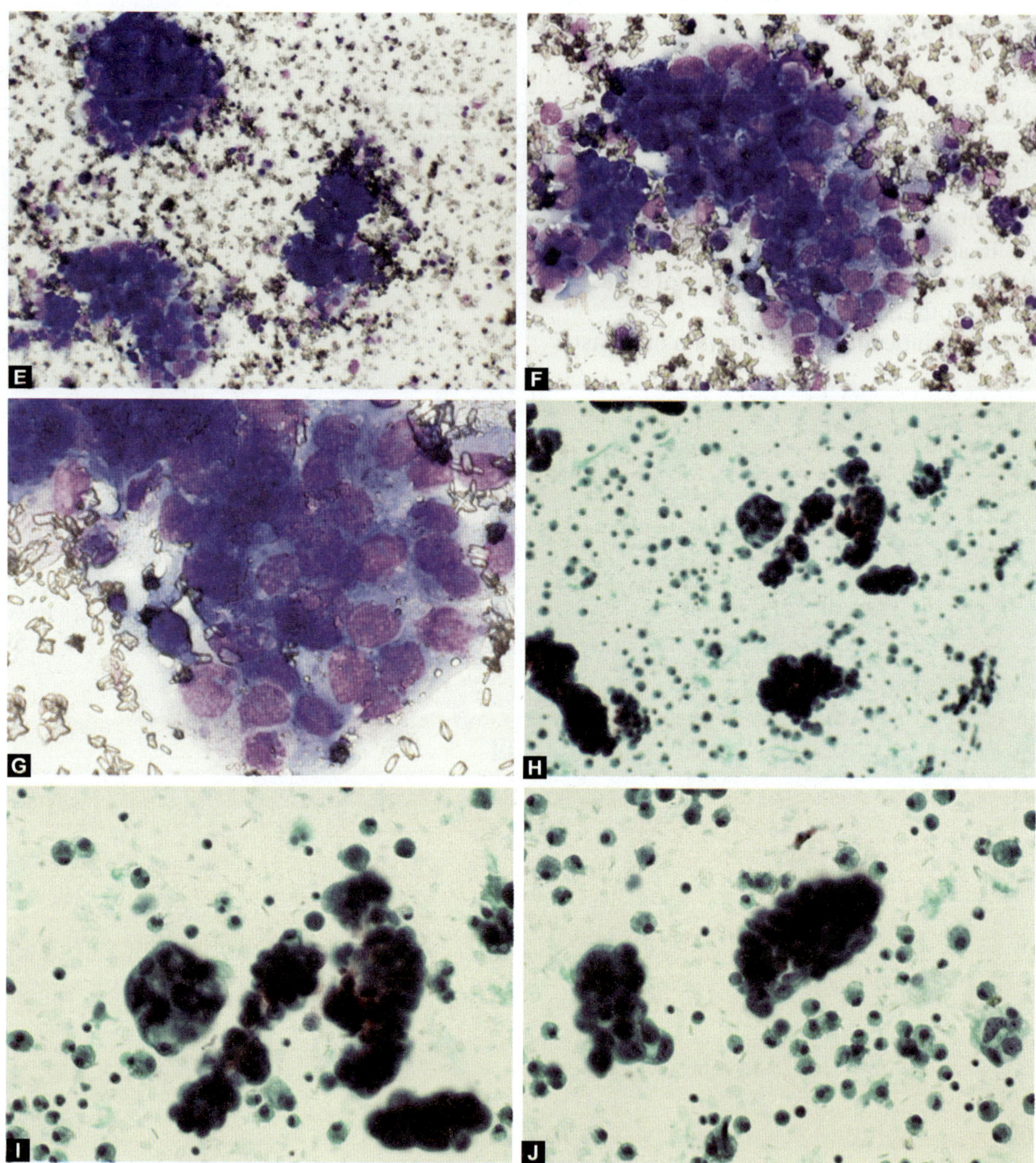

Figures 3.26E to J

Problem 3.27

A 67-year-old male presented with massive left sided pleural effusion. His lung was collapsed and it was difficult to detect any space occupying lesion in the lung on radiograph. He had increased frequency of micturation and dysuria for last 2 years. His bowel habit was normal. Detailed immunocytochemistry was done from the cell block preparation of effusion cytology sample. The most likely diagnosis from Figure 3.27 is:

A. Metastatic adenocarcinoma from large intestine
B. Metastatic adenocarcinoma from lung
C. Reactive mesothelial cells
D. Metastatic prostatic carcinoma

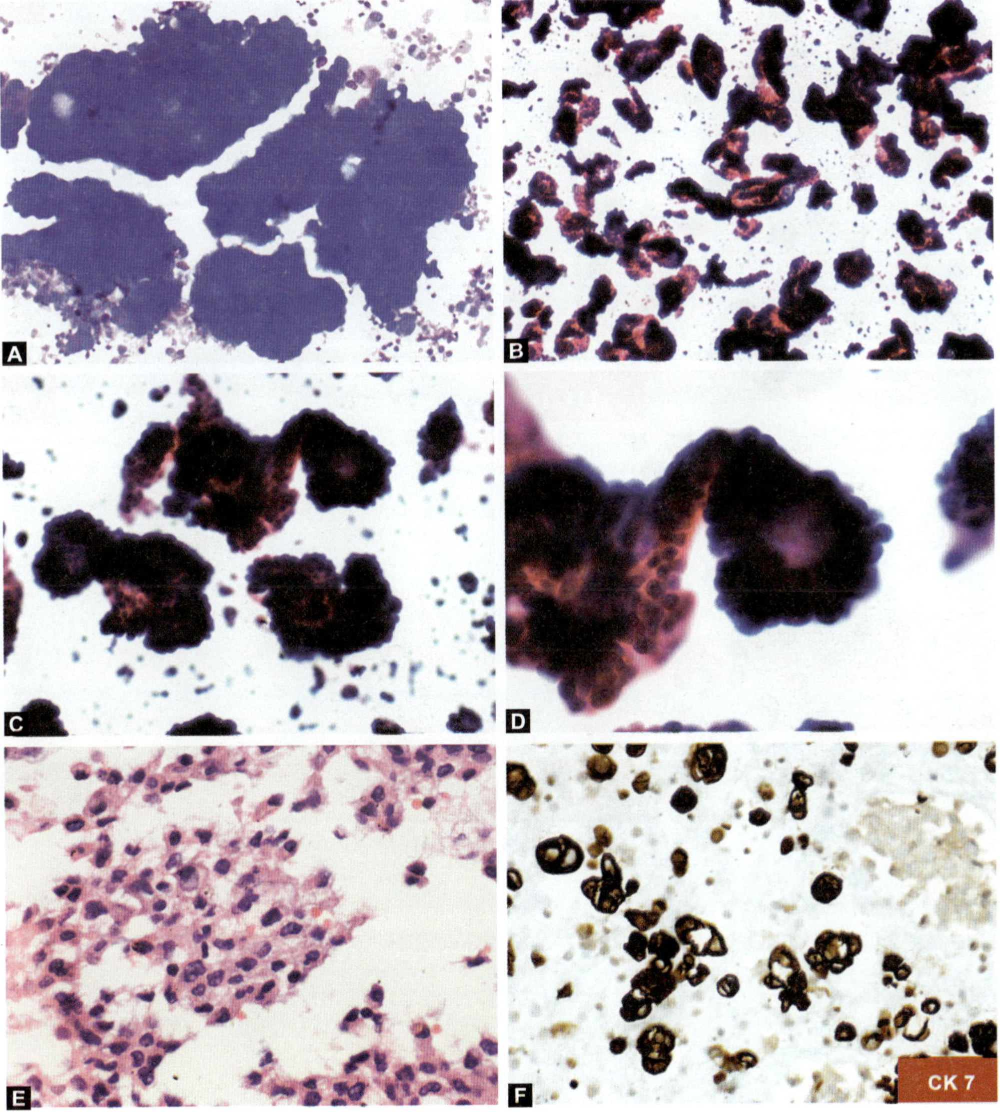

Figures 3.27A to F

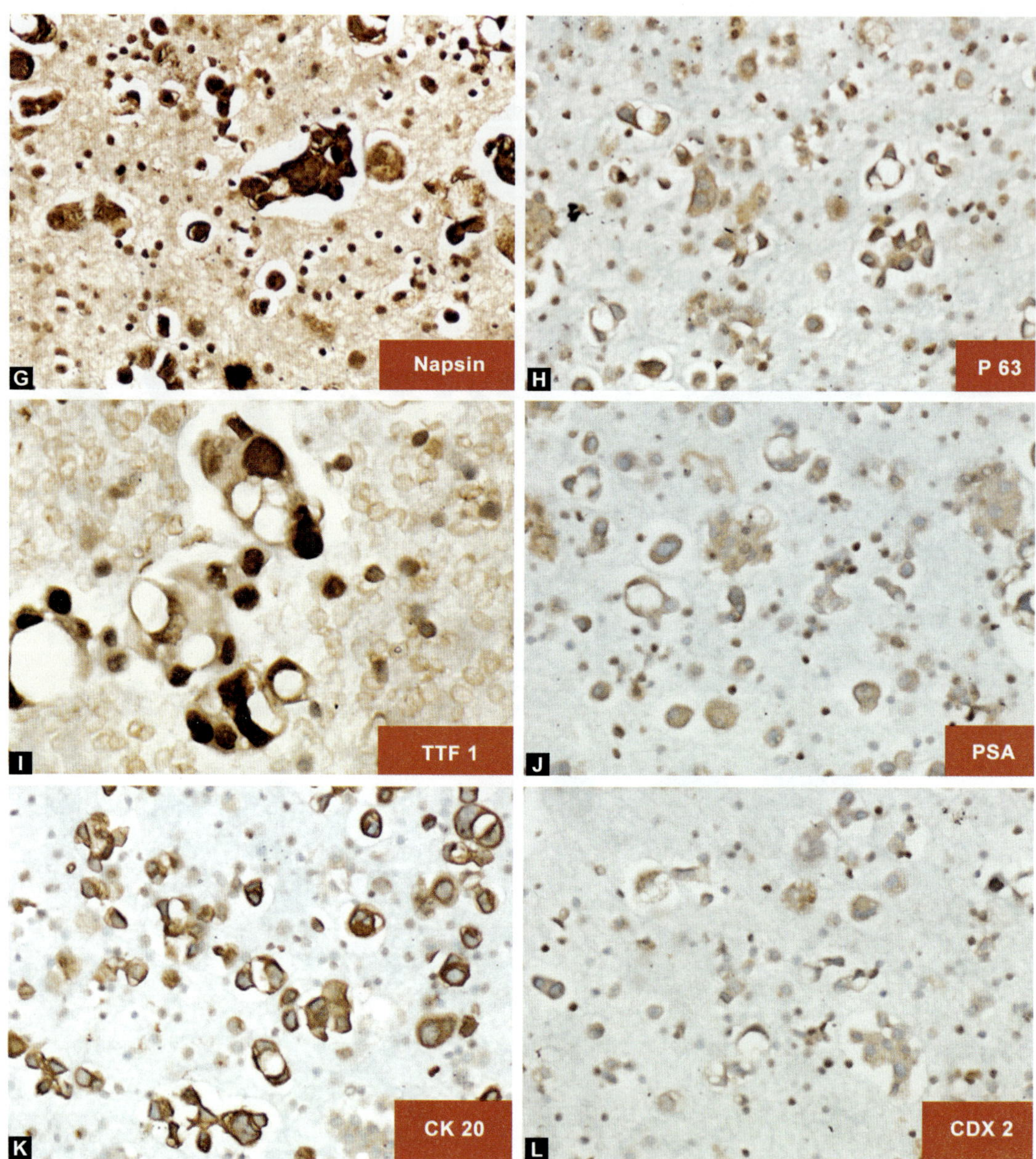

Figures 3.27G to L

Problem 3.28

A 51-year-old male presented with projectile vomiting and weight loss for 2 months. He also had peritoneal effusion. The most likely diagnosis from effusion cytology smear (Fig. 3.28) is:

A. Reactive mesothelial cells
B. Tuberculous effusion
C. Signet ring cell carcinoma
D. Mesothelioma

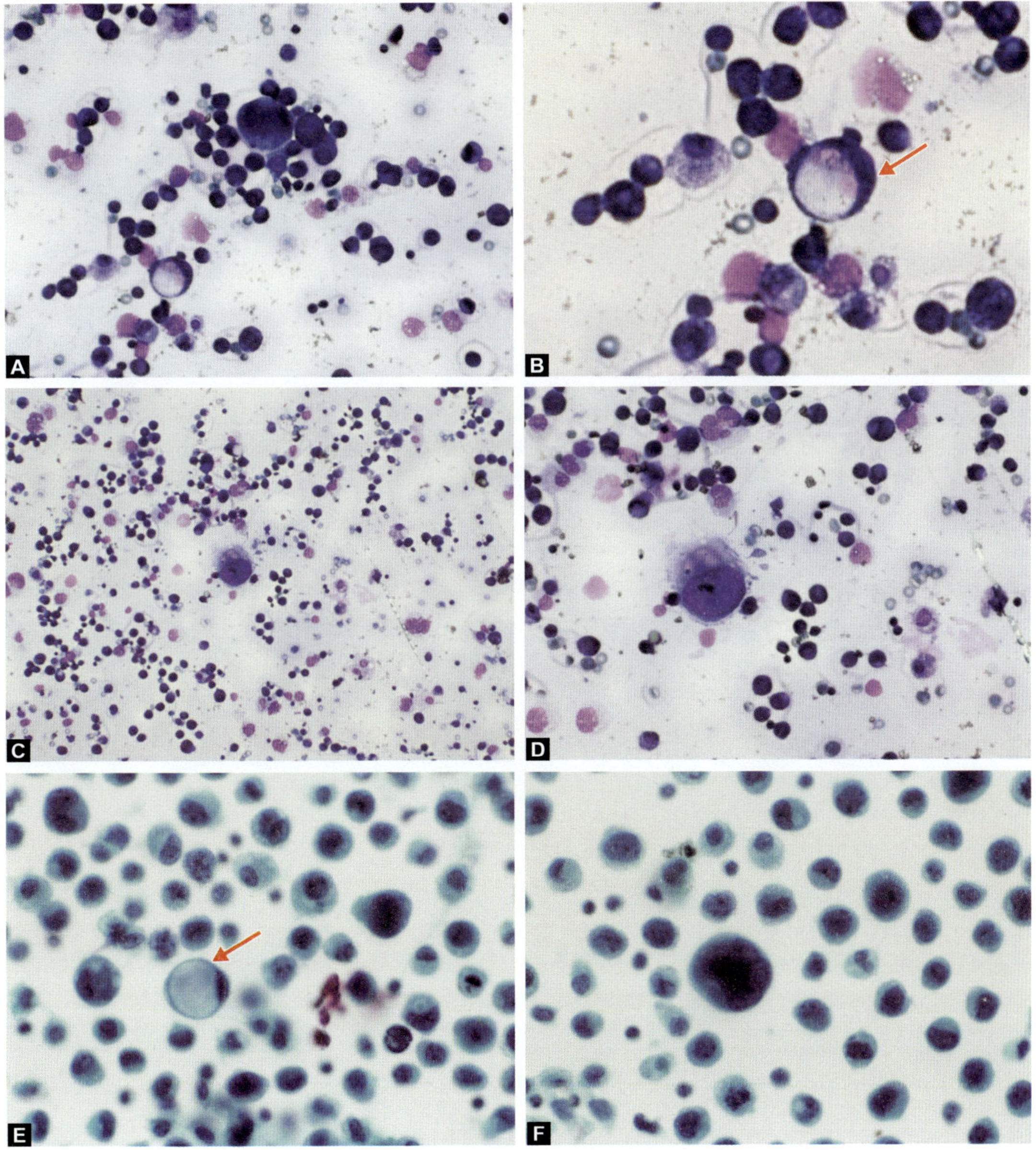

Figures 3.28A to F

Problem 3.29

A 51-year-old female presented with ascites and adnexal mass. The ascitic tap yielded thick mucinous material. The most likely diagnosis from Figure 3.29 is:

A. Pseudomyxoma peritonei
B. Reactive mesothelial cells
C. Metastatic adenocarcinoma
D. Mesothelioma

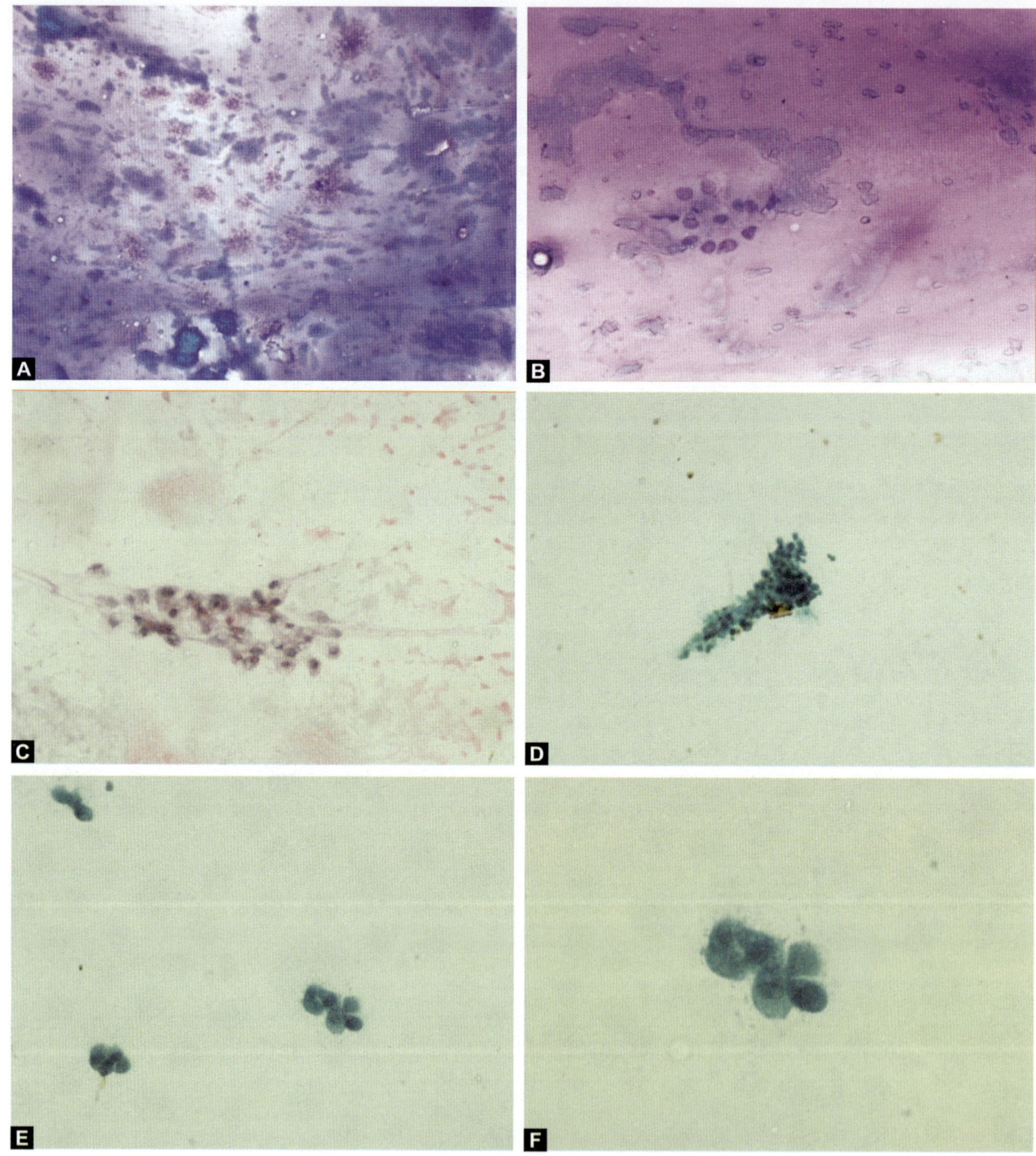

Figures 3.29A to F

Problem 3.30

A 41-year-old female presented with 10 cm diameter right adnexal mass along with ascites. Her CA 125 level in the serum was 70 IU/mL. The peritoneal washing cytology was done at the time of removal of the adnexal mass. The microphotographs of the histopathology section of the mass were shown in Figures 3.30A to D. The microphotographs of the cytology smear of peritoneal wash are shown in Figures 3.30E to J. The most likely diagnosis from Figure 3.30 is:

A. Peritoneal implants of ovarian tumor
B. Reactive mesothelial cells
C. Mesothelioma
D. Lymphocyte rich effusion

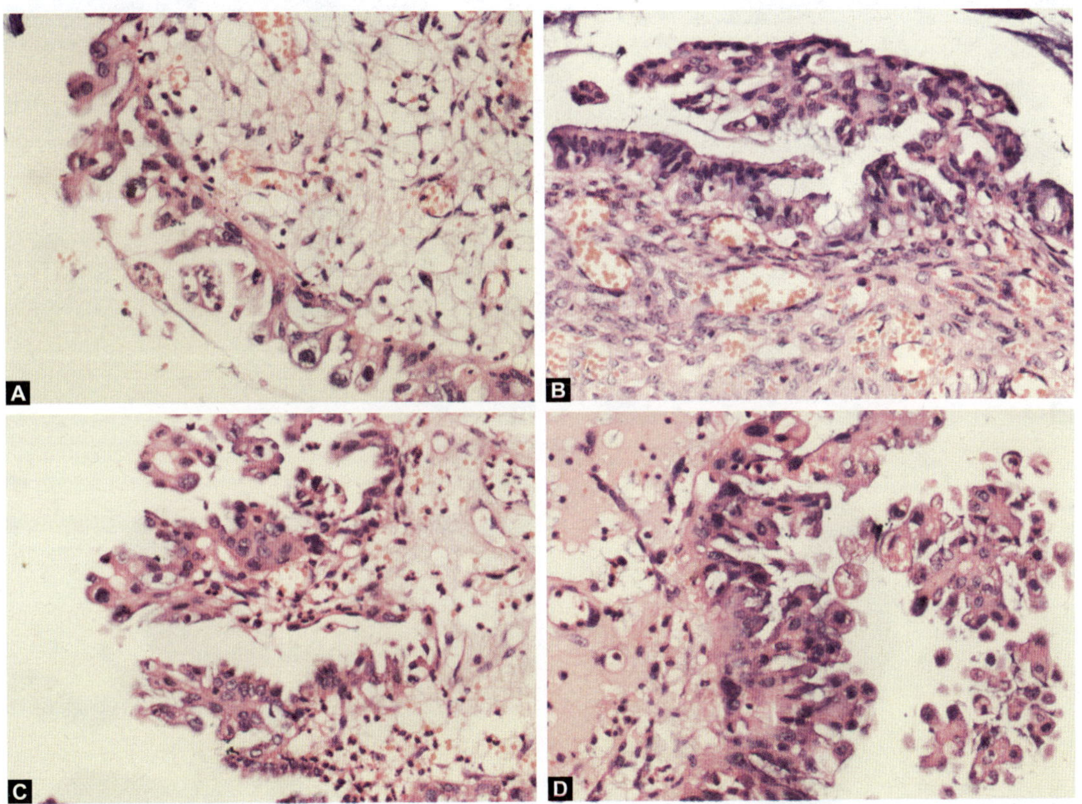

Figures 3.30A to D

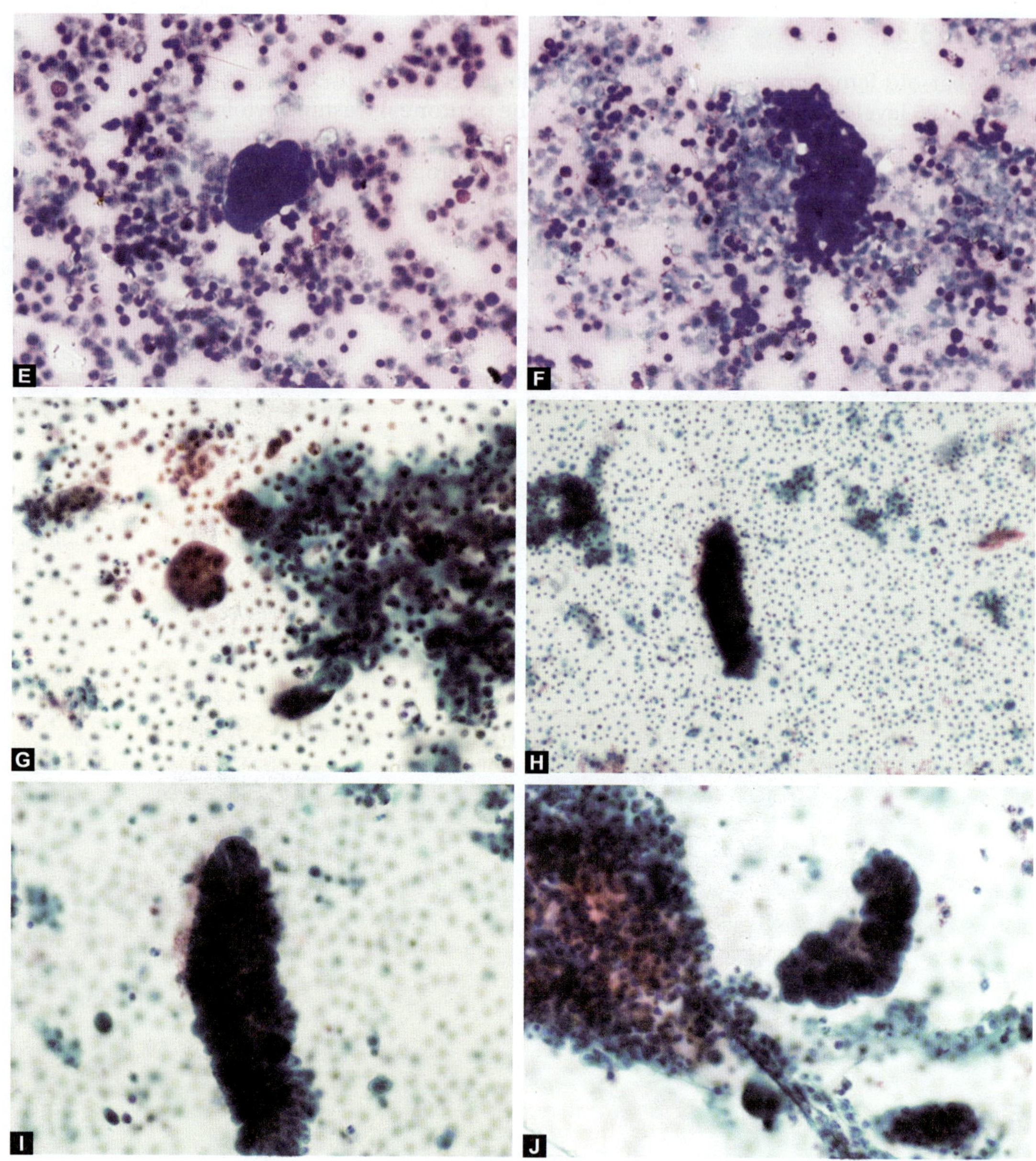

Figures 3.30E to J

Problem 3.31

A 39-year-old female presented with massive ascites along with lower abdominal mass. The most likely diagnosis from the Figure 3.31 is:

A. Metastatic adenocarcinoma from ovary
B. Reactive mesothelial cells only
C. Lymphocyte rich effusion
D. Mesothelioma

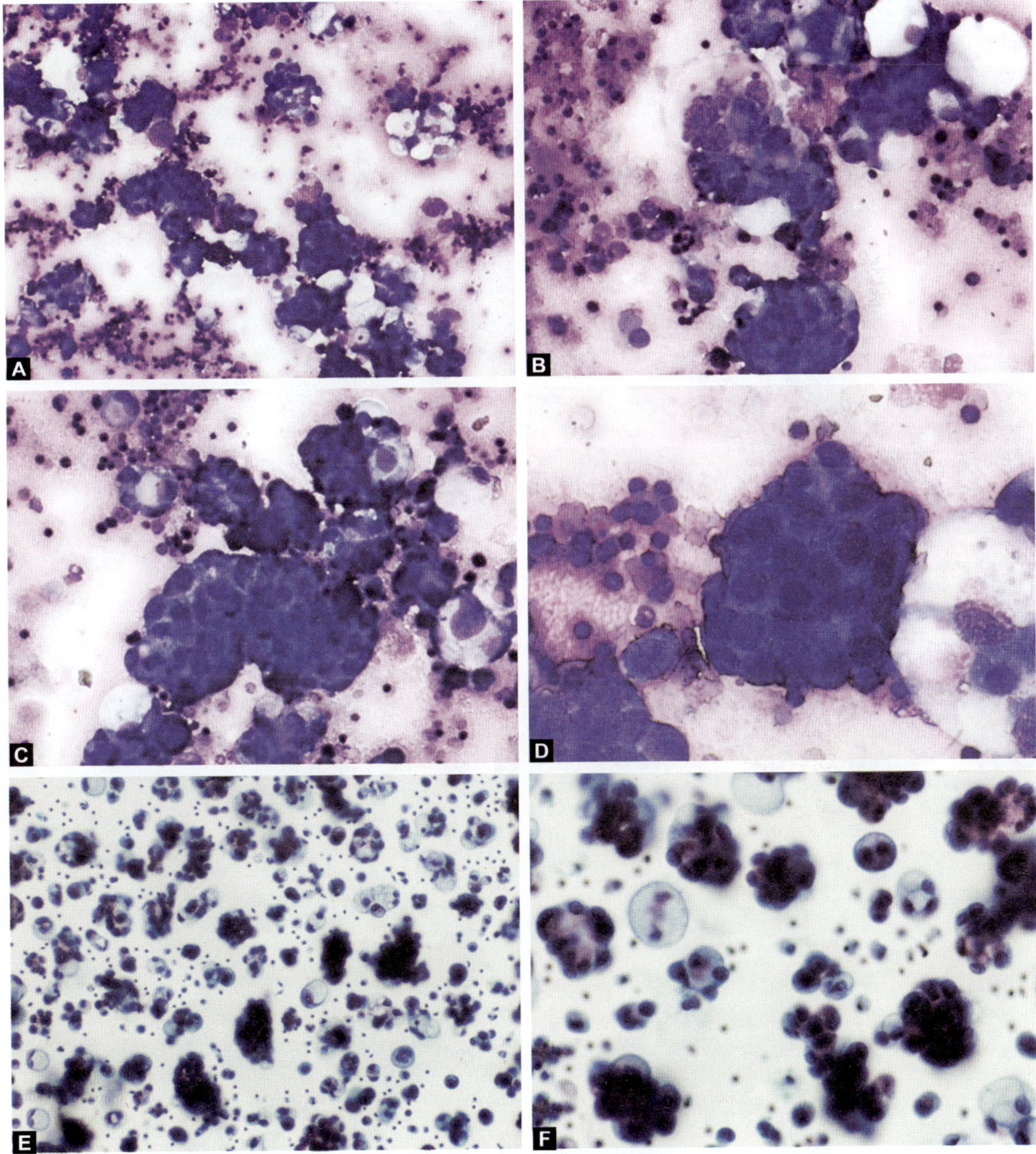

Figures 3.31A to F

Problem 3.32

A 45-year-old female presented with massive peritoneal effusion along with pelvic mass. Her CA 125 level is 759 IU/mL. The most likely diagnosis from Figure 3.32 is:

A. Metastatic serous carcinoma of ovary
B. Mesothelioma
C. Reactive mesothelial cells
D. Pseudomyxoma ovary

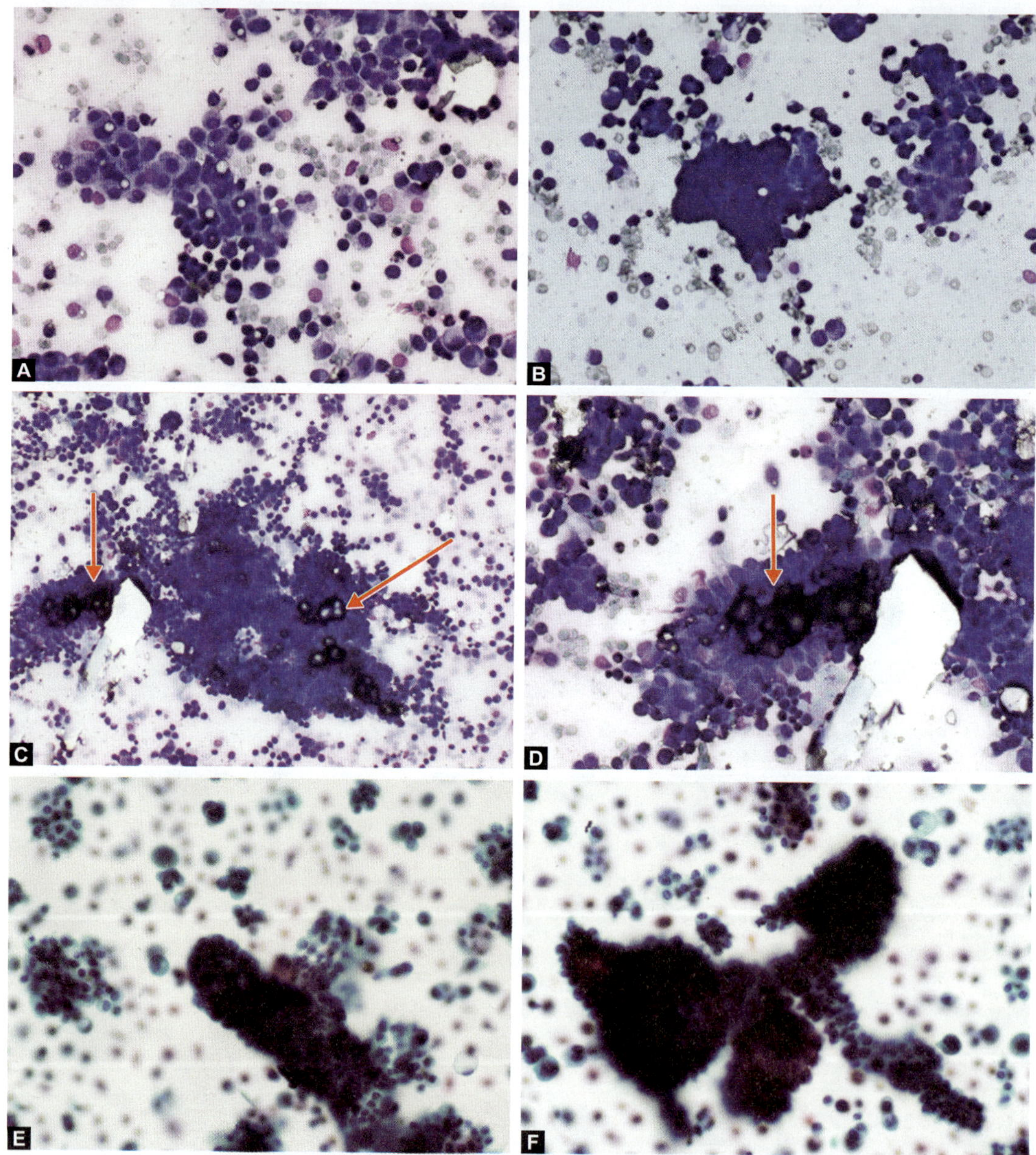

Figures 3.32A to F

Problem 3.33

A 52-year-old female patient was diagnosed as acute myeloid leukemia. The patient had completed her chemotherapy regimen 6 months back and now she presented with pleural effusion of right side. The most likely diagnosis from Figure 3.33 is:

A. Reactive mesothelial cells only
B. Acute leukemia
C. Metastatic carcinoma
D. Histoplasma infection

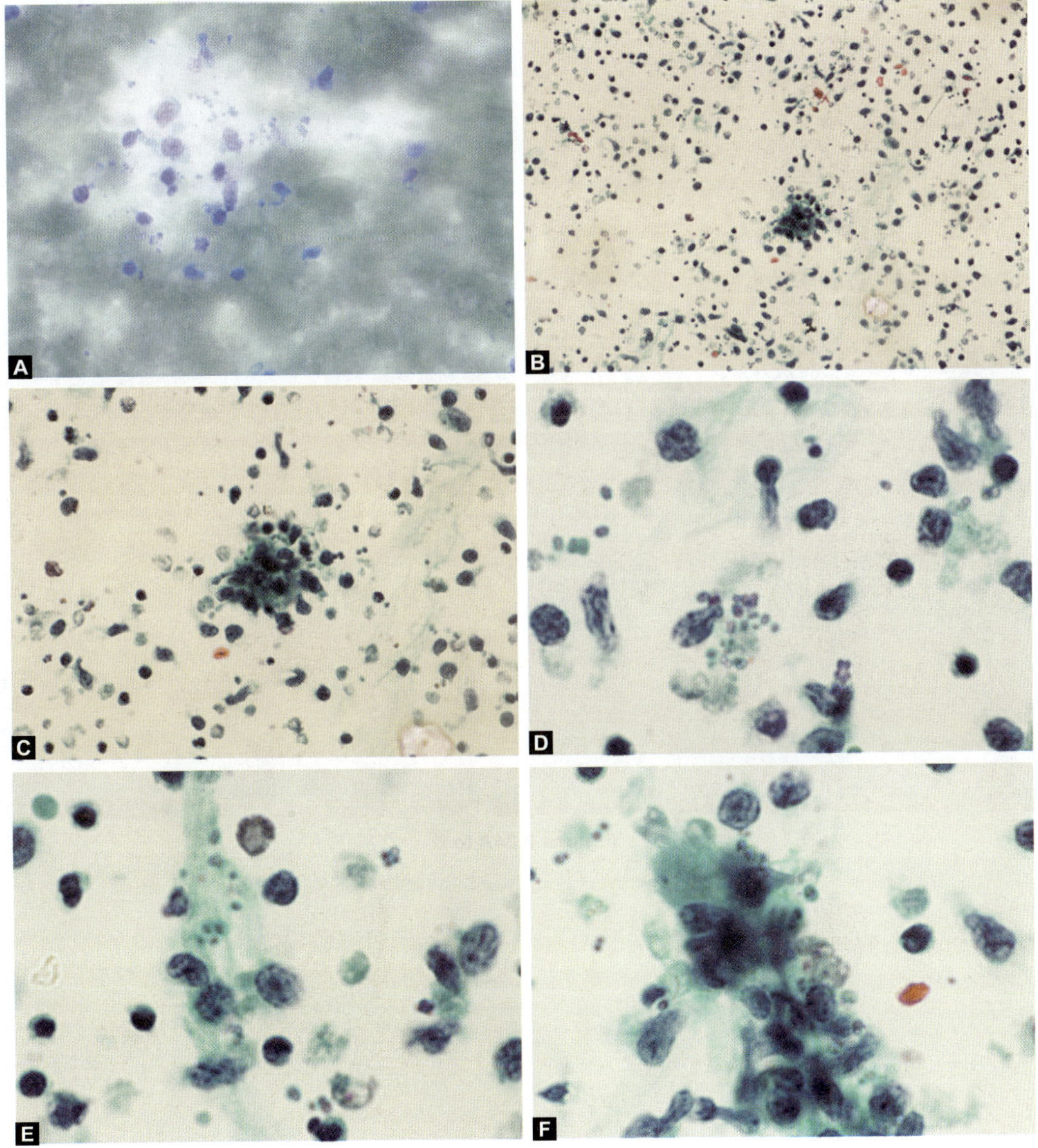

Figures 3.33A to F

Problem 3.34

A 16-year-old female was operated for an adnexal mass in the lower abdomen. Histopathology of the mass is shown in Figures 3.34A to D. After 2 years the patient developed right sided pleural effusion. The cytological examination of the pleural fluid was done. The most likely diagnosis from Figures 3.34E to J is:

A. Reactive mesothelial cells only
B. Non-Hodgkin lymphoma
C. Infiltration by yolk sac tumor
D. Dysgerminoma

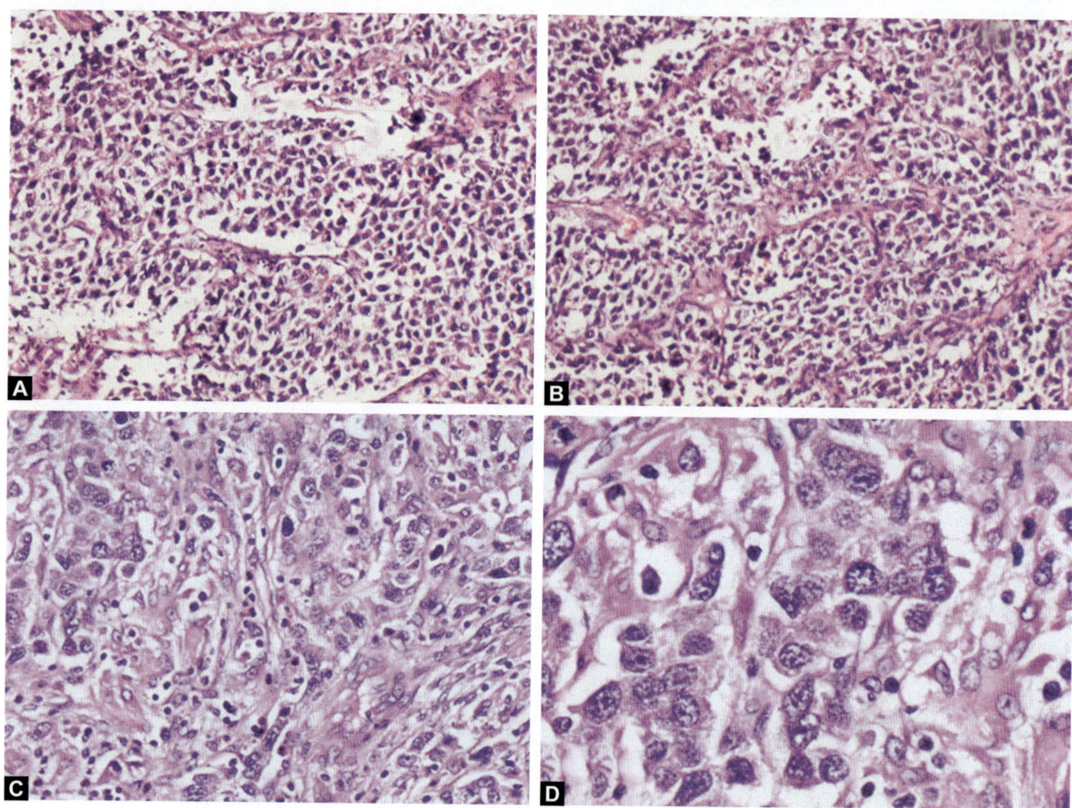

Figures 3.34A to D

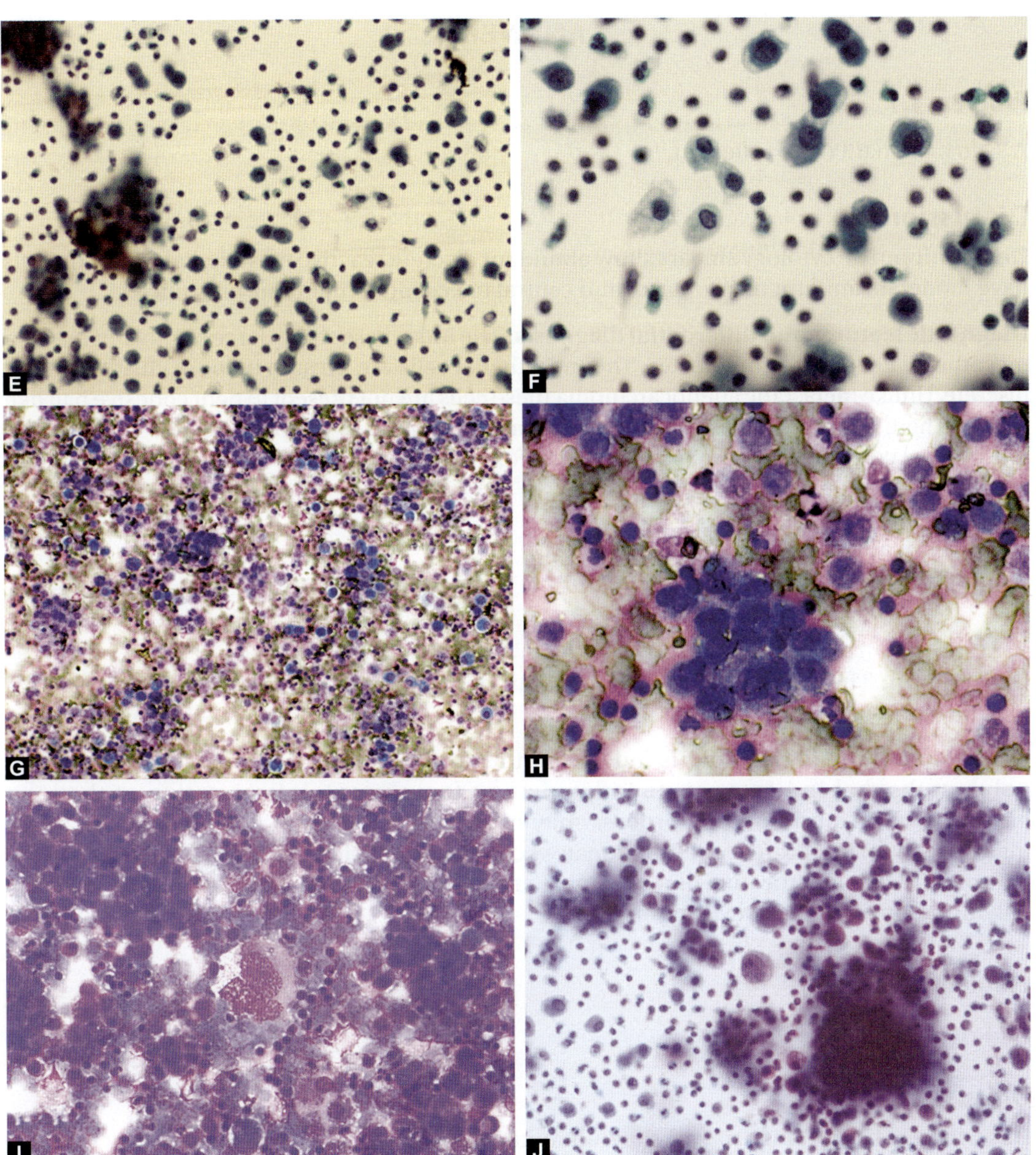

Figures 3.34E to J

ANSWERS

Problem 3.1

A. No malignant cell in effusion

Discussion

Description: The cytology smears show abundant discrete reactive mesothelial cells. The cells have moderate amount of cytoplasm and central to eccentric monomorphic nuclei.

Diagnostic features and differential diagnosis: The index case shows large number of reactive mesothelial cells and lymphocytes. The mesothelial cells are present in loose clusters and in discrete. Unlike metastatic adenocarcinoma these cells are not present in tight cohesive ball like cluster. There is no excess lymphocytes in the smear to label it as "lymphocyte rich effusion". The mesothelial cells are benign looking and therefore the diagnosis of mesothelioma is excluded.

Problem 3.2

A. Acute inflammation (empyema)

Discussion

Description: The cytology smears show abundant polymorphs and occasional reactive mesothelial cells.

Discussion of diagnosis: In this index case the effusion cytology smear shows abundant polymorphs that indicates acute inflammation (empyema). Inflammation of the lung parenchyma in a pneumonic process may extend to the pleural surface causing acute inflammation. The nature of the inflammatory cells in the effusion fluid mainly depends on the nature of infection in the lung parenchyma. Early stage of bacterial infection usually provokes polymorphonuclear leukocyte infiltration.

Problem 3.3

A. Lymphocyte rich effusion

Discussion

Description: The cytology smears show abundant mature looking lymphocytes and occasional mesothelial cells.

Discussion on diagnosis: The index case shows abundant lymphoid cells along with reactive mesothelial cells. The lymphoid cells have scanty cytoplasm with round monomorphic nuclei. Chromatin is regular. Lymphocyte rich effusion is a broad term and it encompasses all the conditions that cause predominant lymphocyte rich effusion cytology. The common malignant etiology of lymphocyte rich effusion is NHL and benign etiology is tuberculous effusion.

Problem 3.4

D. Eosinophilic effusion

Discussion

Description: The cytology smear shows abundant eosinophils along with occasional reactive mesothelial cells. No parasites or malignant cells seen.

Discussion on diagnosis: Eosinophilic pleural effusions are defined as a pleural effusion that contains at least 10% eosinophils. In this index case the presence of abundant eosinophils along with reactive mesothelial cells suggests the diagnosis of eosinophilic effusion. Hemothorax and pneumothorax are the most common causes of eosinophilic effusion (Table 3.2). Repeated thoracocentesis may also cause eosinophilic effusion because the procedure may introduce a small amount of air into the pleural cavity that may generate eosinophilic reaction. About one third of the cases no exact etiologic cause is identified. Most of the eosinophilic effusions are self-limiting and resolve spontaneously.

Malignant cells may be associated with eosinophilia however; there were no evidences of malignancy in this case.

Table 3.2: Common causes of eosinophilic effusion

Pleural effusion	*Pericardial*	*Peritoneal*
• Pneumothorax • Cancer • Post surgical • Traumatic • Parasitic • Pancreatitis • Pneumonic • Tuberculosis • Allergic • Idiopathic	• Pulmonary eosinophilia • Hypersensitivity reaction	• Allergic • Peritoneal dialysis • Parasitic

Problem 3.5

C. Reactive mesothelial cells

Discussion

Description: The cytology smears show abundant reactive mesothelial cells. Occasional discrete cells are enlarged with large mildly pleomorphic nuclei.

Discussion on diagnosis: The smear shows predominantly discrete cells. Unlike metastatic carcinoma no ball like clusters are seen. The predominant cell population in the smear is monomorphic and only occasional large mesothelial cells are seen. The mesothelial cells in effusion often proliferate due to rich source of nutrients and therefore occasional atypical looking mesothelial cells in benign effusion is not uncommon.

Problem 3.6

B. Candida infection

Discussion

Description: The cytology smear shows many reactive mesothelial cells along with intracellular and also extracellular candidal spore.

Discussion on diagnosis: The patient is immunocompromised due to chronic liver disease. Therefore it is not unusual to have infection in the peritoneal effusion. Candida is an opportunistic infective organism and is often seen in the immunocompromised patient. Positive fungal culture is necessary to label such cases as infection rather than harmless colonization of the fungus. The candidal infection in ascitic fluid has been described in cirrhotic patients by Choi et al.[1] Out of 21 patients, ten patients had spontaneous candidal infection in ascitic fluid along with cirrhosis of liver. All these patients died within one year period.

Reference

1. Choi SH, Soo Kim Y, Chung JW, et al. Clinical significance of untreated *Candida* species isolated from ascites in cirrhotic patients. Scand J Infect Dis. 2004;36:649-55.

Problem 3.7

C. Aspergillus infection

Discussion

Description: The cytology smear shows abundant polymorphs, reactive mesothelial cells and long thin fungal hyphae with acute angle branching.

Discussion on diagnosis: The organism with long thin septate having acute angle branching is consistent with Aspergillus fungi. Aspergillus fungal infection usually occurs in the immunocompromised patient. In the index case the patient is renal transplant recipient and hence immunocompromised.

Problem 3.8

A. Metastatic adenocarcinoma

Discussion

Description: The cytology smears show multiple ball like three-dimensional clusters of cells along with background mesothelial cells. The cells have moderate amount of vacuolated cytoplasm and large hyperchromatic pleomorphic nuclei.

Diagnostic points: The following features suggest the diagnosis of metastatic adenocarcinoma:
- The large three-dimensional ball like clusters
- Large vacuoles in the cell that displaced the nucleus towards periphery of the cell
- Moderately pleomorphic nuclei
- Large prominent nucleoli.

These features are not specific for ovarian adenocarcinoma and may be seen in other metastatic adenocarcinoma also. However, the presences of adnexal mass in an elderly female along with these cytological features indicate the diagnosis of ovarian adenocarcinoma. The index case showed omental metastasis of serous adenocarcinoma of ovary.

Differential diagnosis: The distinct two populations of cells consisting of mesothelial cells and other foreign type of cells eliminate the possibility of mesothelioma. Moreover the malignant cells do not resemble mesothelial cells. The reactive mesothelial cells may often resemble malignant cells due to nuclear enlargement, multinucleation and nuclear pleomorphism. However, three-dimensional ball like clusters are absent in reactive mesothelial cell proliferation. The distinguishing features of these two conditions have been highlighted in the Table 3.3.

Table 3.3: Differentiating points between reactive mesothelial cells and adenocarcinoma

Cytological features	*Reactive mesothelial cells*	*Adenocarcinoma*
Ball like spherical clusters	Absent	Present
Two cell population	Absent	Present
Cell with high N/C ratio	Absent	Present
Large pleomorphic nucleoli	Absent	Present
Irregular nuclear margin	Absent	Present
Sharply demarcated cytoplasmic margin	Absent	Present
Cytochemistry		
Mucicarmine	Negative	Positive
PAS	Positive	Positive
Immunocytochemistry		
CEA	Negative	Positive
Ber-Ep4	Negative	Positive
Leu M1	Negative	Positive
B72.3	Negative	Positive

Problem 3.9

C. Plasma cell neoplasm

Discussion

Description: The cytology smears show abundant plasma cells admixed with occasional mesothelial cells. The plasma cells have single to double nuclei with peripheral clumped chromatin.

Diagnostic features: In the index case large number of plasma cells in the effusion cytology along with many immature plasma cells with prominent nucleoli suggests the diagnosis of infiltration by plasma cell neoplasm.

Disease description: The incidence of plasma cell tumor involving the pleural cavity is very uncommon and it is less than 1 to 2% of all myeloma cases. The plasma cell infiltration in the pleural cavity may occur due to direct involvement from the thoracic bone or by the infiltration of the plasma cells in the lymph node causing obstruction and poor lymphatic drainage. The thorough clinical history (such as back pain in the index case) along with ancillary techniques particularly bone marrow examination, serum electrophoresis and flow cytometric immunophenotyping of the plasma cell are helpful in diagnosis of such case. Immunocytochemistry, particularly CD 38 and CD 138 marker of plasma cell, is helpful for final confirmation of plasma cells in effusion fluid.

Differential diagnosis: The various other conditions may also show increased plasma cell in the effusion fluid such as viral infection, tuberculosis and Hodgkin lymphoma. In these conditions the number of plasma cells in the effusion fluid is much less and the cells are mainly mature plasma cells.

Further Reading

1. Deshpande AH, Munshi MM. Pleural effusion as an initial manifestation of multiple myeloma. Acta Cytol. 2000; 44:103-4.

Problem 3.10

D. Yolk sac tumor

Discussion

Description: The cytology smears show large number of malignant cells predominantly in small loose clusters. The individual cells have moderate amount of cytoplasm and large pleomorphic nuclei with multiple prominent nucleoli.

Diagnostic points: In this index case the patient had a large ovarian mass (20 cm in diameter) along with high serum α-fetoprotein level. The subsequent histopathology also showed features of yolk sac tumor (both Shiller Duval bodies and hyaline globules present). The cytomorphology of the ascitic fluid is consistent with yolk sac tumor. The cells are large and pleomorphic. The nucleoli are prominent.

In ascitic fluid examination the direct diagnosis of yolk sac tumor is difficult. Serum α-fetoprotein level or immunocytochemistry of α-fetoprotein are helpful.

Further Reading

1. Murugan P, Siddaraju N, Sridhar E, Soundararaghavan J, Habeebullah S. Unusual ovarian malignancies in ascitic fluid: a report of 2 cases. Acta Cytol. 2010;54(4):611-7.
2. Petrakakou E, Grapsa D, Stergiou ME, Mikou P, Tsarpalis D, et al. Ascitic fluid cytology of yolk sac tumor of the ovary: a case report. Acta Cytol. 2009;53(6):7.

Problem 3.11

A. Non-Hodgkin lymphoma

Discussion

Description: The FNAC smear shows discrete immature lymphoid cells with scanty vacuolated cytoplasm and monomorphic nuclei (Figs 3.11A and B). The nuclear chromatin is fine with prominent nucleoli. Similar population of cells are also seen in the effusion fluid (Figs 3.11C to F).

Diagnostic features: In the index case the FNAC of the lymph node shows monomorphic population of the immature lymphoid cells. These cells show round nuclei with vacuolated cytoplasm having fine nuclear chromatin and occasional prominent nucleoli. The cytomorphology is that of NHL. The effusion cytology smear shows cells with similar morphology and is consistent with NHL.

Differential diagnosis: The FNAC features, clinical history and cytology of effusion fluid exclude the possibility of dysgerminoma. The lymphoid cells are immature in look and the possibility of benign lymphoid cells is not considered.

Problem 3.12

C. Mesothelioma

Discussion

Description: The cytology smear shows abundant monomorphic mesothelial cells with abundant cytoplasm and monomorphic round nuclei.

Diagnostic points: In the index case the features suggestive of mesothelioma are:

- Abundant monomorphic discrete mesothelial cells and only one cell population
- The cells with abundant cytoplasm with monomorphic nuclei
- The typical history of rapidly filling pleural effusion
- Diffuse thickened lung pleura.

All such cases should have calretinin immunostaining to confirm the mesothelial origin of the cells.

Problem 3.13

B. B-cell non-Hodgkin lymphoma possibly Burkitt's lymphoma

Discussion

Cytology description: The cytology smear shows abundant atypical lymphoid cells. The individual cells have scanty vacuolated cytoplasm with mildly pleomorphic round nuclei having prominent nucleoli.

Diagnostic points: In the index case smear shows abundant atypical lymphoid cells. The cells have scanty cytoplasm with mildly pleomorphic nuclei having fine chromatin. Lymphoid nature of the cell is established by CD 45 positive staining. The predominant lambda positive staining indicates light chain restriction and favors lymphoma. In addition, CD 19 positivity indicates B-cell origin of the lymphoid cells. Negative TdT staining excludes the possibility of lymphoblastic lymphoma.

Problem 3.14

C. Infiltration by dysgerminoma

Discussion

The histopathology section shows round to oval tumor cells arranged in small lobules separated by fibrous septae that are infiltrated by lymphocytes. The individual tumor cells are round with fine chromatin and single prominent nucleoli. The histological features are that of dysgerminoma.

The fluid cytology smear shows discrete round to oval tumor cells with scanty cytoplasm. These cells are scanty in number and admixed with large number of mesothelial cells. The typical histology along with the presence of these tumor cells eliminate the possibilities of any other diagnosis.

Further Reading

1. Hajdu SI, Nolan MA. Exfoliative cytology of malignant germ cell tumors. Acta Cytol. 1975;19(3):255-60.

Problem 3.15

D. Infiltration by soft tissue Ewings' tumor/peripheral neuroectodermal tumor

Discussion

Cytology description and diagnosis: The fine needle aspiration cytology smear of the chest wall mass shows discrete round cells with scanty cytoplasm and fine granular chromatin. The cells show Mic 2 (CD 99) and focal CD 56 positivity. The tumor cells are negative for CD 45 (leucocyte common antigen, LCA) and cytokeratin (CK). The overall cytomorphology and immunocytochemistry strongly favor the diagnosis of soft tissue Ewings' tumor/peripheral neuroectodermal tumor. The effusion cytology smear shows similar cells as that of FNAC smear.

Differential diagnosis: The negative CD 45 immunostain eliminates the possibility of non-Hodgkin lymphoma. The loose clusters of tumor cells along with absent lymphoglandular bodies also exclude such possibility. The possibilities of reactive mesothelial cells is remote in this index case.

Further Reading

1. Oliveira MJ, de Almeida LP, Wengerkievicz AC, Siqueira SA, Antonangelo L. From conventional fluid cytology to unusual histological diagnosis: report of four cases. Diagn Cytopathol. 2013;41(4):348-53.

Problem 3.16

C. Pseudomyxoma peritonei

Discussion

Description: The cytology smear shows abundant mucinous material along with reactive mesothelial cells and clusters of epithelial cells with moderate amount of cytoplasm and round monomorphic nuclei.

Diagnostic features in the index case: The following features suggest the diagnosis of pseudomyxoma peritonei (PMP):

- The gross description of thick gelatinous material
- Cytology smear shows thick mucinous material
- Occasional clusters and discrete benign looking epithelial cells entrapped within the mucus material.

Disease description: PMP is a poorly understood low grade neoplastic condition characterized by mucinous ascites and mucinous implants diffusely involving the peritoneal surfaces. The condition is most commonly noted in appendicular mucinous tumor followed by adenocarcinoma of ovary and colon.

The PMP is classified as disseminated peritoneal adenomucinosis (DPAM) and peritoneal mucinous carcinomatosis (PMCA). DPAM is characterized by superficial peritoneal implants composed of abundant extracellular mucin containing scanty simple to focally proliferative mucinous epithelium with little cytological atypia or mitotic activity. DPAM has a benign and indolent clinical course. PMCA is characterized by invasive peritoneal lesions with more abundant mucinous epithelium with the architectural and cytological features of carcinoma. In cytology smear it is difficult to distinguish these two conditions.

Further Reading

1. Ronnett BM, Zahn CM, Kurman RJ, Kass ME, Sugarbaker PH, Shmookler BM. Disseminated peritoneal adenomucinosis and peritoneal mucinous carcinomatosis. A clinicopathologic analysis of 109 cases with emphasis on distinguishing pathologic features, site of origin, prognosis, and relationship to "pseudomyxoma peritonei". Am J Surg Pathol. 1995;19(12):1390-408.

Problem 3.17

A. Metastatic renal cell carcinoma

Discussion

Description: The cytology smear shows multiple small clusters of cells with abundant cytoplasm and centrally placed large nuclei having single prominent nucleoli.

Diagnostic features in the index case: In this index case the following features suggest the diagnosis of renal cell carcinoma in the ascitic fluid:

- History of right upper abdominal mass with hematuria
- Tight cohesive clusters and loose cluster of cells
- Cells with abundant clear cytoplasm.

Differential diagnosis: At time it is difficult to differentiate renal cell carcinoma versus metastatic carcinoma from gastrointestinal tract. Typical history along with cells containing abundant cytoplasm are helpful features in that context.

Problem 3.18

C. Metastatic signet ring carcinoma

Discussion

Description: The cytology smear shows abundant malignant cells admixed with reactive mesothelial cells. The individual cells show abundant cytoplasm and eccentrically placed enlarged nuclei giving signet ring appearance.

Diagnostic features of the index case: The following features suggest the diagnosis of metastatic signet ring carcinoma:

- Dispersed tumor cells
- Cells with abundant cytoplasm and pushed peripheral nuclei giving a signet ring appearance
- Moderate nuclear atypia.

This was a case of gastric carcinoma.

Differential diagnosis: Occasionally reactive mesothelial cells may show signet ring type of cells. However, nuclei of such cells do not show any atypia and the amount of such cells are also scanty. The signet ring cells from gastric carcinoma are usually positive for Alcian blue and mucicarmine staining as the cells contain mucinous material.

Problem 3.19

B. Metastatic small cell carcinoma from lung

Discussion

Description: The cytology smears show multiple loose clusters of malignant cells. The cells are predominantly arranged in small rows and the nuclei are compressed with each other. At places the nuclei are compressed and show "cell in cell" arrangement (Fig. 3.19D). The nuclei of the cells are enlarged and hyperchromatic with irregular margin and condensed chromatin.

Diagnostic features of the index case: In this index case the following features suggest the diagnosis of metastatic small cell carcinoma:

- The small rows of tumor cells arranged like Indian file pattern
- Cell to cell compression giving a vertebrae like appearance
- Small round cells with scanty cytoplasm and hyperchromatic nuclei with no nucleoli
- The clinical history of dry cough, chest pain and hemoptysis suggest the possibility of lung origin of the tumor.

Differential diagnosis: Metastatic adenocarcinoma from lung usually shows ball like tight clusters of cells. The cells usually contain abundant vacuolated cytoplasm.

Problem 3.20

D. Metastatic renal cell carcinoma

Discussion

Description: The cytology smears show many loose clusters of malignant cells. The cells have abundant vacuolated cytoplasm with centrally placed moderately enlarged nuclei. The nuclear chromatin is fine with prominent nucleoli. Occasional signet ring like cells are also seen (Fig. 3.20F).

Diagnostic features: In this index case the following features indicate the diagnosis of renal cell carcinoma in the ascetic fluid:

- Discrete and loose cluster of cells
- Cells with abundant clear cytoplasm
- Prominent nucleoli.

Differential diagnosis: Small cell carcinoma usually shows small cells with scanty cytoplasm and hyperchromatic nuclei. The reactive mesothelial cells do not show such large pleomorphic nuclei. Mesothelioma is a remote possibility in this index case.

Problem 3.21

A. Metastatic squamous cell carcinoma

Discussion

Description: The cytology smears show many discrete cells of oval to polyhedral in shape. The cells have enlarged hyperchromatic nuclei. Occasional tad pole like cells are also seen (Figs 3.21E and F).

Diagnostic features of the index case: In this index case the following features suggest the diagnosis of metastatic squamous cell carcinoma:

- Discrete cell population
- Oval to polyhedral cells with centrally placed dark hyperchromatic nuclei
- Elongated tadpole like cells.

Differential diagnosis: The metastatic adenocarcinoma cases usually show tight ball like clusters and individual cells show abundant vacuolated cytoplasm. Nuclear characteristic is also different in adenocarcinoma as the cells show prominent nucleoli.

Problem 3.22

D. Metastatic adenocarcinoma

Discussion

Description: The FNAC smears (Figs 3.22A to C) show multiple tight cohesive clusters of malignant cells with papillae like arrangement. The individual cells show moderate amount of vacuolated cytoplasm and round moderately pleomorphic nuclei. The features are that of a non-small cell carcinoma consistent with adenocarcinoma.

The effusion cytology smears (Figs 3.22D to I) show abundant small ball like three-dimensional clusters of cells along with background mesothelial cells. The individual cells show moderate amount of vacuolated cytoplasm and central to eccentric nuclei with moderate pleomorphism.

Diagnostic points in the index case: The FNAC smears show features of non-small cell carcinoma consistent with adenocarcinoma. Effusion cytology smears show features of adenocarcinoma. Considering the clinical history and FNAC of lung lesion the most likely diagnosis is adenocarcinoma metastasis from lung. Immunocytochemistry on cell block of the effusion such as TTF-1 is sufficient to prove the primary lung origin of the tumor.

Problem 3.23

A. Metastatic lobular carcinoma of breast

Discussion

Description: The histopathology section (Figs 3.23A to C) shows a malignant tumor and the cells are predominantly arranged in rows separated by fibrous stroma. The individual cells are round with scanty cytoplasm having hyperchromatic nuclei. The histological features are consistent with invasive lobular carcinoma of breast.

The cytology smears (Figs 3.23D to I) show abundant discrete malignant cells with round nuclei and scanty cytoplasm. The individual cells show round mildly pleomorphic nuclei with condensed chromatin and inconspicuous nucleoli.

Diagnostic features in the index case: The histopathology of the breast tumor is that of lobular carcinoma. In the index case the following features suggest the diagnosis of lobular carcinoma of breast:

- Discrete small round cells
- Cells with scanty cytoplasm
- Nuclei show condensed chromatin with inconspicuous nucleoli
- The individual cell morphology of the effusion cytology is consistent with the histopathology.

Differential diagnosis: The absence of lymphoglandular bodies and previous history of breast carcinoma eliminate the possibility of non-Hodgkin lymphoma. The reactive mesothelial cells show more cytoplasm and often window like gap in between the cells. The mature lymphocytes show scanty deep blue cytoplasm with round nuclei and condensed chromatin. The mature lymphocytes are smaller than the cells of lobular carcinoma of breast.

Further Reading

1. Danner DE, Gmelich JT. A comparative study of tumor cells from metastatic carcinoma of the breast in effusions. Acta Cytol. 1975;19(6):509-18.
2. Hemachandran M, Dey P. Indian file pattern of adenocarcinoma cells in effusions. Acta Cytol. 2004;48(3):385-6.

Problem 3.24

D. Metastatic squamous cell carcinoma

Discussion

Description: The cytology smear shows small clusters and discrete malignant cells. The cells show moderate amount of cytoplasm and enlarged pleomorphic hyperchromatic nuclei. Occasional fibre like squamous cells are also seen (Fig. 3.24E).

Diagnostic features: In the index case the following features favor the diagnosis of metastatic squamous cell carcinoma:

- Predominantly discrete and loose clusters of cell population
- Oval to polyhedral cells with hyperchromatic nuclei
- Many polyhedral cells with orangeophilic cytoplasm indicating squamoid differentiation
- Typical fiber-like cells.

The patient had squamous cell carcinoma in the esophagus with infiltration in the mediastinal lymph node and pleural cavity.

Differential diagnosis: The typical squamoid differentiation in the malignant cells eliminates the possibility of adenocarcinoma. The smears show abundant polymorphs in the background and therefore one may consider the diagnosis of empyema. However the presence of malignant cells eliminates this possibility.

Problem 3.25

A. Lymphoma (recurrence)

Discussion

Description: The cytology smears show abundant discrete immature lymphoid cells admixed with reactive mesothelial cells. The lymphoid cells show deep blue vacuolated cytoplasm with central large nuclei having fine chromatin and prominent nucleoli.

Diagnostic points in the index case: The diagnostic points in the index case include:
- History of acute leukemia
- Immature lymphoid cells.

Further Reading

1. Das DK. Serous effusions in malignant lymphomas: A Review. Diagn Cytopathol. 2006;34:335-47.

Problem 3.26

B. Metastatic ductal carcinoma of breast

Discussion

Description: The histopathology sections (Figs 3.26A to D) show islands of malignant ductal cells with moderate nuclear enlargement and pleomorphism. The tumor cells infiltrated into the adjacent stroma.

The cytology smears (Figs 3.26E to J) show multiple ball like clusters of malignant cells admixed with reactive mesothelial cells. The individual cells have enlarged nuclei with fine nuclear chromatin and prominent nucleoli.

Diagnostic features: The cytomorphology of the malignant cells in the effusion fluid is consistent with the cells of infiltrating duct carcinoma of breast. Without the history of carcinoma of breast the effusion cytology smears indicate only the presence of metastatic carcinoma. Antic T et al[1] described three different type of pattern of breast carcinoma in effusion fluid cytology:
- Single cell like pattern
- Cannon ball like pattern
- Mesothelial cell like pattern

The present case shows cannon ball like pattern of infiltration which is the most common pattern.

Reference

1. Antic T, Gong Y, Sneige N. Tumor type and single-cell/mesothelial-like cell pattern of breast carcinoma metastases in pleural and peritoneal effusions. Diagn Cytopathol. 2012;40(4):311-5.

Problem 3.27

B. Metastatic adenocarcinoma from lung

Discussion

Description: The cytology smears (Figs 3.27A to D) show multiple tight cohesive clusters of malignant cells. The individual cells show scanty to moderate cytoplasm and hyperchromatic nuclei. The cell block section (Fig. 3.27E) shows sheet of tumor cells with vacuolated cytoplasm and moderately pleomorphic nuclei. The tumor cells are positive for CK 7, napsin and TTF-1 that indicate the primary origin of lung adenocarcinoma.

Problem 3.28

C. Signet ring cell carcinoma

Discussion: The microphotographs of the cytology smear show abundant malignant cells admixed with many reactive mesothelial cells. The individual cells show abundant cytoplasm and eccentrically placed large nuclei that simulate signet ring appearance.

Diagnostic features in the index case: The following features suggest the diagnosis of signet ring carcinoma:

- ❑ Dispersed population of malignant cells
- ❑ Cells with abundant cytoplasm and eccentrically placed nuclei giving a signet ring appearance
- ❑ Moderate nuclear atypia.

This patient was proved as a case of gastric adenocarcinoma.

Problem 3.29

A. Pseudomyxoma peronei

Discussion

Description: The microphotograph of the cytology smear shows occasional reactive mesothelial cells and clusters of epithelial cells in abundant deep blue mucinous material. The individual epithelial cells show moderate amount of cytoplasm and round monomorphic nuclei.

Diagnostic features in the index case: The following features suggest the diagnosis of pseudomyxoma peronei:

- ❑ The gross description of thick mucinous material
- ❑ Deep blue mucinous material
- ❑ Occasional clusters and discrete benign looking epithelial cells entrapped within the mucus material.

Detailed description of pseudomyxoma peritonei: See case 3.16.

Problem 3.30

A. Peritoneal implants of ovarian tumor

Description: The histology section shows ovarian tumor lined by cuboidal to columnar cells having moderate nuclear pleomorphism. No stromal invasion is noted. The features are consistent with borderline serous adenocarcinoma.

The cytology smear shows clusters and papillae like structures of tumor cells along with abundant reactive mesothelial cells. The individual cells are round moderately pleomorphic nuclei with scanty cytoplasm.

Diagnostic features:
- Multiple ball like clusters and papillae
- Regular outline of papillary group of cells
- Nuclear hyperchromasia and pleomorphism.

Disease description: This index case was a patient of borderline serous adenocarcinoma with peritoneal involvement. Borderline tumor may show peritoneal implants and this implant is not considered as metastasis. It is very difficult to distinguish peritoneal implants from the metastatic adenocarcinoma in effusion cytology. In fact, the histology of the primary ovarian tumor is the gold standard to determine the nature of ovarian tumor and histology of the omentum may be required to confirm the peritoneal implant.

Further Reading

1. Bell DA, Weinstock MA, Scully RE. Peritoneal implants of ovarian serous borderline tumors. Histologic features and prognosis. Cancer. 1988;62(10):2212-22.
2. Johnson TL, Kumar NB, Hopkins M, Hughes JD. Cytologic features of ovarian tumors of low malignant potential in peritoneal fluids. Acta Cytol. 1988;32(4):513-8.

Problem 3.31

A. Metastatic adenocarcinoma from ovary

Discussion

Description: The cytology smears show abundant tight three-dimensional ball like clusters along with reactive mesothelial cells. The individual cells show moderately enlarged hyperchromatic nuclei with large prominent nucleoli.

Diagnostic features of the index case: The following features suggest the diagnosis of metastatic adenocarcinoma:
- Distinct two population of cells: Mesothelial cells and malignant cells
- Abundant large three dimensional ball like clusters
- Malignant cells with moderately pleomorphic nuclei
- The tumor cells with large vacuoles in the cell
- Large prominent nucleoli.

Problem 3.32

A. Metastatic serous carcinoma of ovary

Discussion

Description: The cytology smears show multiple tight clusters of malignant cells admixed with reactive mesothelial cells. The individual cells show moderately pleomorphic hyperchromatic nuclei. In addition, the background of the smear shows many psammoma bodies.

Diagnostic features in the index case:
- Three-dimensional clusters of cells
- The cells with moderately pleomorphic hyperchromatic nuclei
- Multiple psammoma bodies.

Psammoma bodes in effusion cytology smears are seen in following conditions:
- Serous adenocarcinoma of ovary
- Borderline adenocarcinoma of ovary
- Mesothelioma.

Problem 3.33

D. Histoplasma infection

Discussion

Description: The cytology smears show large number of histoplasma within the mesothelial cells and also occasionally extracellular.

Disease: Histoplasma infection occurs in immunosuppressed patients particularly in AIDS. The index patient was immunosuppressed due to the chemotherapy effect. The occurrence of this fungi in the effusion fluid is uncommon. The infection has probably occurred secondary to lung infection or systemic infection by vascular spread.

Further Reading

1. Kortsik C, Elmer A, Tamm I. Pleural effusion due to histoplasma capsulatum and idiopathic CD4 lymphocytopenia. Respiration. 2003;70(1):118-22.

Problem 3.34

D. Dysgerminoma

Discussion

Description: The histopathology section of the ovarian mass (Figs 3.34A to D) shows nests of tumor cells separated by fibrous septae. The individual tumor cells are enlarged with round moderately pleomorphic nuclei having prominent nucleoli.

The cytology smears (Figs 3.34E to J) show abundant reactive mesothelial cells and occasional clusters and discrete tumor cells with similar morphology as that of histology section.

Diagnostic features in the index case: Histopathology of the adnexal mass shows features of dysgerminoma. The pleural effusion cytology smear shows many discrete and also loose clusters of tumor cells along with reactive mesothelial cells and lymphocytes. The cytomorphology of the tumor cell is consistent with histopathology. The cells are round to oval with round nuclei having fine chromatin and prominent nucleoli.

Differential diagnosis: On effusion cytology it is difficult to differentiate dysgerminoma from non-Hodgkin lymphoma. However typical history and primary histopathology along with complete absence of lymphoglandular bodies eliminate any possibility of lymphoma. Yolk sac tumor shows cells with large prominent nucleoli.

Further Reading

1. Hajdu SI, Nolan MA. Exfoliative cytology of malignant germ cell tumors. Acta Cytol. 1975;19(3):255-60.

CHAPTER 4

Respiratory Cytology

Introduction

The cytological examination of the respiratory tract is very helpful in the detection of both infective organisms and various neoplastic lesions. Carcinoma of the lung is the most common cancer in the world. The common complaints of the patient are progressive shortness of breath, chest pain, cough, dyspnea, and hemoptysis. World Health Organization (WHO) classified lung cancer as mentioned in the Box 4.1.[1]

Box 4.1: WHO classification of lung tumors, 2015[1]

- Epithelial tumors
  - Adenocarcinoma
    - Lepidic adenocarcinoma
    - Acinar adenocarcinoma
    - Papillary adenocarcinoma
    - Solid adenocarcinoma
    - Invasive mucinous adenocarcinoma
    - Minimally invasive adenocarcinoma
    - Colloid adenocarcinoma
  - Squamous cell carcinoma
    - Keratinizing squamous cell carcinoma
    - Nonkeratinizing squamous cell carcinoma
    - Basaloid squamous cell carcinoma
  - Neuroendocrine tumors
    - Small cell carcinoma
    - Large cell neuroendocrine carcinoma
    - Carcinoid tumors
    - Typical carcinoid
    - Atypical carcinoid
  - Large cell carcinoma
  - Adenosquamous carcinoma
  - Pleomorphic carcinoma
  - Giant cell carcinoma
  - Carcinosarcoma
  - Pulmonary blastoma

Contd...

Contd...

- Salivary gland type tumor
  - Adenoid cystic carcinoma
  - Mucoepidermoid carcinoma
  - Pleomorphic adenoma
- Papillomas: Squamous cell papilloma and glandular papilloma
- Mesenchymal tumor
- Lymphohistiocytic tumor
- Tumors of ectopic origin
- Metastatic tumors

Squamous Cell Carcinoma[2]

The tumor shows polyhedral cells with eosinophilic cytoplasm. Cytoplasmic orangeophilia indicates intracellular keratin. The nuclei are round, hyperchromatic and moderately pleomorphic. Many fiber cells and tadpole cells are also seen. The tumor often shows background necrosis and granular debris.

Immunocytochemistry

Positive for CK 5/6, p40 and p 63. The tumor cells are negative for TTF and napsin and CK 7.

Genetic Profile

Squamous cell carcinoma shows gene amplification of the following chromosomes

- 3q : SOX2, TP 63
- 7p: EGFR

The frequent gene mutations are seen in: TP 53, CDKN2A, and PTEN.

Adenocarcinoma[2]

Adenocarcinomas of the lung are the most common subtype in females. The various subtypes of adenocarcinoma includes: Lepidic adenocarcinoma, acinar adenocarcinoma, papillary adenocarcinoma, solid adenocarcinoma, invasive mucinous adenocarcinoma, minimally invasive adenocarcinoma and colloid adenocarcinoma. The cytological features of adenocarcinoma depend on the subtypes. Overall, the cells show glandular or acinar arrangement along with honeycomb like sheets and discrete cells. Many three-dimensional cell balls of malignant cells are also seen. The background mucinous material may also be noted in a small number of cases. The cytoplasm of the cell is moderate to abundant and often contains large vacuoles filled with mucin. Large vacuoles may push the nucleus to the periphery of the cell causing indented margin of the nucleus. The nuclei are vesicular having prominent nucleoli.

Immunocytochemistry

Adenocarcinomas of lungs are positive for CK 7, napsin A, and TTF-1. The tumor cells are negative for p40, p63 and CK 5/6.

Genetic Profile

- Epidermal growth factor receptor (EGFR) mutation: Point mutation of codon 858 and in frame deletions in exon 19 (90% cases) are seen. EGFR mutation is specific for adenocarcinoma and it also acts as a prognostic factor. It is a predictor of response to EGFR tyrosine kinase inhibitor treatment. EGFR mutation is related with TTF-1 positivity.

- KRAS mutation
- Anaplastic lymphoma kinase (ALK) rearrangement: ALK rearrangement is predictive of response to ALK inhibitor treatment, however it is not a prognostic factor.

Small Cell Carcinoma[2]

Small cell carcinoma (SCC) accounts for 20% of lung cancer and 90% of these tumors are centrally located. SCC is classified as:

- Pure small cell carcinoma
- Combined small cell carcinoma with other carcinomas such as squamous cell carcinoma, adenocarcinoma, large cell carcinoma, etc.

The cytology smears of SCC show a predominant discrete population of the small round cells about one and half times larger than mature lymphocytes. The tumor cells show nuclear threading and crushing artifact due to the fragile nature of the cells. The nuclei are relatively monomorphic hyperchromatic with indistinct nucleoli. Nuclear molding is characteristic of small cell carcinoma.

Immunocytochemistry

The small cell carcinoma is positive for CD 56, chromogranin, synaptophysin, cytokeratin and TTF-1.

Carcinoid[2]

Carcinoid tumor is classified as (a) typical carcinoid and (b) atypical carcinoid.

Typical carcinoid: Cytology smears of the carcinoid tumor usually show dissociated cells. Occasional rosette like structures may be seen. Individual cells have small to moderate amount of cytoplasm with fine reddish granules. The nuclei are relatively monomorphic with finely stippled chromatin giving the characteristic salt and pepper appearance.

Atypical carcinoid: Cells of atypical carcinoid show large nuclei with moderate pleomorphism. The individual cells have moderate to abundant reddish granular cytoplasm.

Immunocytochemistry

Carcinoid tumors are positive for chromogranin A, synaptophysin and CD 56. The tumor cells are also positive for pancytokeratin. The carcinoid tumors are usually negative for TTF-1.

Large Cell Neuroendocrine Carcinoma

These tumors are difficult to recognize as they have overlapping cytological features with neuroendocrine tumors and adenocarcinomas. The cells are large with moderate amount of cytoplasm, monomorphic hyperchromatic nuclei having irregular nuclear margin and prominent nucleoli.

The differentiating features of SCC, typical and atypical carcinoid and large cell neuroendocrine carcinomas are highlighted in the Table 4.1.

Table 4.1: Distinguishing features of various neuroendocrine tumors of lung

Features	*Small cell carcinoma*	*Carcinoid*	*Atypical carcinoid*	*Large cell neuroendocrine carcinoma*
Mitosis	High	High	Frequent	High
Necrosis	Present	No	Occasionally	Present
Nucleus	Hyperchromatic	Salt and pepper chromatin	Salt and pepper chromatin	Salt and pepper chromatin
TTF-1	Positive	Negative	Negative	Positive
CD 56	Positive	Positive	Positive	Positive
Chromogranin	Positive	Positive	Positive	Positive

Large Cell Anaplastic Tumor[2]

The large cell carcinoma (LCC) accounts for 10% of all lung carcinomas. These are undifferentiated large cell carcinomas without any particular evidence of differentiation towards squamous or adenocarcinoma. Majority of large cell carcinomas are located in the peripheral part of lung. WHO recommended that this specific entity should be labelled as non-small cell carcinoma, not otherwise specified on cytology smear. On cytology smear the tumor shows discrete and syncytial clusters of cells. No glandular differentiation is seen. The individual cells have moderate to abundant dense cytoplasm with ill-defined margins. Nuclei show marked pleomorphism, nuclear margin irregularity, irregular coarse chromatin and multiple enlarged prominent nucleoli. Many large mono/multinucleated tumor giant cells are also noted. The characteristic feature of large cell carcinoma is sticking of polymorphs with the malignant cells. Giant cell carcinoma is one of the varieties of the undifferentiated large cell carcinoma. Here, the smear shows many large multinuclear and mononuclear tumor giant cells along with heavy polymorphonuclear leucocytes infiltration.

References

1. Travis WD, Brambilla E, Burke AP, Marx A, Nicholson AG. WHO Classification of tumors of the lung, pleura, thymus and heart. Lyon: International Agency for Research on Cancer, 2015.
2. Dey P. Fine Needle Aspiration Cytology: Interpretation and Diagnostic Difficulties. Second Edition, Jaypee Brothers Medical Publishers (P) Ltd, New Delhi, 2014.

QUESTIONS

Problem 4.1

Match the following (Fig. 4.1)

A	1. Hyperplastic bronchial cells
B	2. Curschmann's spiral
C	3. Ciliated columnar cells
D	4. Alveolar macrophages
E	5. Atypical cells

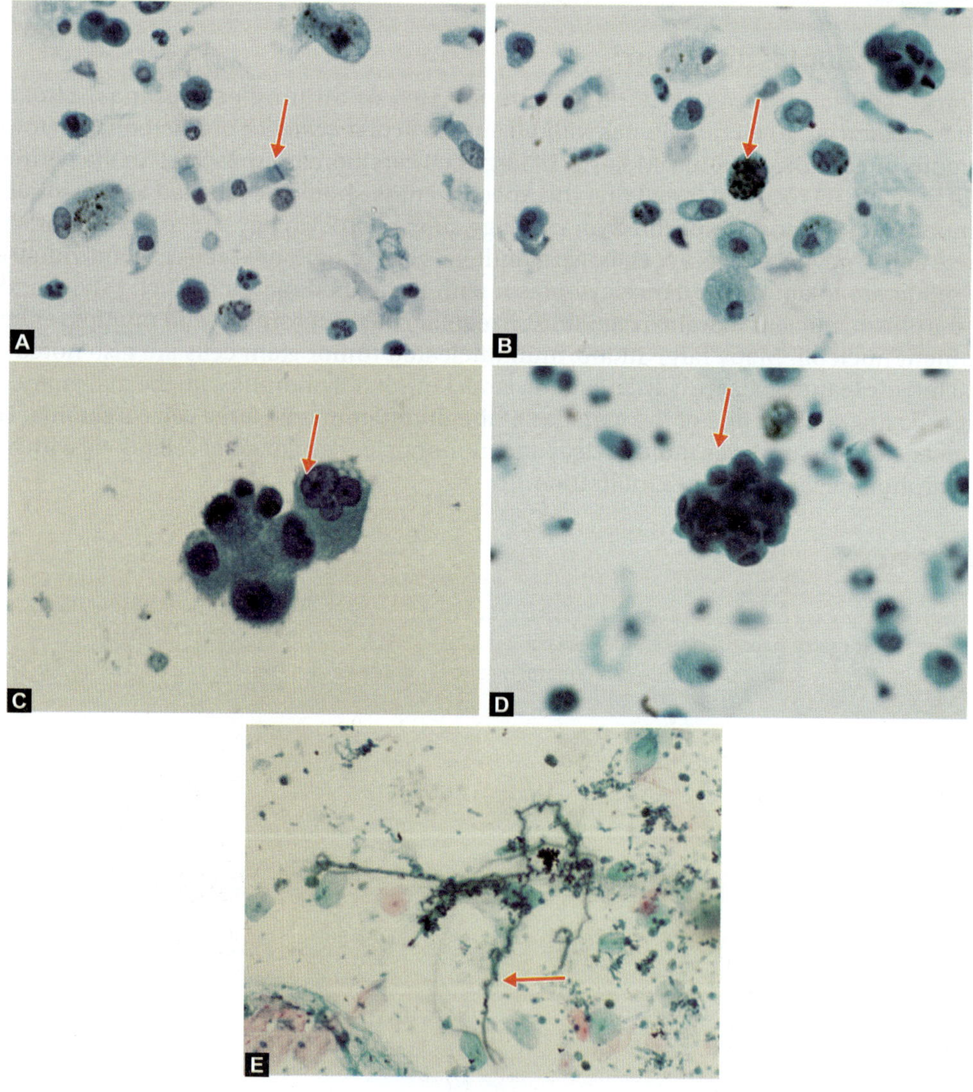

Figures 4.1A to E

Problem 4.2

A 41-year-old female had cough, fever and chest pain for 3 months. She had pallor and loss of weight. Chest radiograph showed upper left lobe consolidation. Bronchoalveolar lavage cytology was done and the microphotographs of the SurePath preparation are displayed in the Figure 4.2. The most likely diagnosis is:

A. Squamous cell carcinoma
B. Reactive bronchial cells
C. Small cell carcinoma
D. Adenocarcinoma

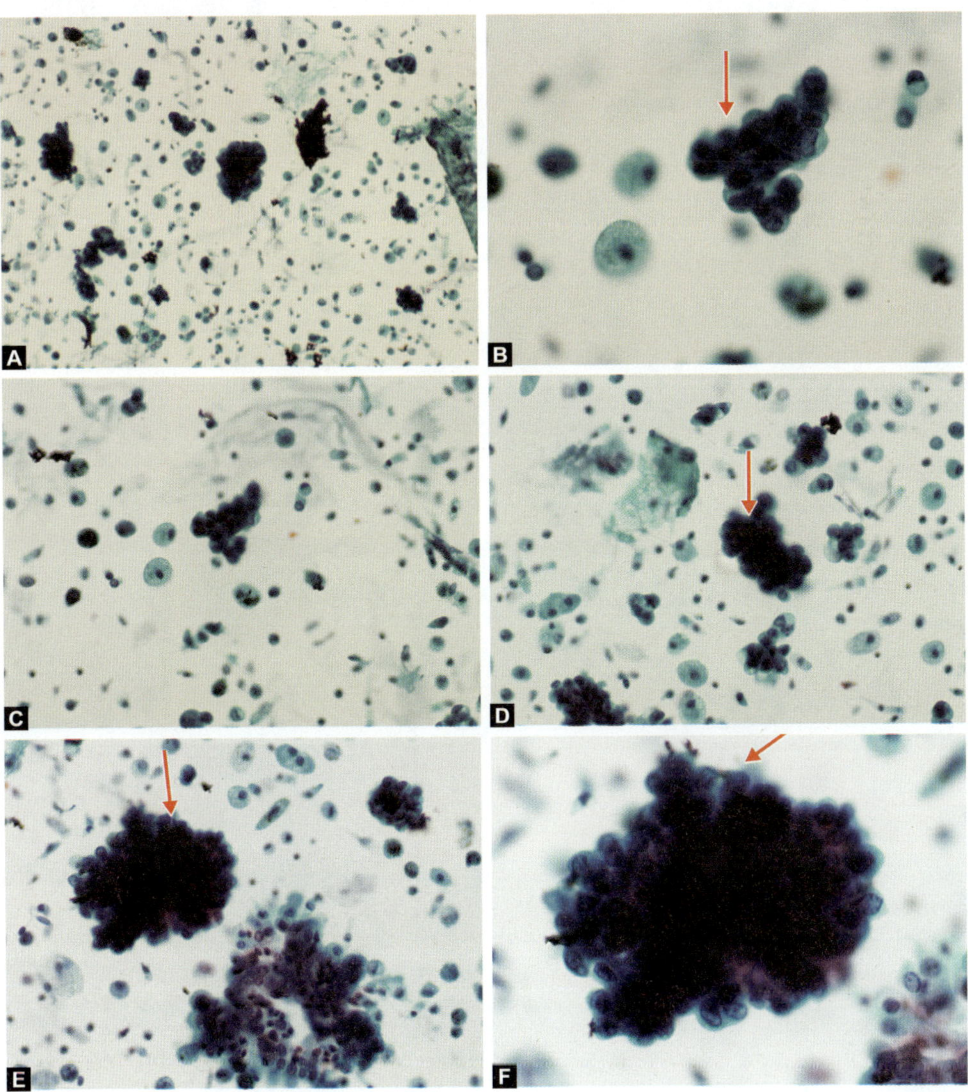

Figures 4.2A to F

Problem 4.3

A 39-year-old male patient recipient of renal transplant patient presented with respiratory distress and lung consolidation. The most likely diagnosis from Figure 4.3 is:

A. Carcinoma
B. Viral infection
C. Aspergillosis
D. Acute inflammation

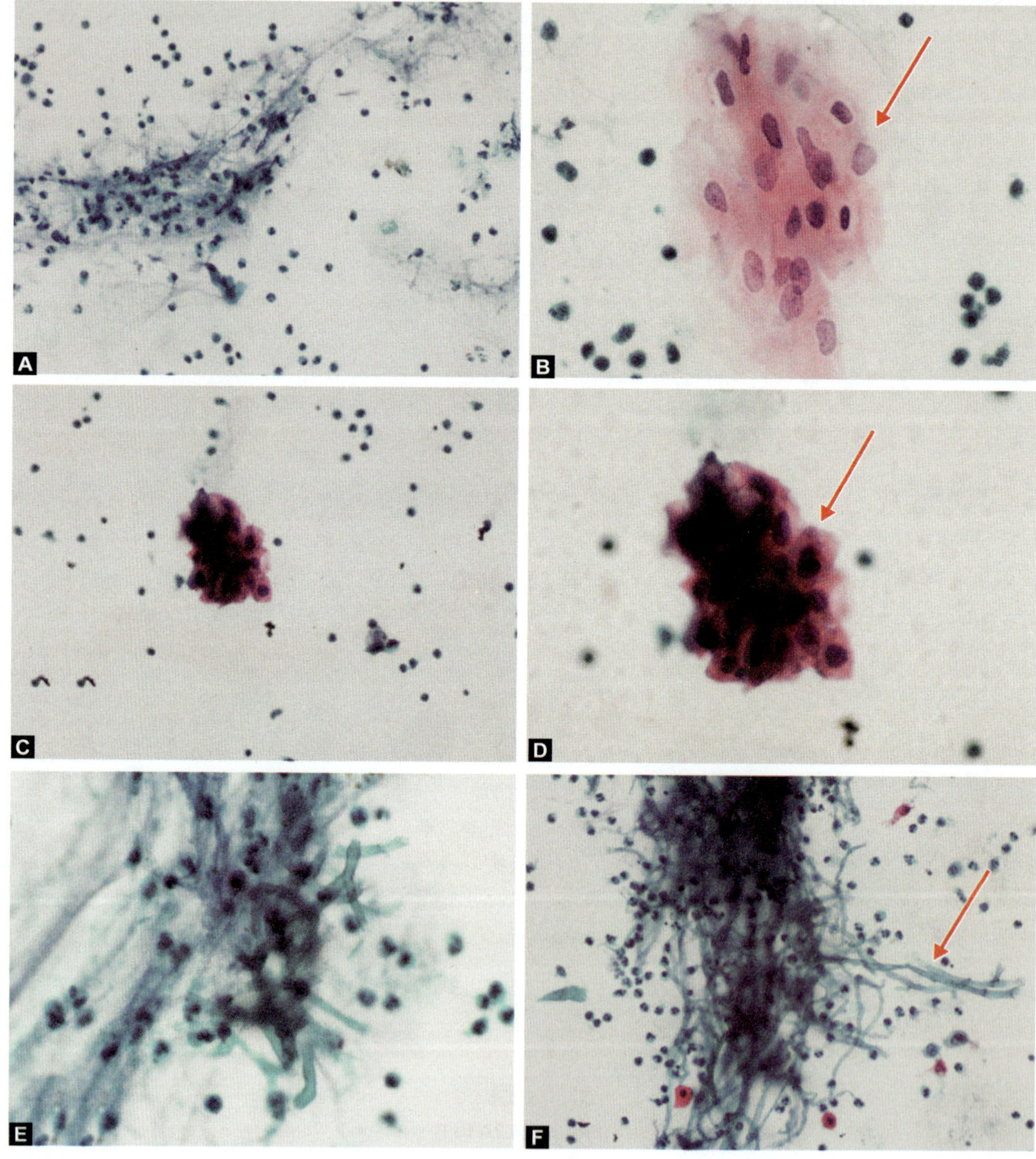

Figures 4.3A to F

Problem 4.4

A 46-year-old male presented with cough, respiratory distress for 3 months. The bronchoalveolar lavage was done and microphotographs are shown in Figure 4.4. The most likely diagnosis is:

A. Adenocarcinoma
B. Hyperplastic bronchial cells
C. Squamous cell carcinoma
D. Small cell carcinoma

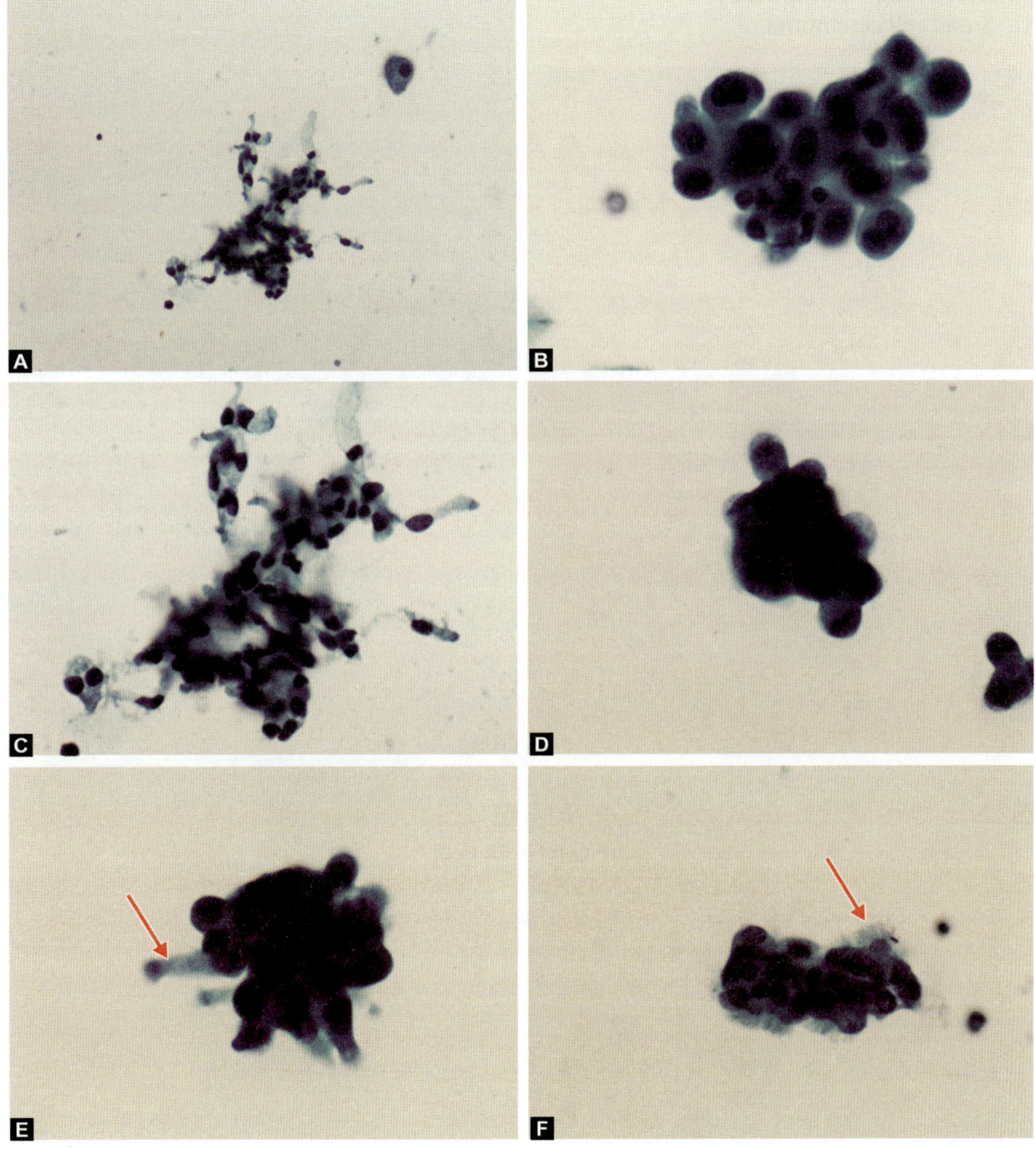

Figures 4.4A to F

Problem 4.5

A 55-year-old male presented with dry cough, hemoptysis and chest pain for 3 months. Chest radiograph showed bilateral upper lobe ground glass opacities. His mediastinal lymph node was enlarged. Endoscopic ultrasound guided transbronchial needle aspiration (EUS TBNA) was done. The most likely diagnosis from Figure 4.5 is:

A. Granulomatous inflammation
B. Reactive lymphoid hyperplasia
C. Non-Hodgkin lymphoma
D. Small cell carcinoma

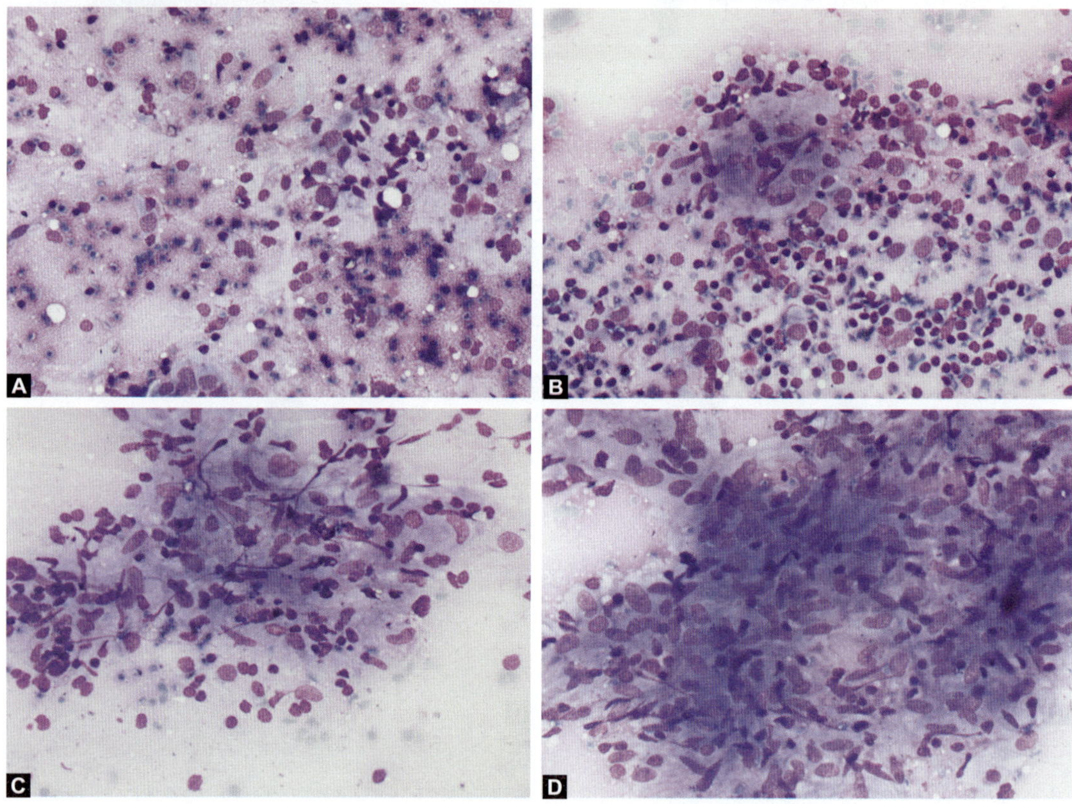

Figures 4.5A to D

Problem 4.6

A 37-year-old female had renal transplantation 2 year back. Presently the patient developed fever, and jaundice along with pruritus for 4 months. On chest radiograph the patient had bilateral ground glass opacity. Her bronchoalveolar lavage (BAL) was done. The most likely diagnosis from Figure 4.6 is:

A. Herpes viral infection
B. No abnormality
C. CMV positive
D. Pneumocystis infection

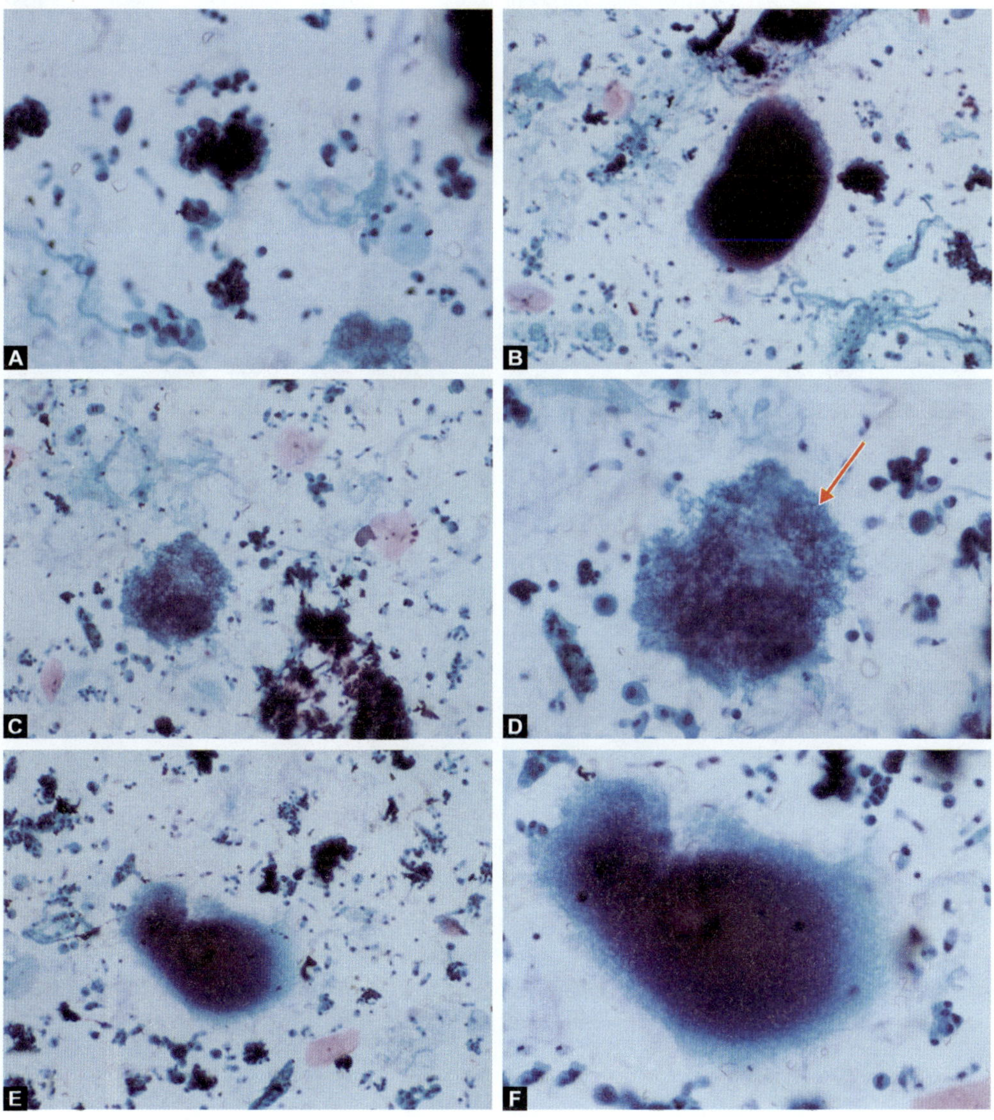

Figures 4.6A to F

Problem 4.7

A 49-year-old male complaints of hemoptysis and chest pain for 2 months. His chest radiograph shows right upper lobe consolidation. BAL was done. The microphotographs of the smears are shown in Figure 4.7. The most likely diagnosis is:

A. Squamous cell carcinoma
B. Reactive bronchial cells
C. Small cell carcinoma
D. Adenocarcinoma

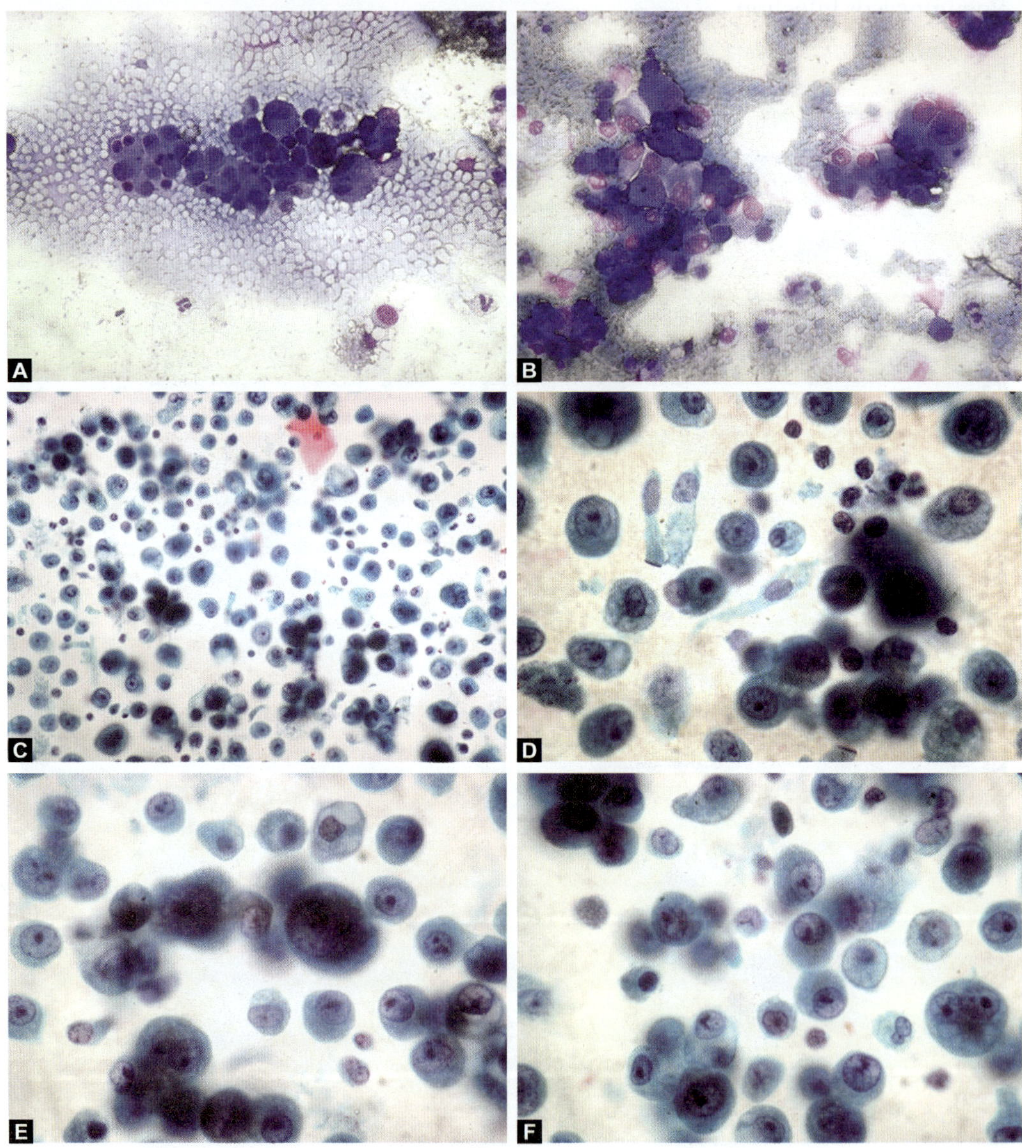

Figures 4.7A to F

Problem 4.8

A 55-year-old male presented with respiratory distress, cough and loss of weight for 2 months. He has lower lung space occupying lesion. The microphotographs of BAL are shown in Figures 4.8A to J. The most likely diagnosis is:

A. Adenocarcinoma
B. Squamous cell carcinoma
C. Small cell carcinoma
D. Reactive atypia

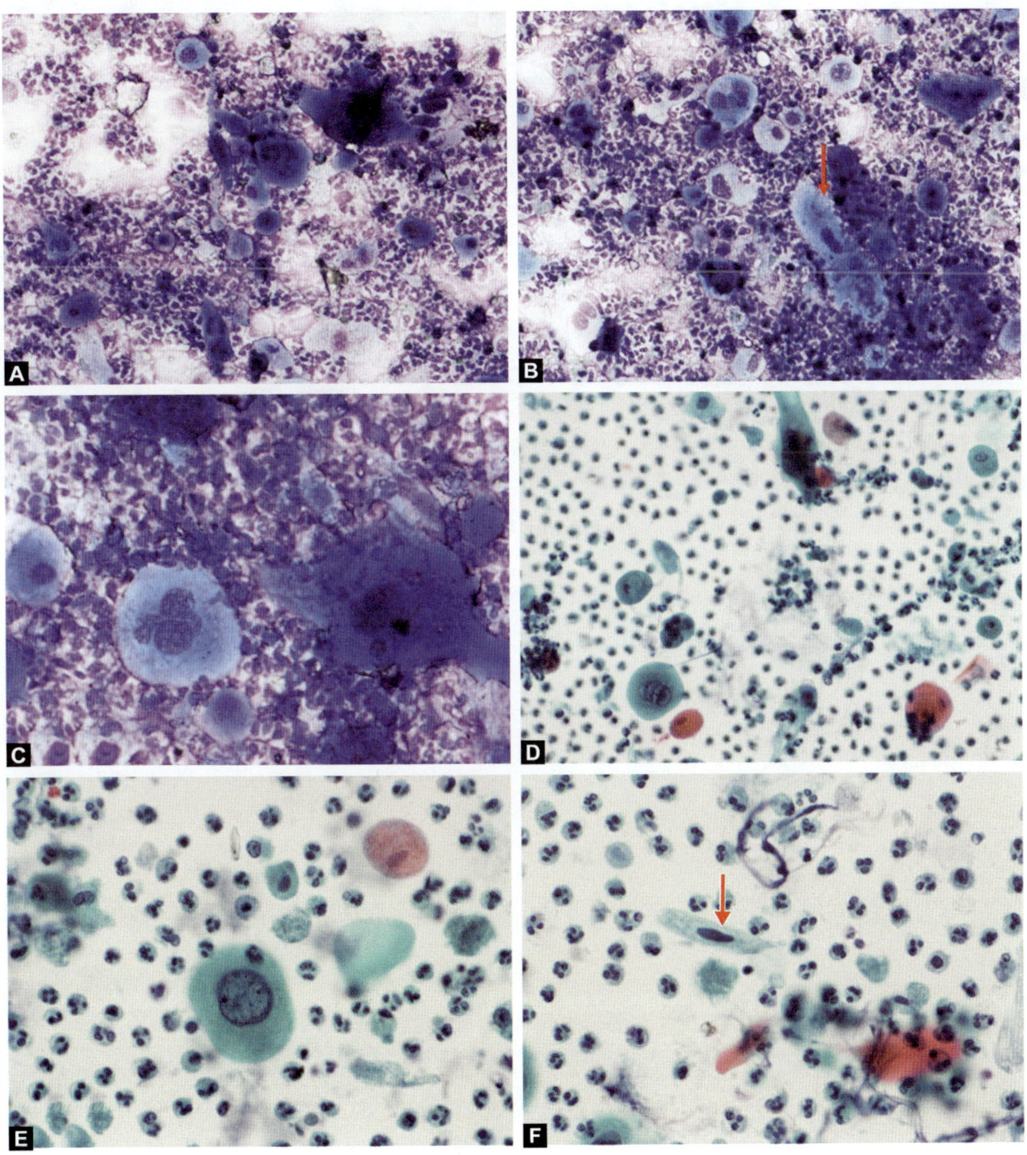

Figures 4.8A to F

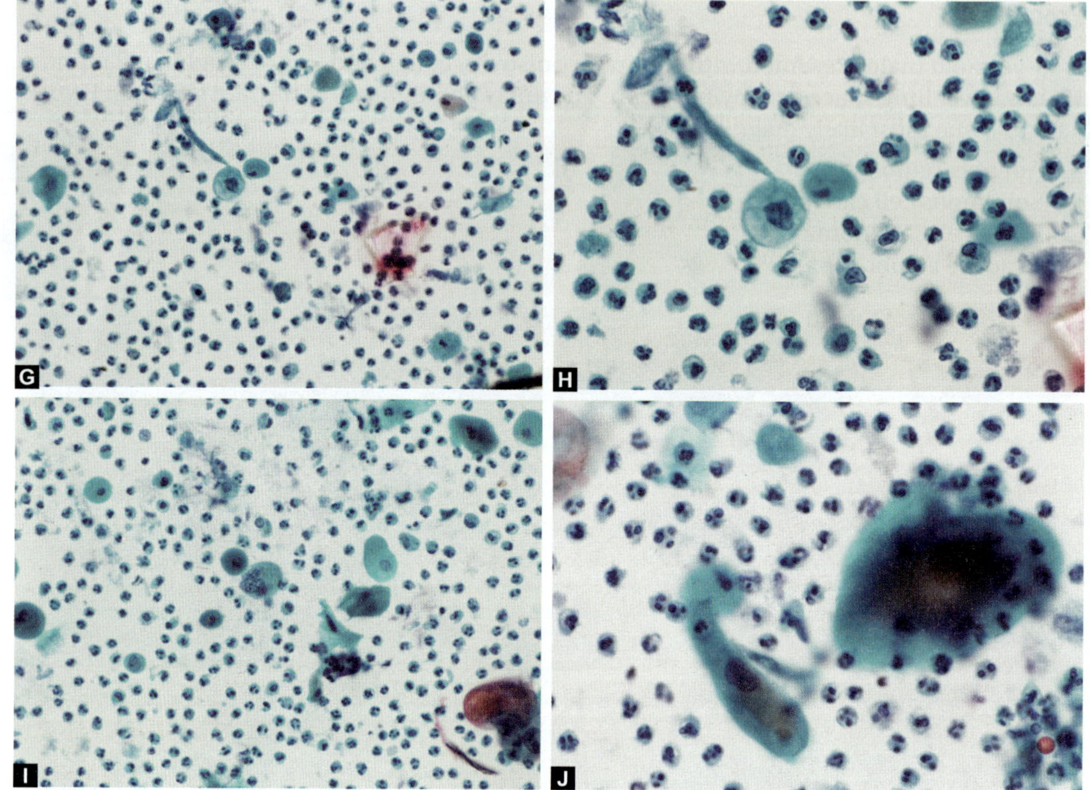

Figures 4.8G to J

Problem 4.9

A 63-year-old female presented with cough and hemoptysis for one month. Her chest radiograph showed consolidated 4 cm diameter space occupying lesion in the left upper lobe of lung. Routine sputum examination was done. The most likely diagnosis from the Figure 4.9 of SurePath preparation of the sputum is:

A. Reactive bronchial cells, no malignancy
B. Squamous cell carcinoma
C. Adenocarcinoma
D. Small cell carcinoma

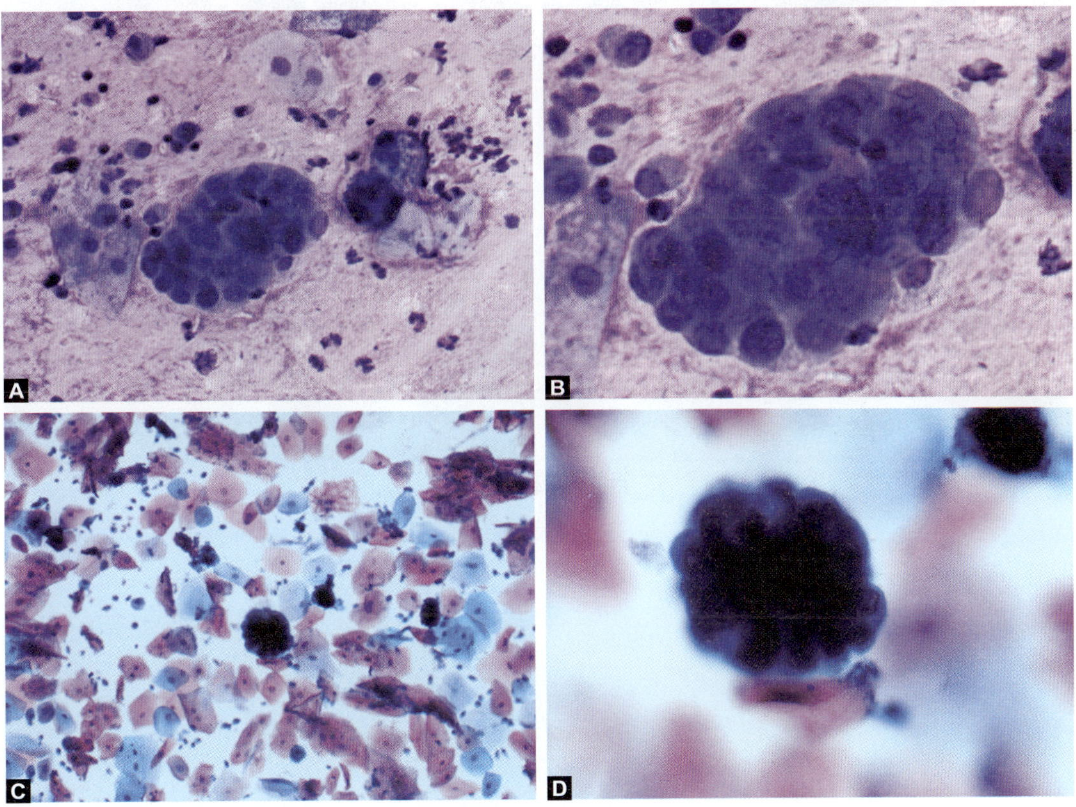

Figures 4.9A to D

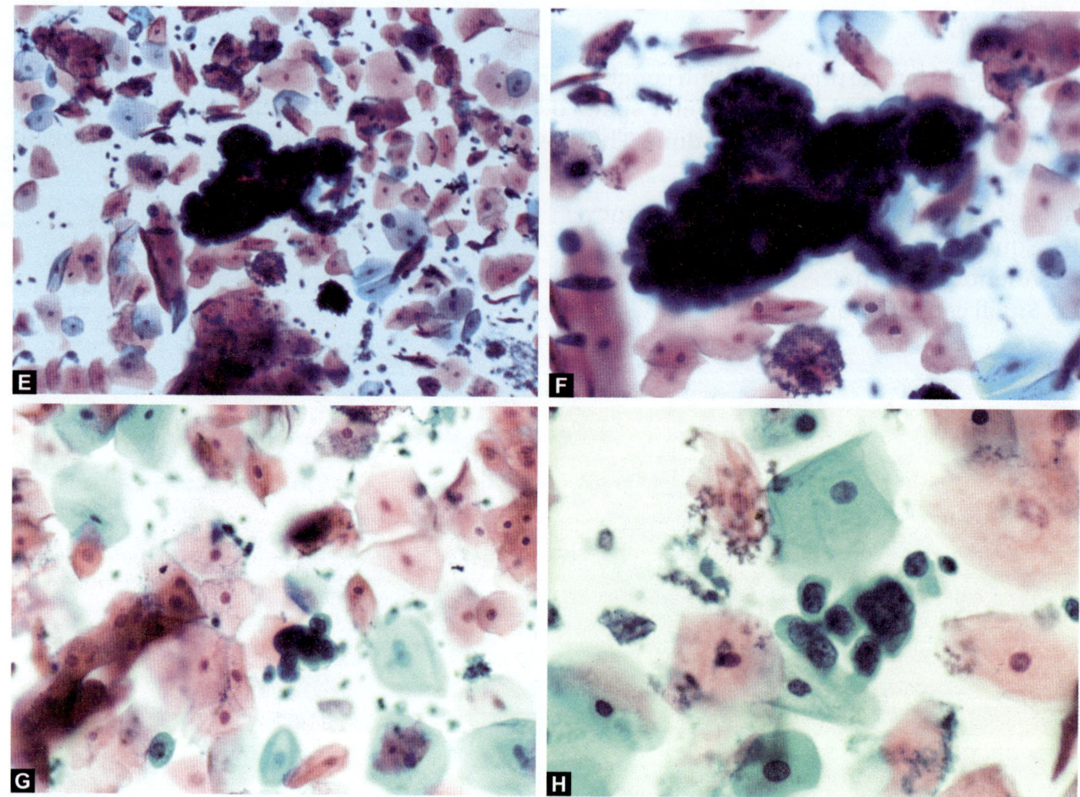

Figures 4.9E to H

Problem 4.10

A 59-year-old male presented with fever, cough and left sided chest pain for 2 months. His chest radiograph showed 2 cm diameter consolidated lesion in upper lobe of left lung. The routine sputum examination was done. The most likely diagnosis from Figure 4.10 is:

A. No malignancy
B. Squamous cell carcinoma
C. Adenocarcinoma
D. Small cell carcinoma

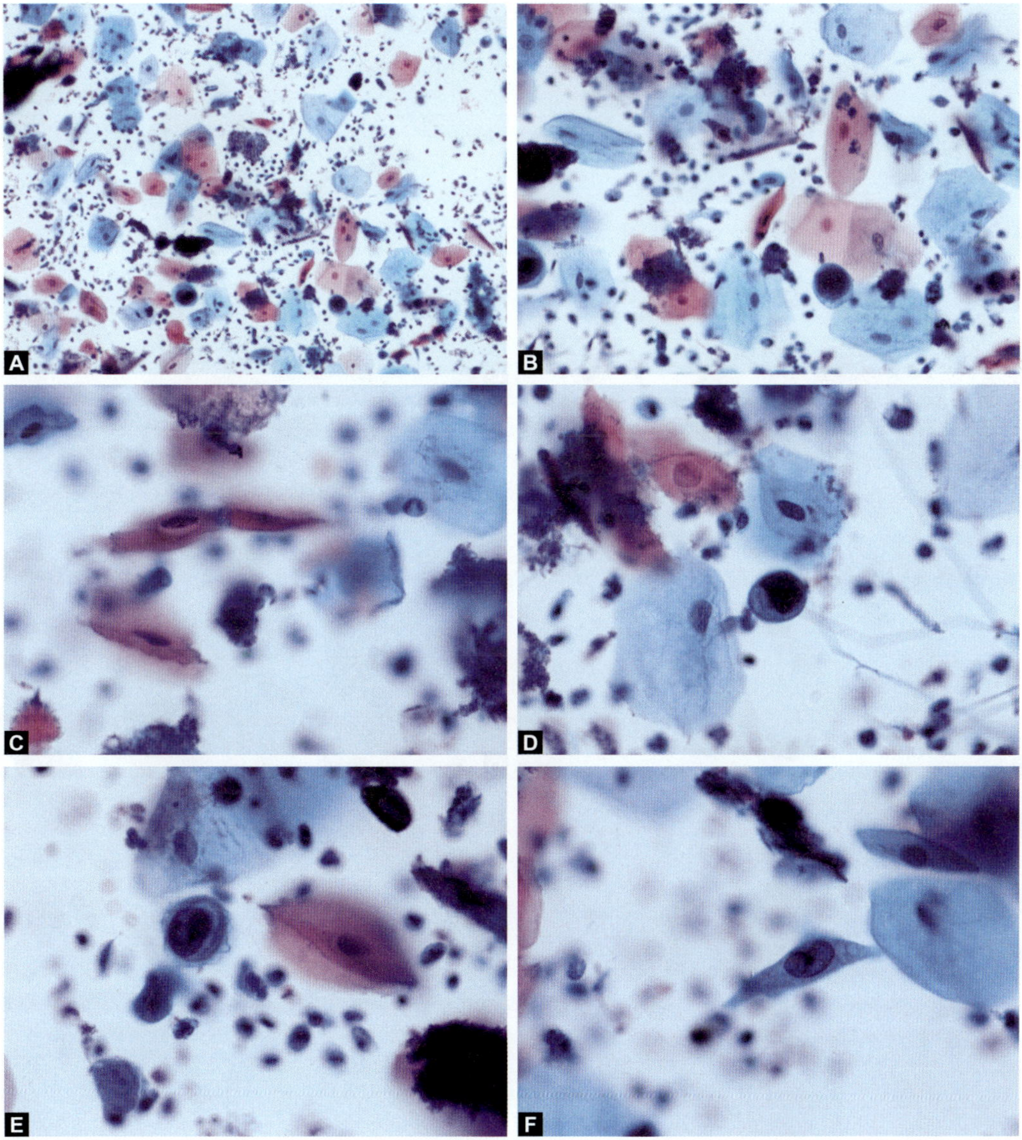

Figures 4.10A to F

Problem 4.11

A 56-year-old man presented with right sided chest pain and dry cough for 21 days. Routine sputum cytological examination (SurePath preparation) was done. The most likely diagnosis form Figure 4.11 is:

A. No evidence of any malignancy
B. Reactive bronchial cells due to viral infection
C. Adenocarcinoma
D. Small cell carcinoma

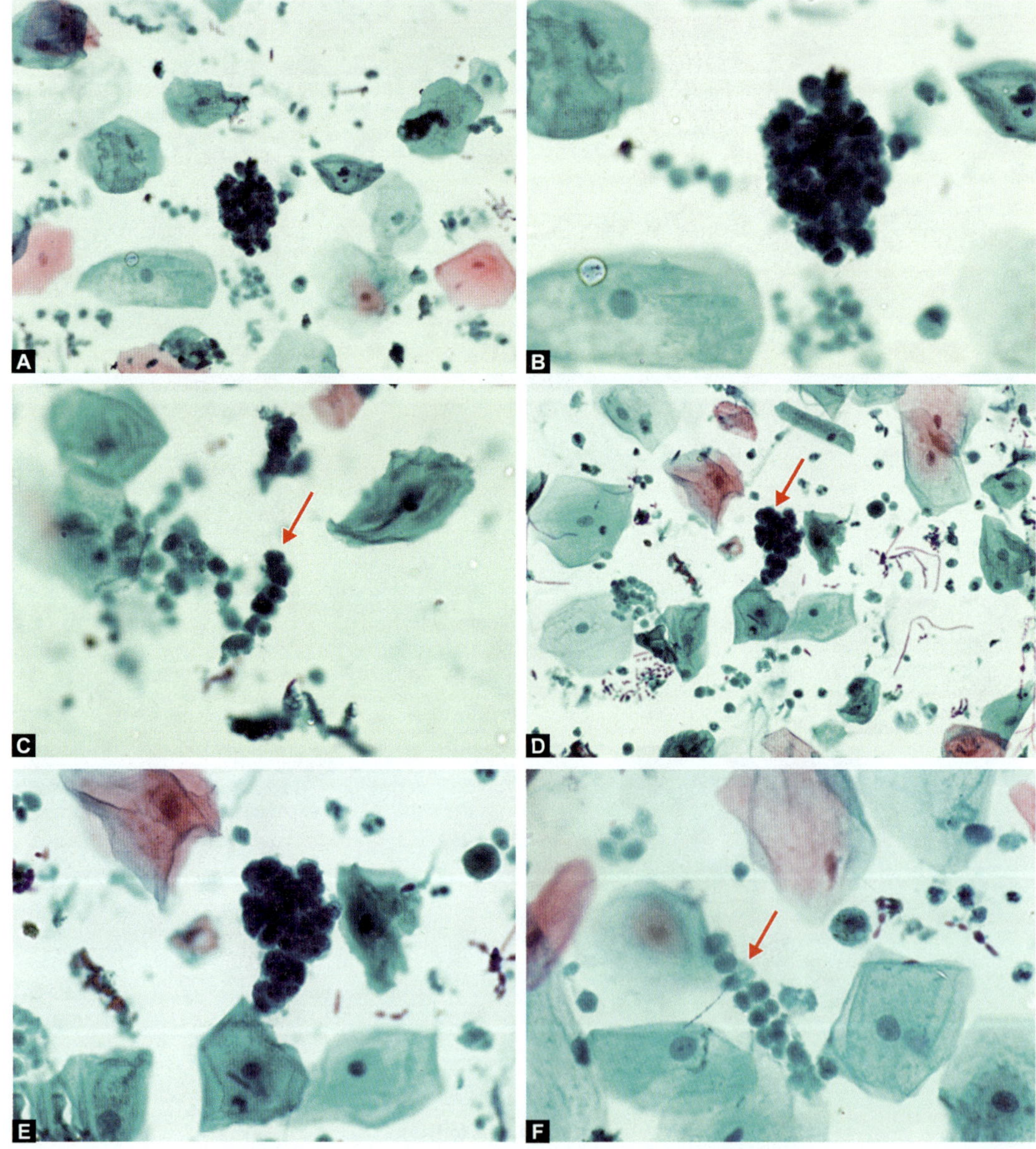

Figures 4.11A to F

Problem 4.12

A 51-year-old male presented with hemoptysis and chest pain in right side for 7 days. Bronchial wash was done from the space occupying solid lesion of right upper lobe. The most likely diagnosis from Figure 4.12 is:

A. Small cell carcinoma
B. Squamous cell carcinoma
C. Adenocarcinoma
D. Granulomatous inflammation

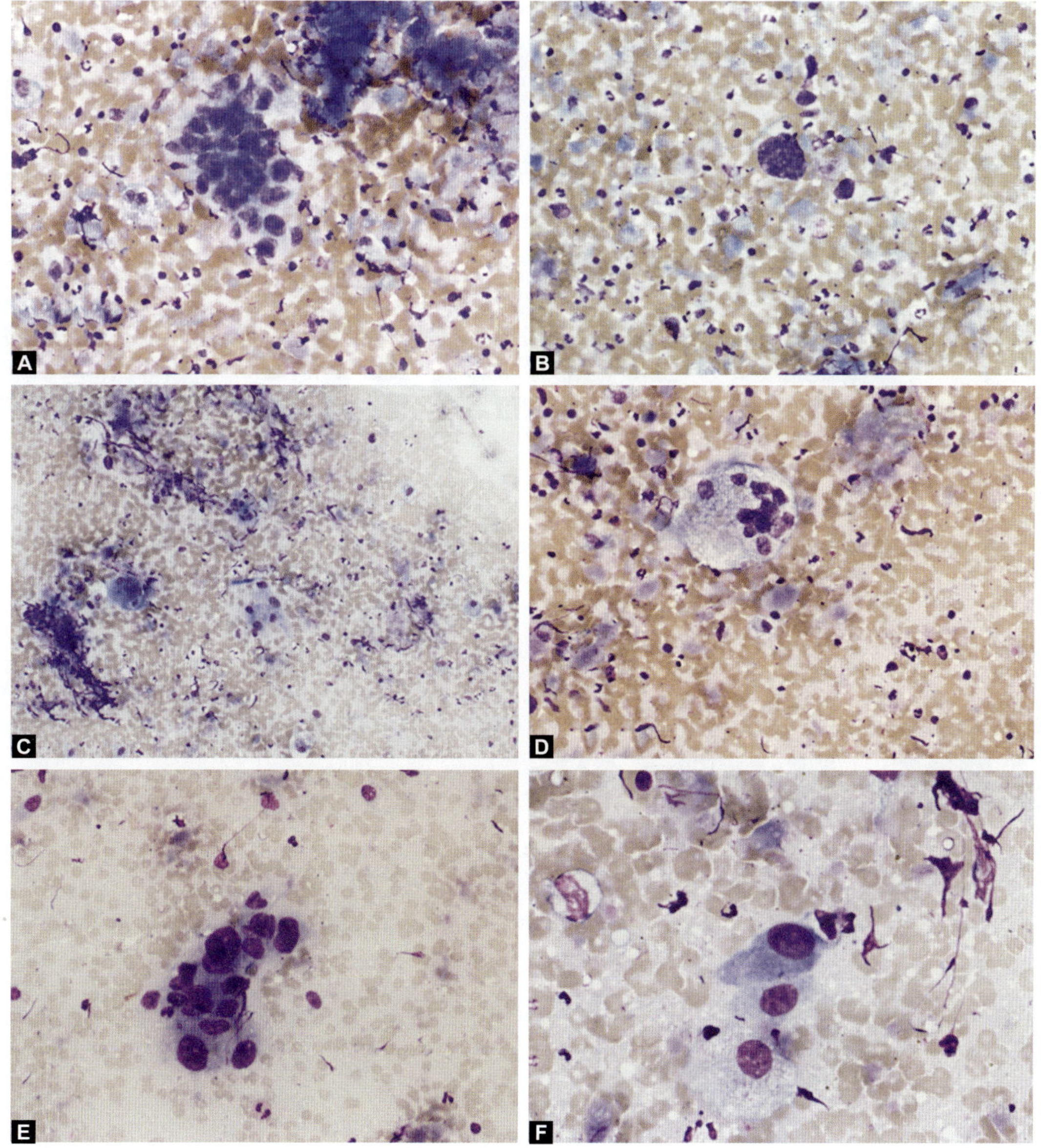

Figures 4.12A to F

Problem 4.13

A 63-year-old male complained of hemoptysis and loss of weight for 6 weeks. The patient had peribronchial 2 cm diameter space occupying lesion in the right upper lobe of lung. Endoscopic ultrasound guided transbronchial needle aspiration (EUS TBNA) was done. The most likely diagnosis from Figure 4.13 is:

A. Adenocarcinoma
B. Squamous cell carcinoma
C. Small cell carcinoma
D. Large cell carcinoma

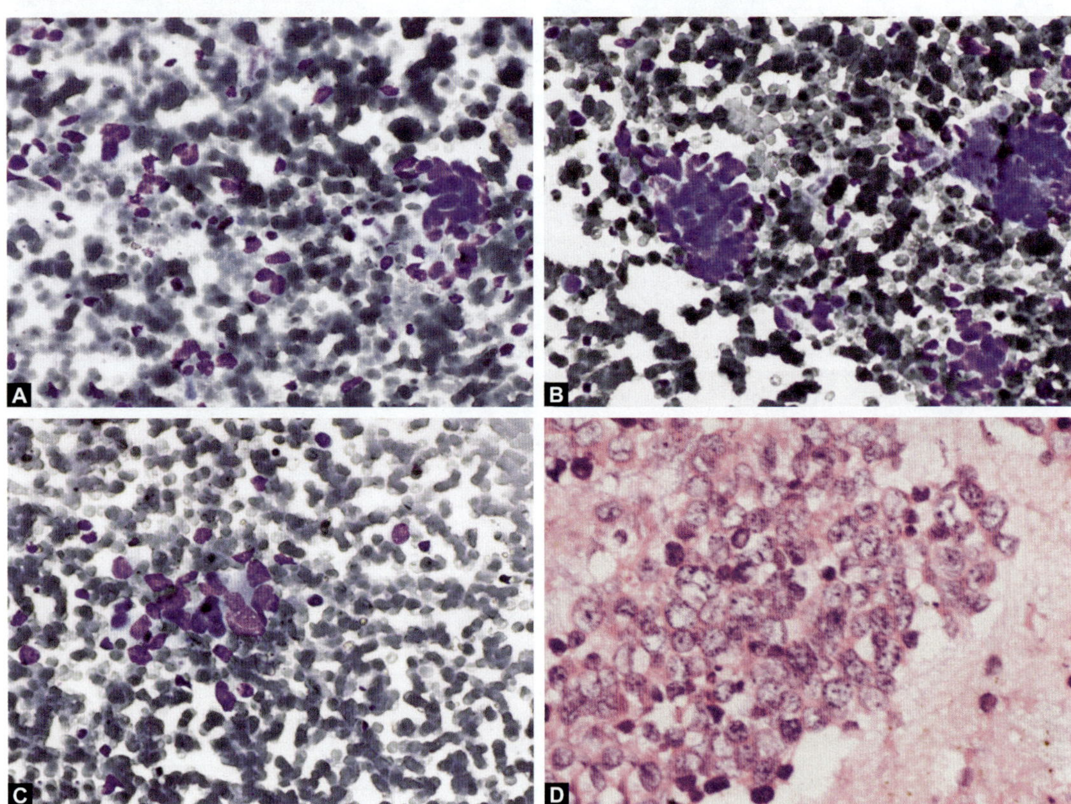

Figures 4.13A to D

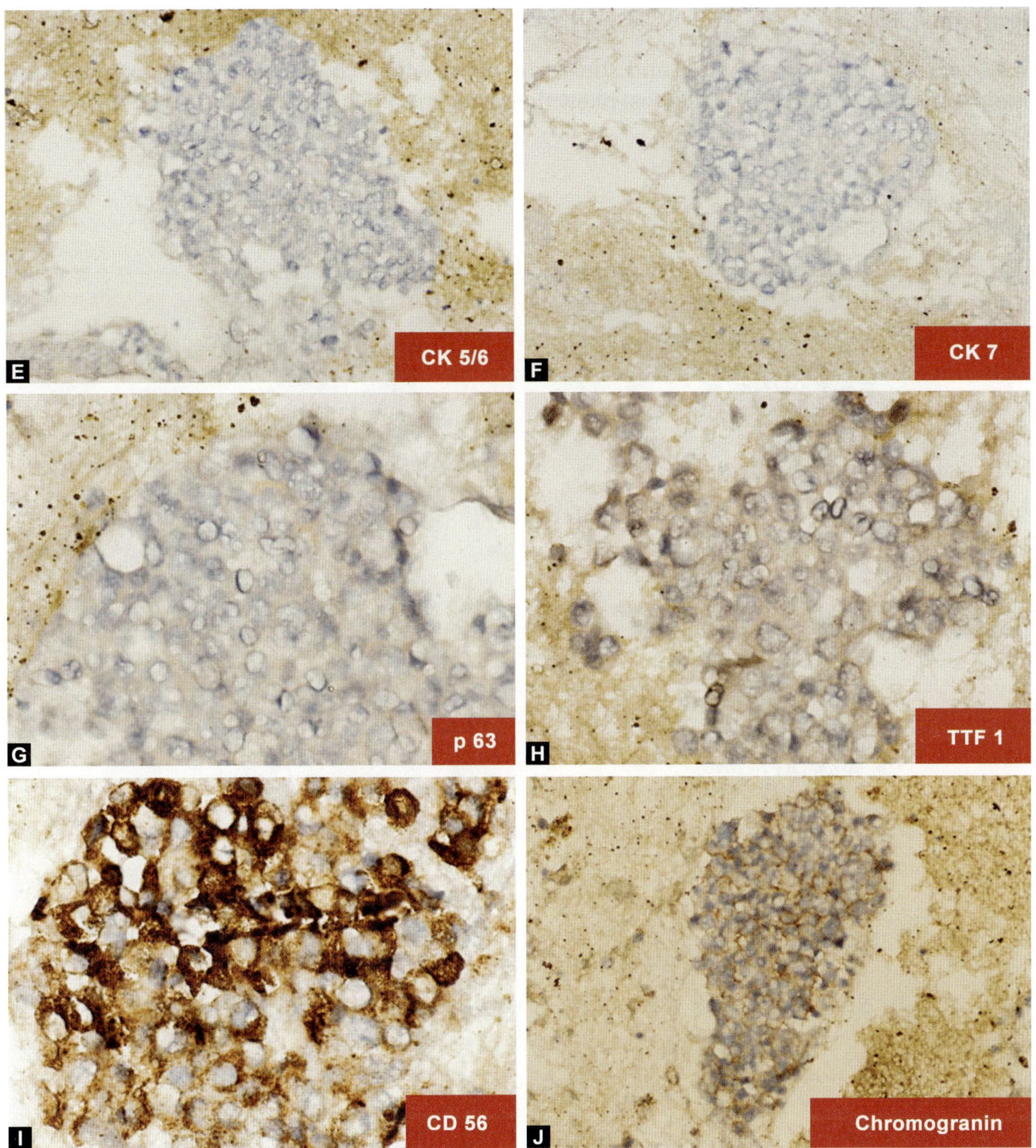

Figures 4.13E to J

Problem 4.14

A 47-year-old male had dyspnea and chest pain along with 4 cm diameter mediastinal lymph node. EUS TBNA was done. The most likely diagnosis from Figure 4.14 is:

a. Adenocarcinoma
b. Squamous cell carcinoma
c. Anaplastic carcinoma
d. Large cell anaplastic carcinoma

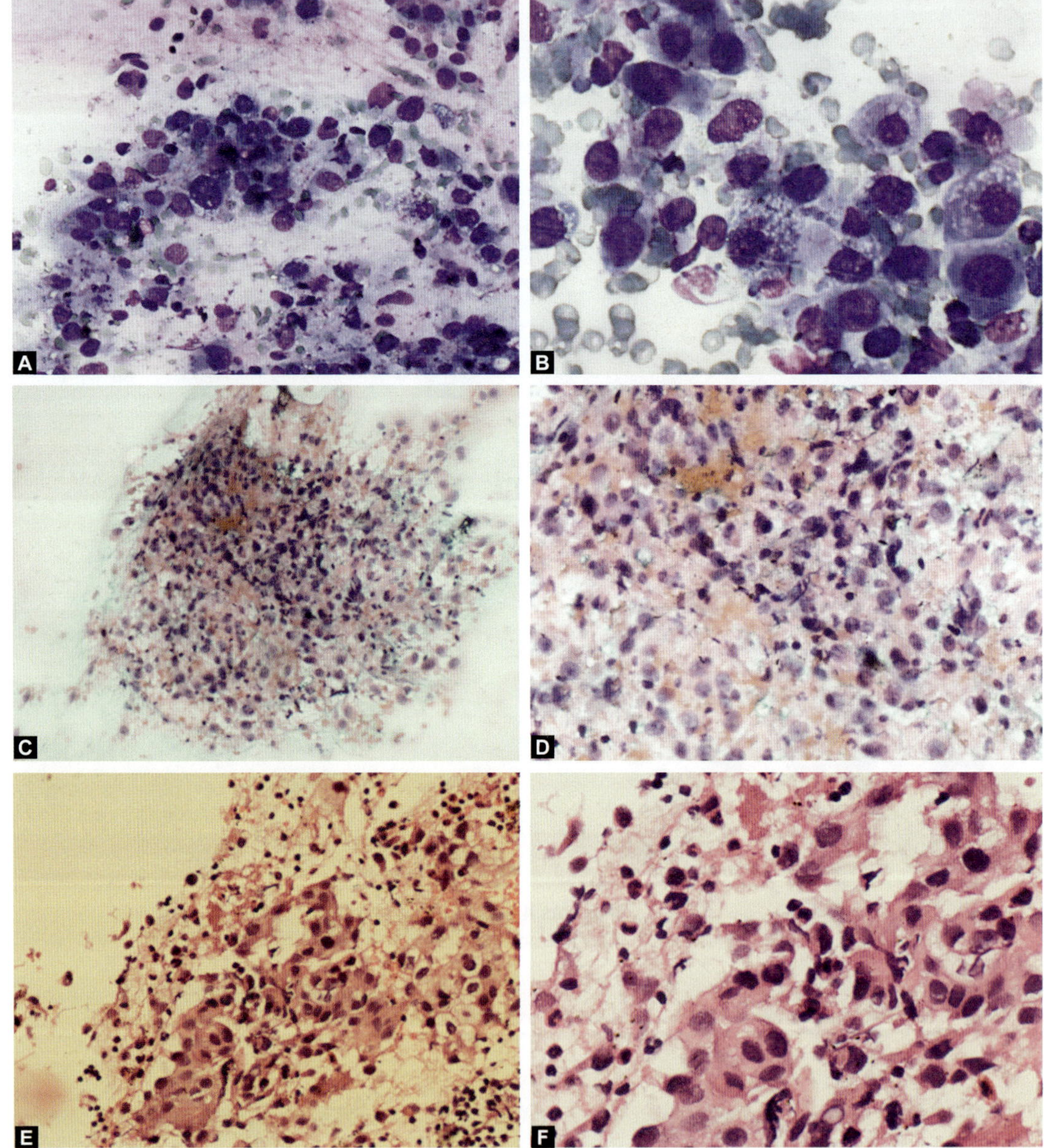

Figures 4.14A to F

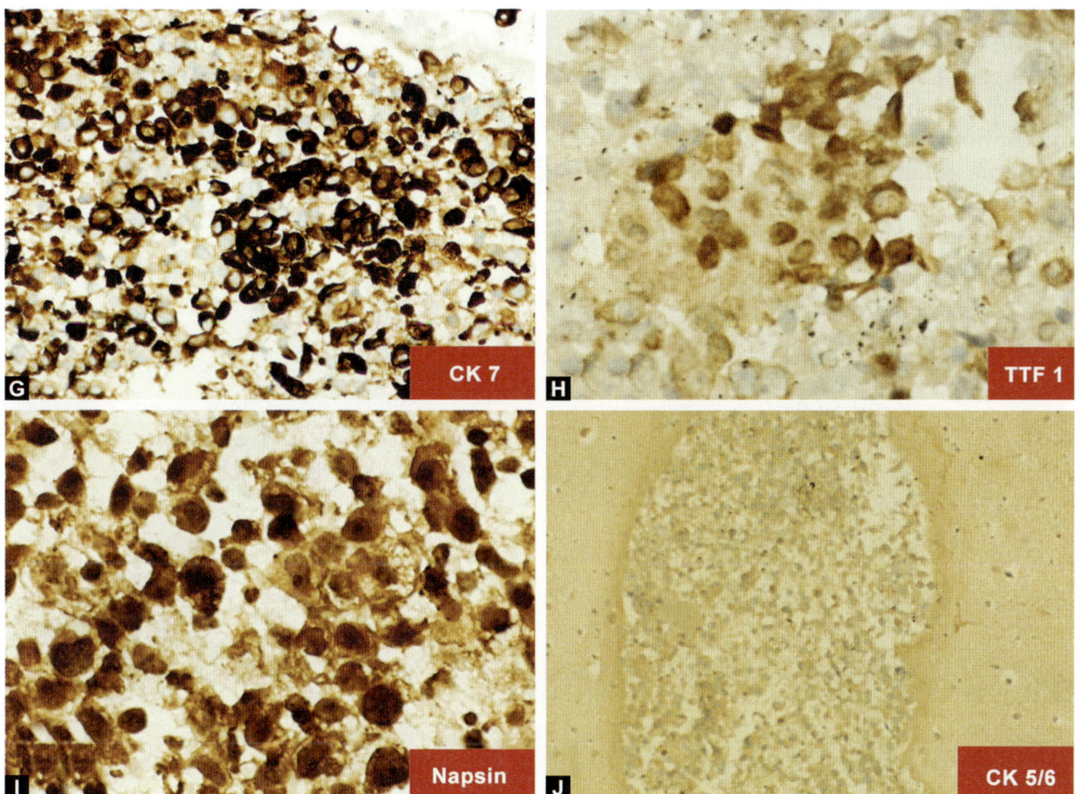

Figures 4.14G to J

Problem 4.15

A 46-year-old male had acute leukemia. He was on remission after chemotherapy. Now for last one week he had respiratory distress. His BAL was done. The most likely diagnosis from the microphotograph (Fig. 4.15) is:

A. Acute inflammation
B. Viral infection
C. Acute leukemia relapse
D. Aspergillosis

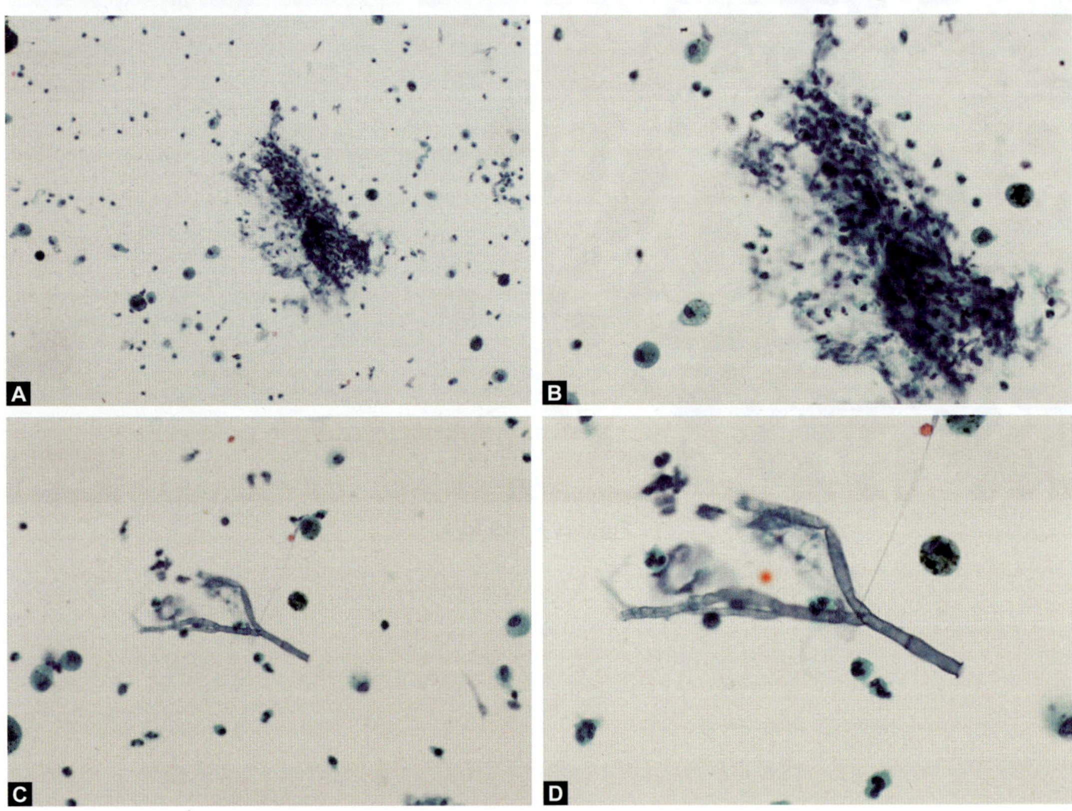

Figures 4.15A to D

Problem 4.16

A 45-year-old female presented with dyspnea, cough and weight loss for 3 months. Her chest radiograph showed nonspecific opacity in the lung. Bronchoalveolar lavage was done. The most likely diagnosis from Figure 4.16 is:

A. Necrosis only
B. Cryptococcal infection
C. Adenocarcinoma
D. Alveolar proteinosis

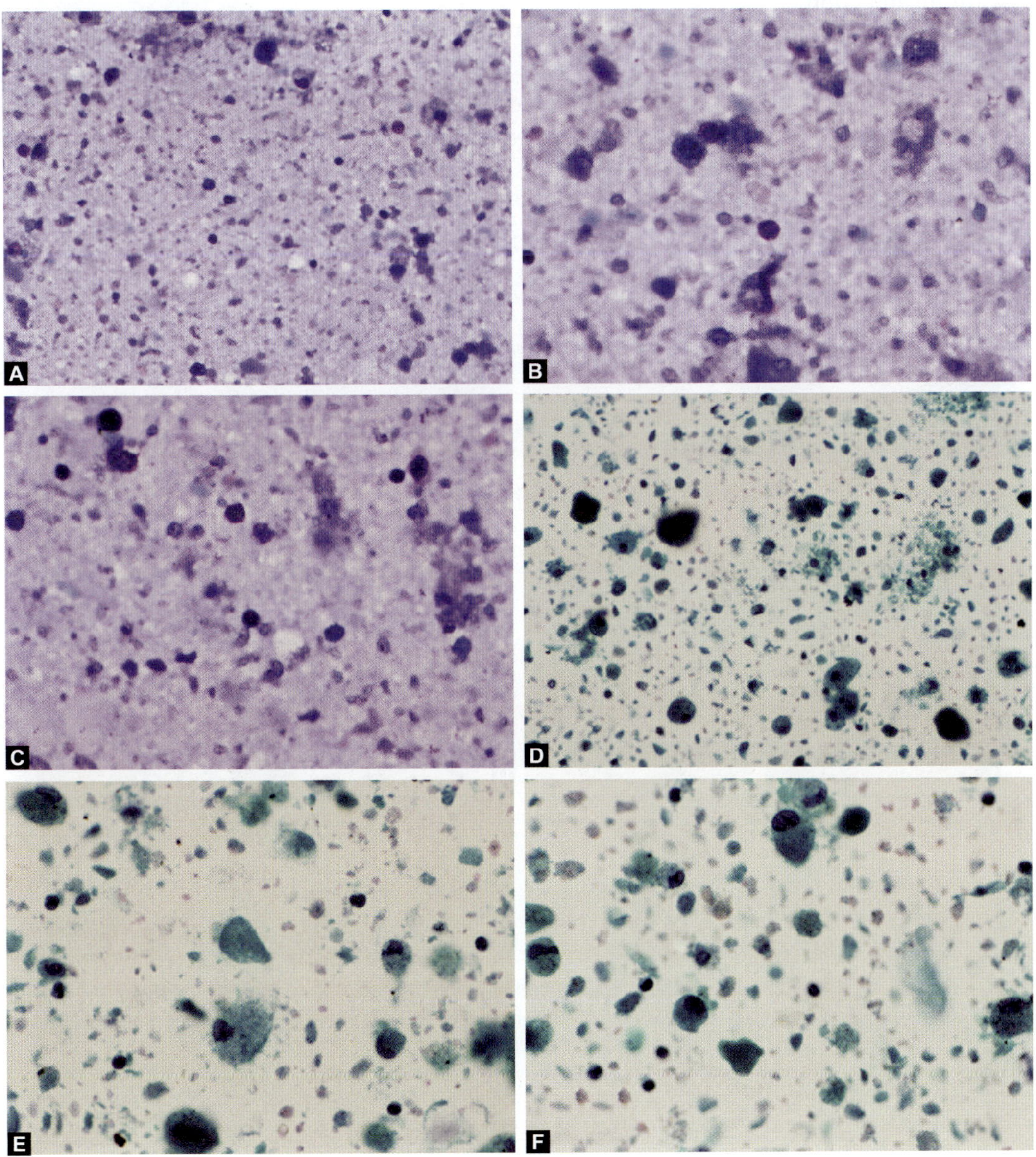

Figures 4.16A to F

Problem 4.17

A 41-year-old male presented with dry cough and hemoptysis. His chest radiograph showed diffuse opacity in right middle lobe of lung. He had history of mediastinal lymphadenopathy which was biopsied before. Microphotographs of the histopathology and immunocytochemistry of the lymph node are shown in Figures 4.17A to F. SurePath smear of the routine sputum examination was done (Figs 4.17G to L). The most likely diagnosis is:

A. Adenocarcinoma
B. Reactive histiocytes
C. Chronic infection
D. Hodgkin lymphoma

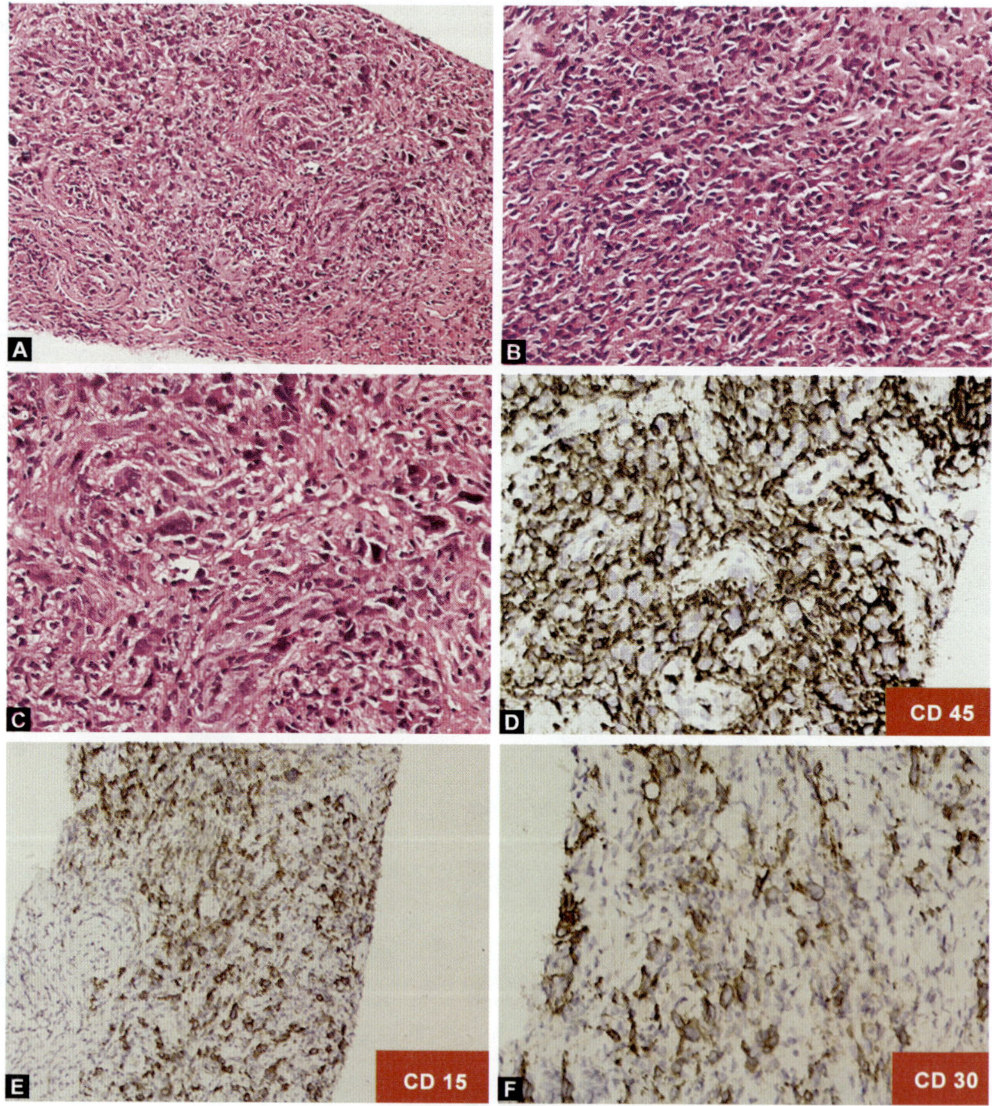

Figures 4.17A to F

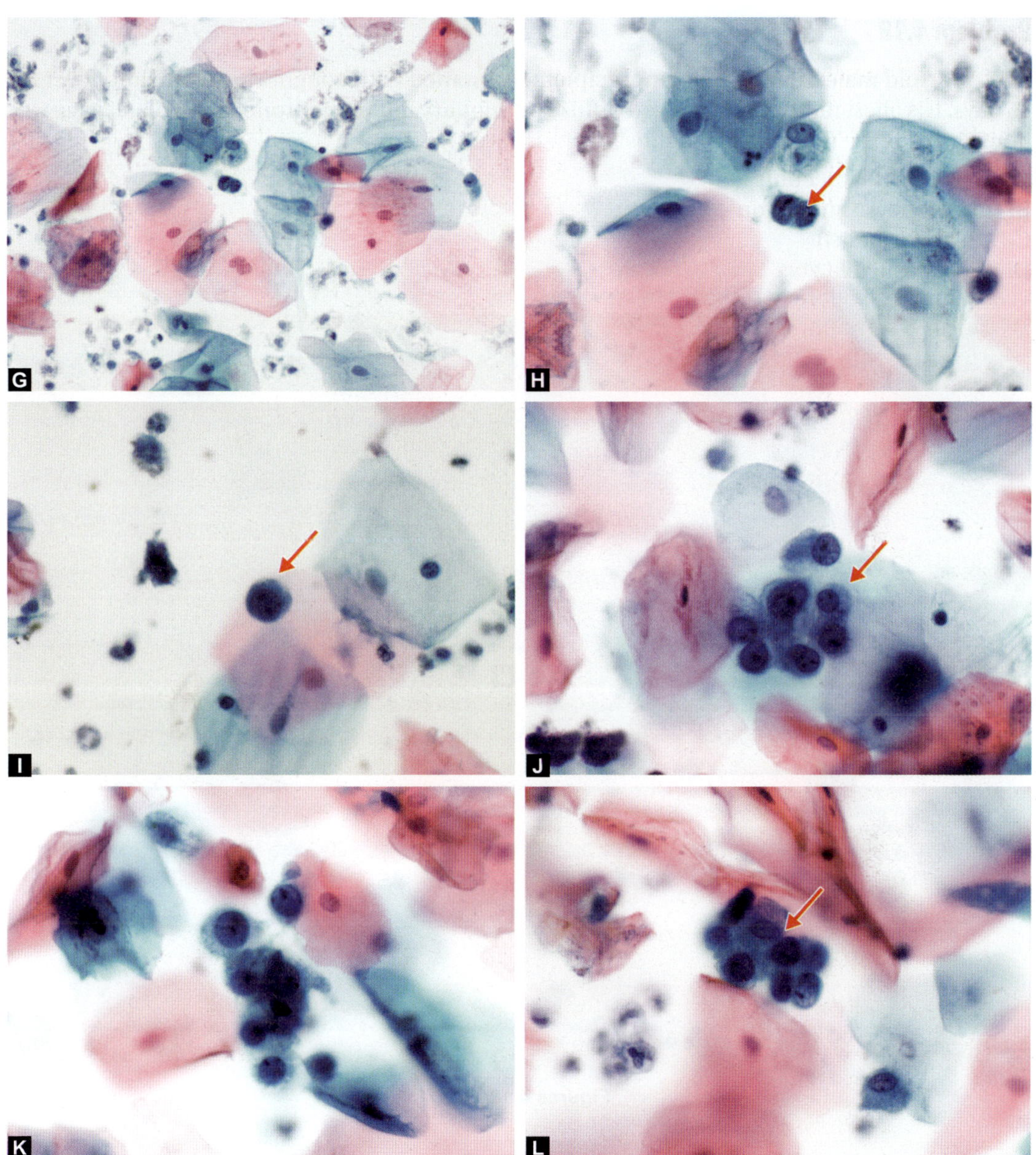

Figures 4.17G to L

Problem 4.18

A 43-year-old male, renal transplant recipient, presented with cough, chest pain and fever for 15 days. His chest radiograph showed diffuse pulmonary infiltrate. Bronchoalveolar lavage was done. What is the most likely diagnosis from Figure 4.18?

A. Herpes viral infection
B. Pneumocystis infection
C. No abnormality
D. CMV positive

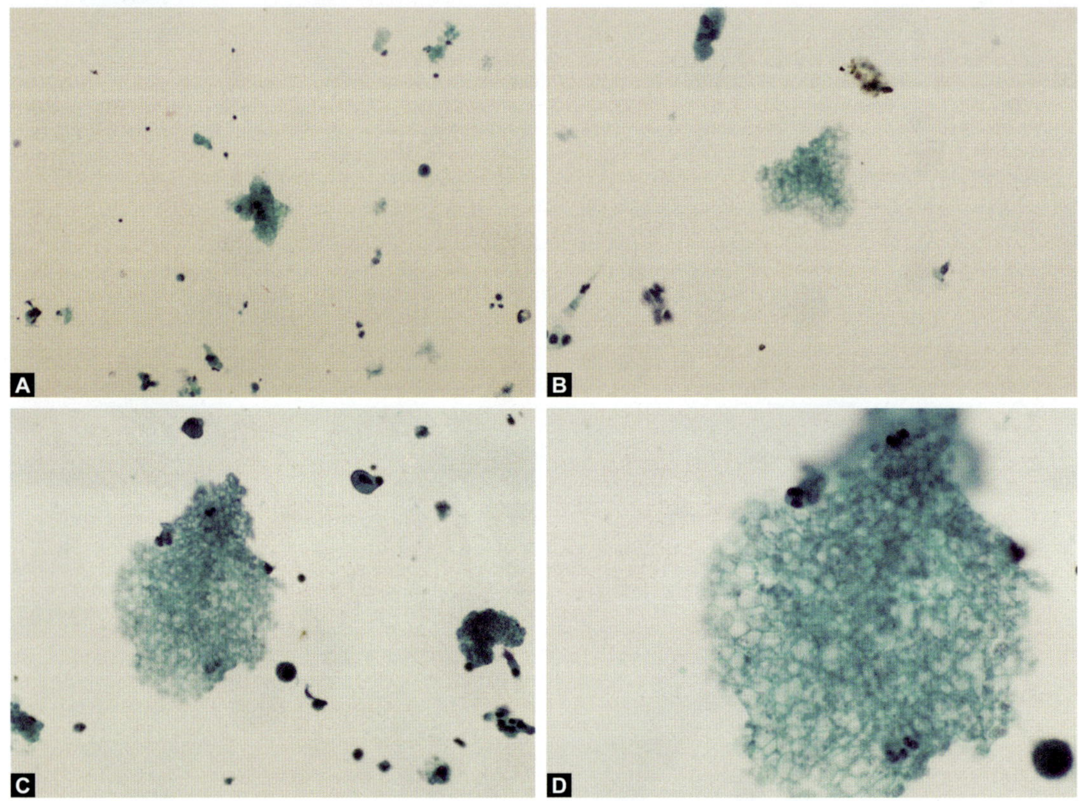

Figures 4.18A to D

Problem 4.19

A 57-year-old male was diagnosed as urothelial cell carcinoma of bladder and treated accordingly 2 year back. Now he presented with cough and hemoptysis. He had right bronchial wall elevated lesion on bronchoscopy. Bronchial brush cytology was done. The most likely diagnosis from Figure 4.19 is:

A. Urothelial cell carcinoma (metastasis)
B. Primary non-small cell carcinoma of lung
C. Hyperplastic bronchial epithelial cells
D. Viral infection

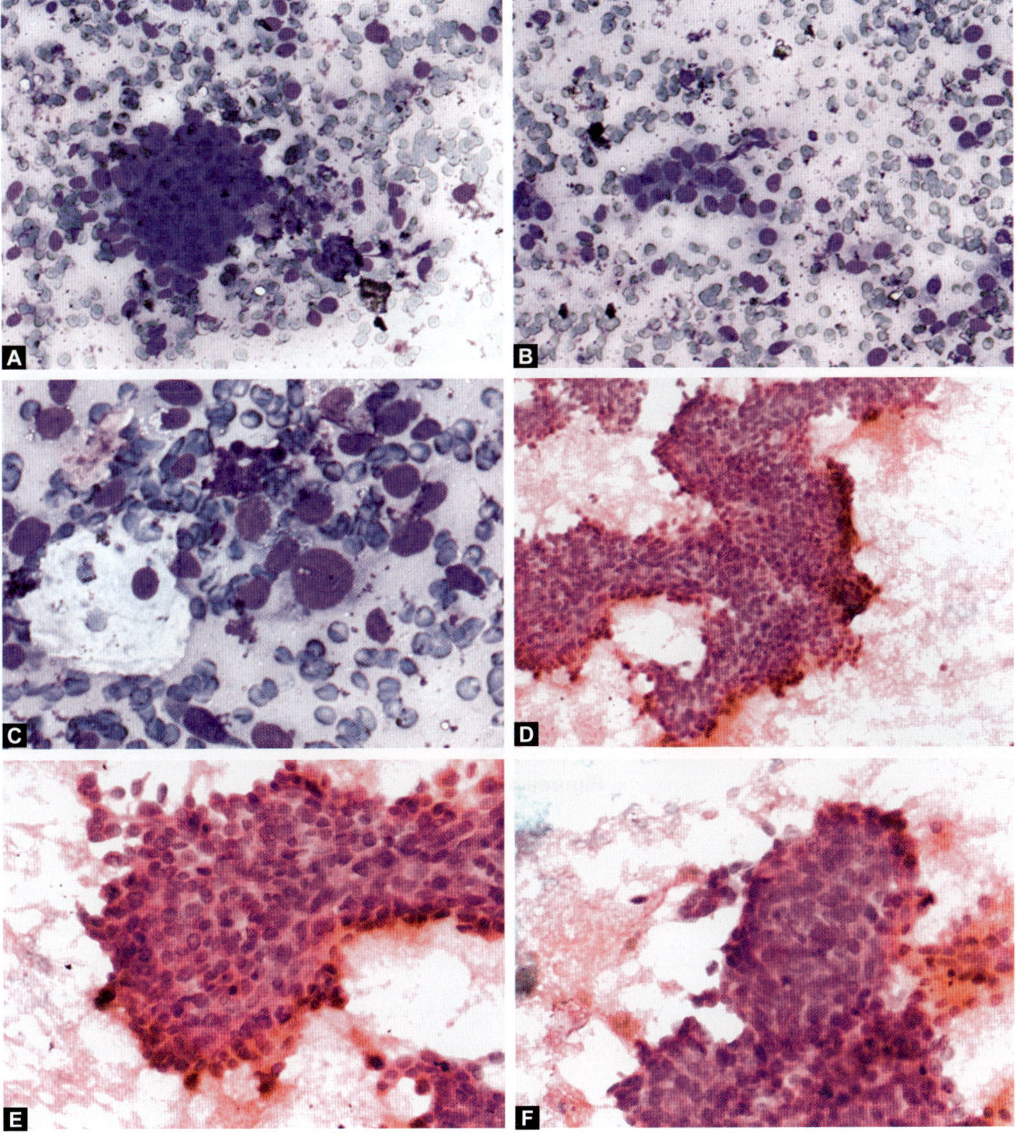

Figures 4.19A to F

Problem 4.20

A 35-year-old male presented with chronic cough, fever and mild chest pain. The patient had history of kidney transplantation 8 months back. His BAL was done. The most likely diagnosis from Figure 4.20 is:

A. Mucormycosis
B. Aspergillosis
C. Acute inflammation only
D. Candidal infection

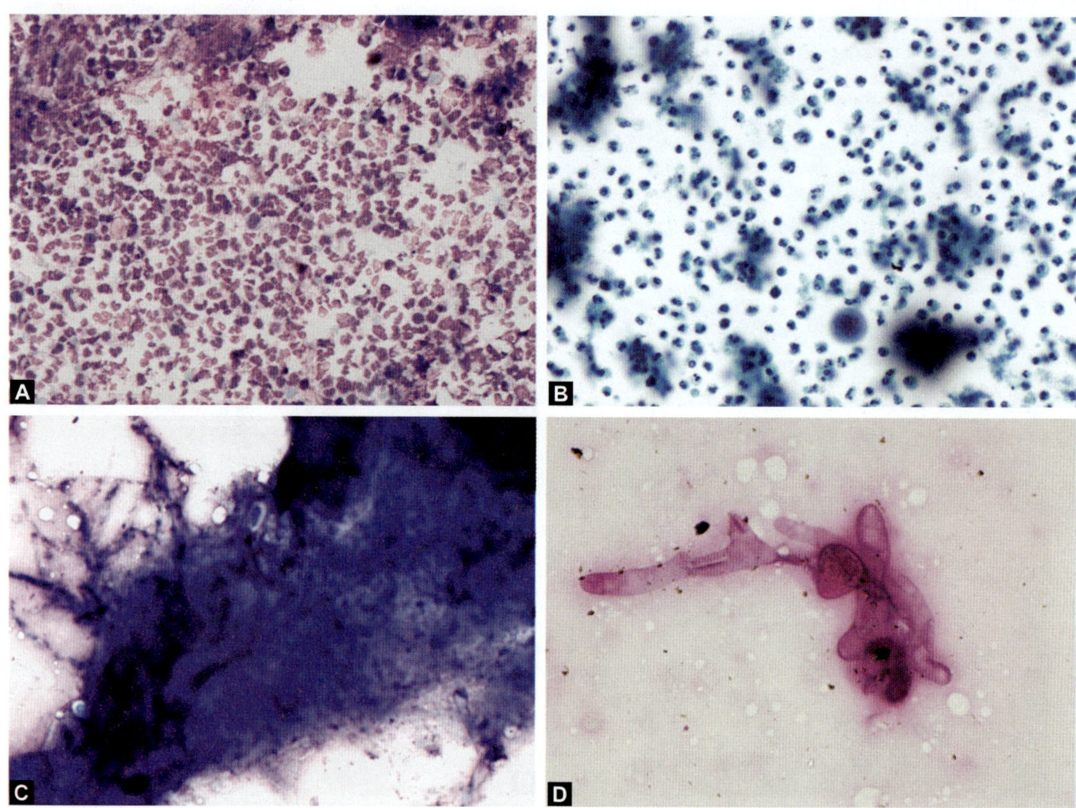

Figures 4.20A to D

Problem 4.21

A 11-year-old boy presented with chest pain, fever and cough. He had multiple 1 to 2 cm solid cystic lesions in the left lung. BAL was done. The most likely diagnosis form Figure 4.21 is:

A. Viral infection
B. Normal cytology
C. Acute inflammation
D. Langerhans cell histiocytosis

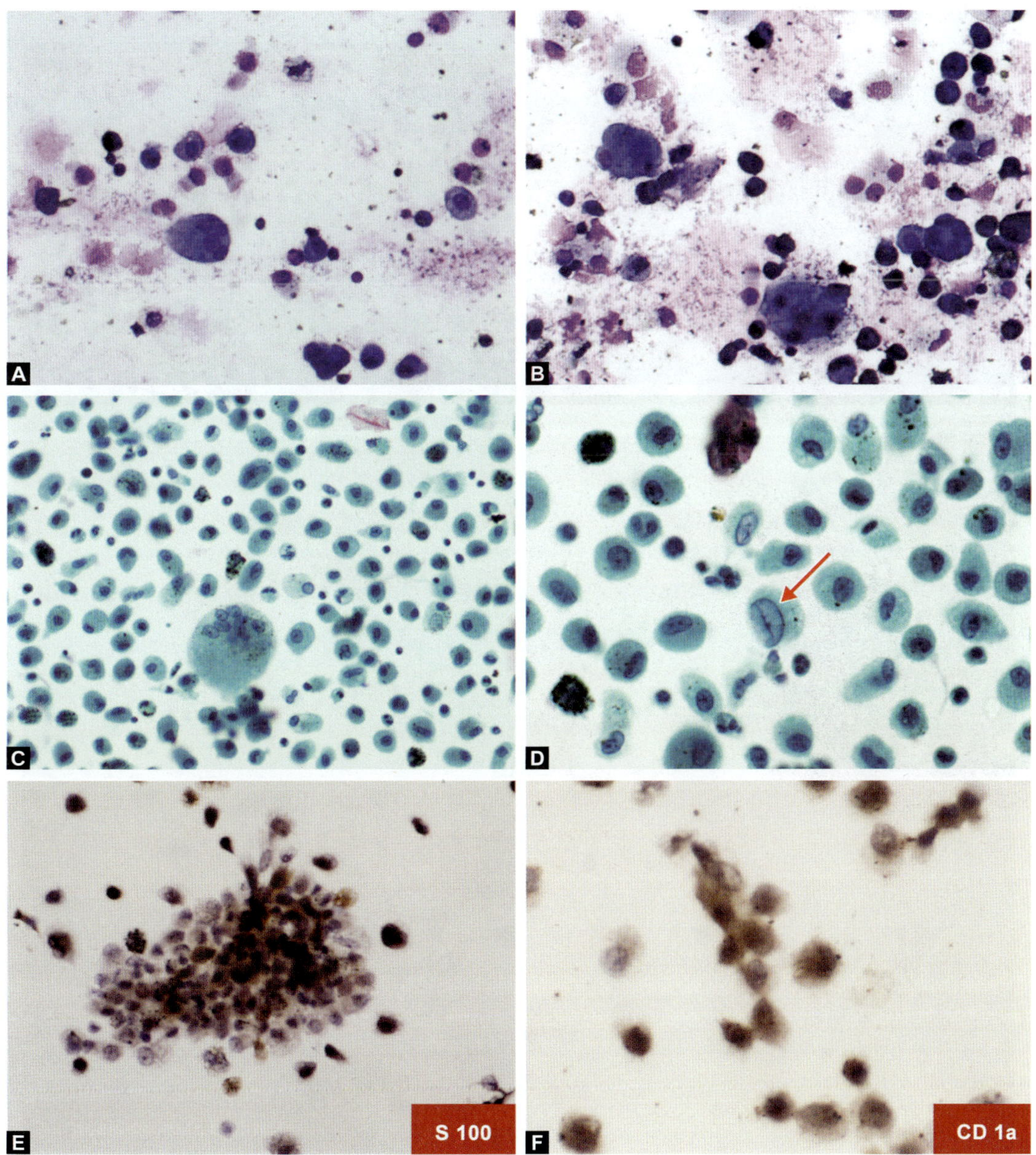

Figures 4.21A to F

Problem 4.22

A 37-year-old female had systemic lupus erythematous and was treated with corticosteroid for 6 months. She had now fever and cough. Her chest radiograph shows diffuse patchy consolidation. Bronchoalveolar lavage was done. The most likely diagnosis from Figure 4.22 is:

A. Candidal infection
B. Aspergillous infection
C. Nocardia infection
D. Mucormycosis

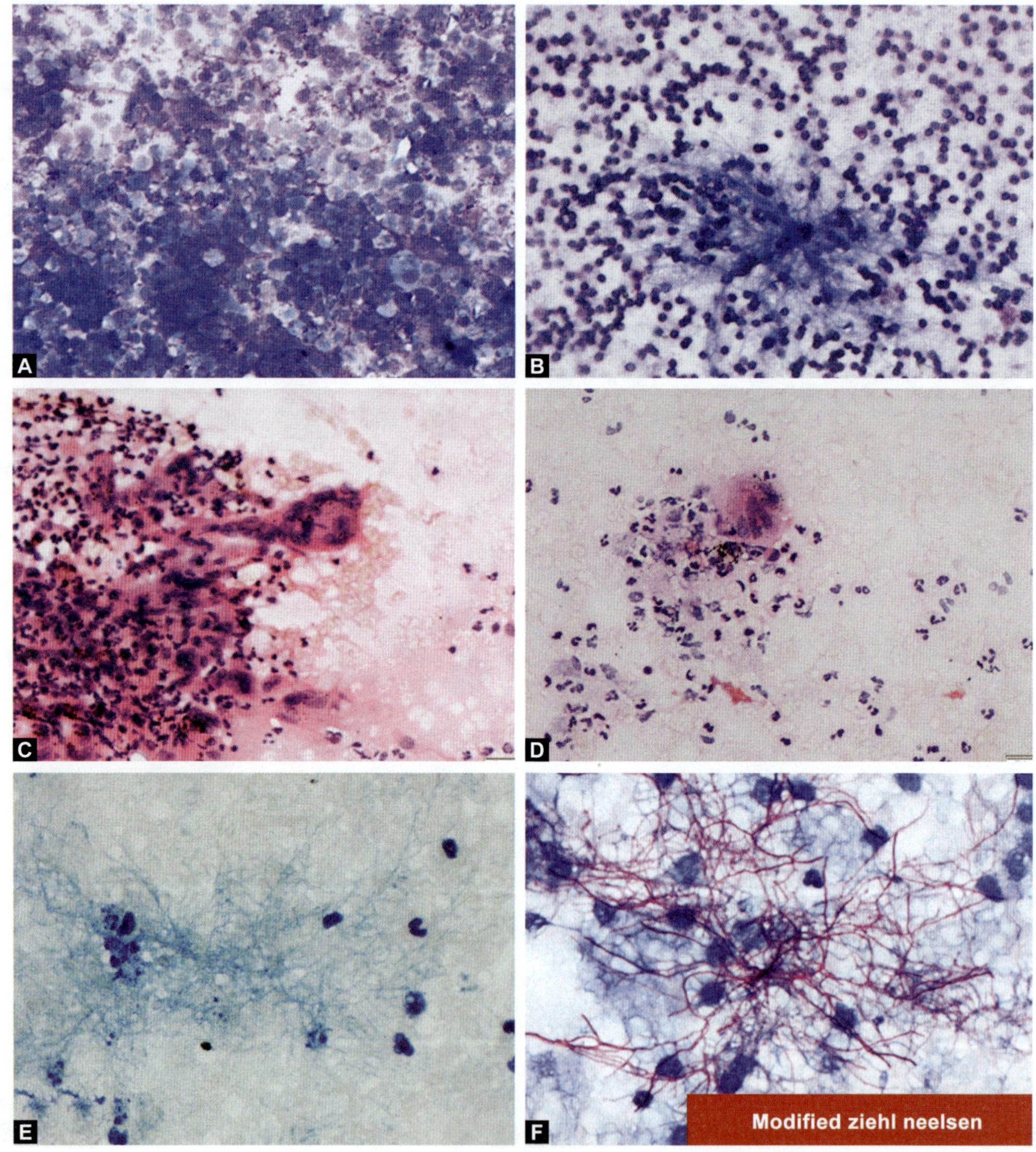

Figures 4.22A to F

Problem 4.23

A 51-year-old male presented with chest pain and fever for 12 days. His chest radiograph showed diffuse opacity in both lung. The patient is a known case of HIV positive case. His bronchoalveolar lavage was done. The microphotographs of SurePath are shown. The most likely diagnosis from Figure 4.23 is:

A. Adenocarcinoma
B. Histoplasma infection
C. Cryptococcal infection
D. Normal cytology

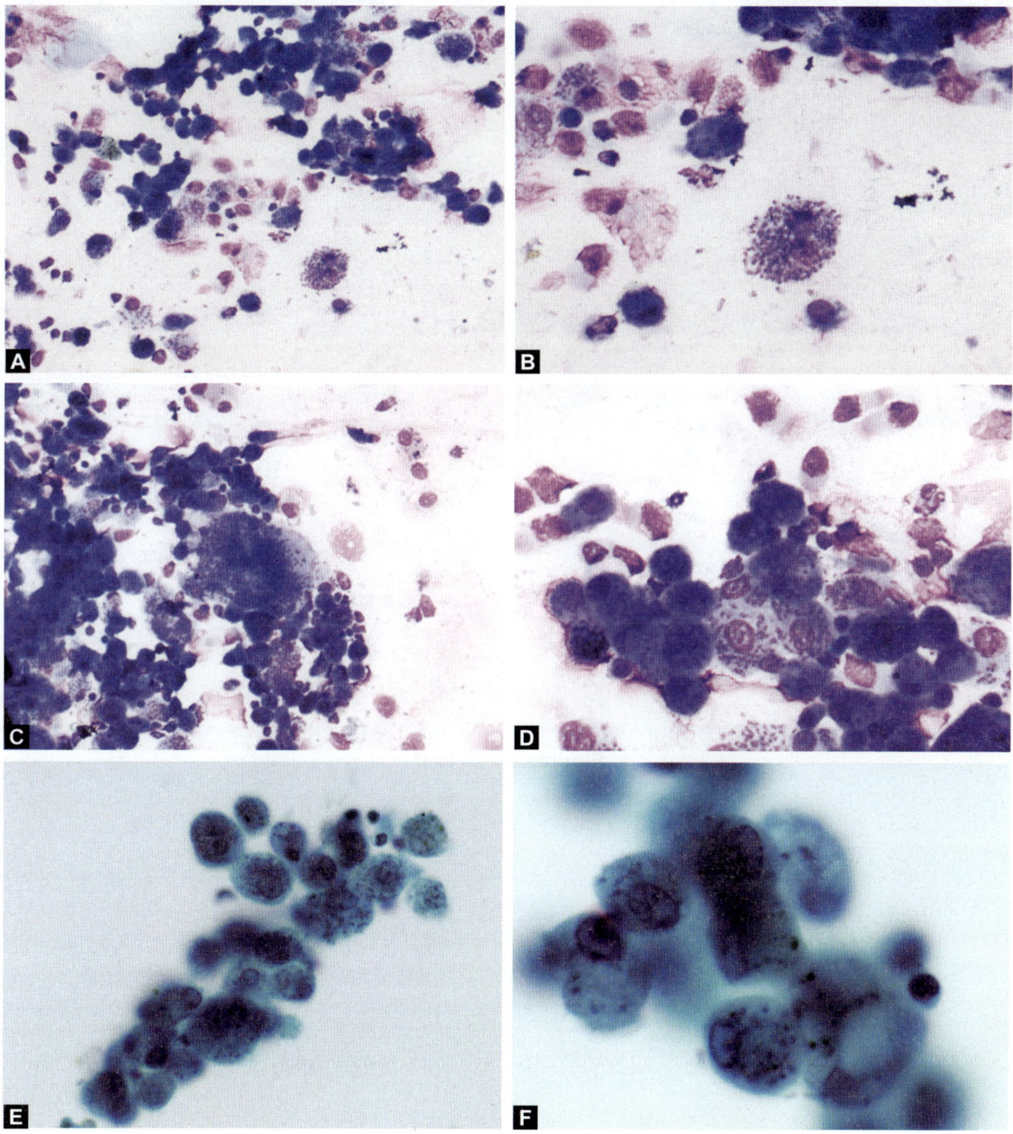

Figures 4.23A to F

Problem 4.24

A 56-year-old female presented with chest pain for 3 months. She had 2.5 cm space occupying lesion in left lung. EUS TBNA of the lesion was done. The most likely diagnosis from Figure 4.24 is:

A. Adenocarcinoma
B. Squamous cell carcinoma
C. Carcinoid
D. Small cell carcinoma

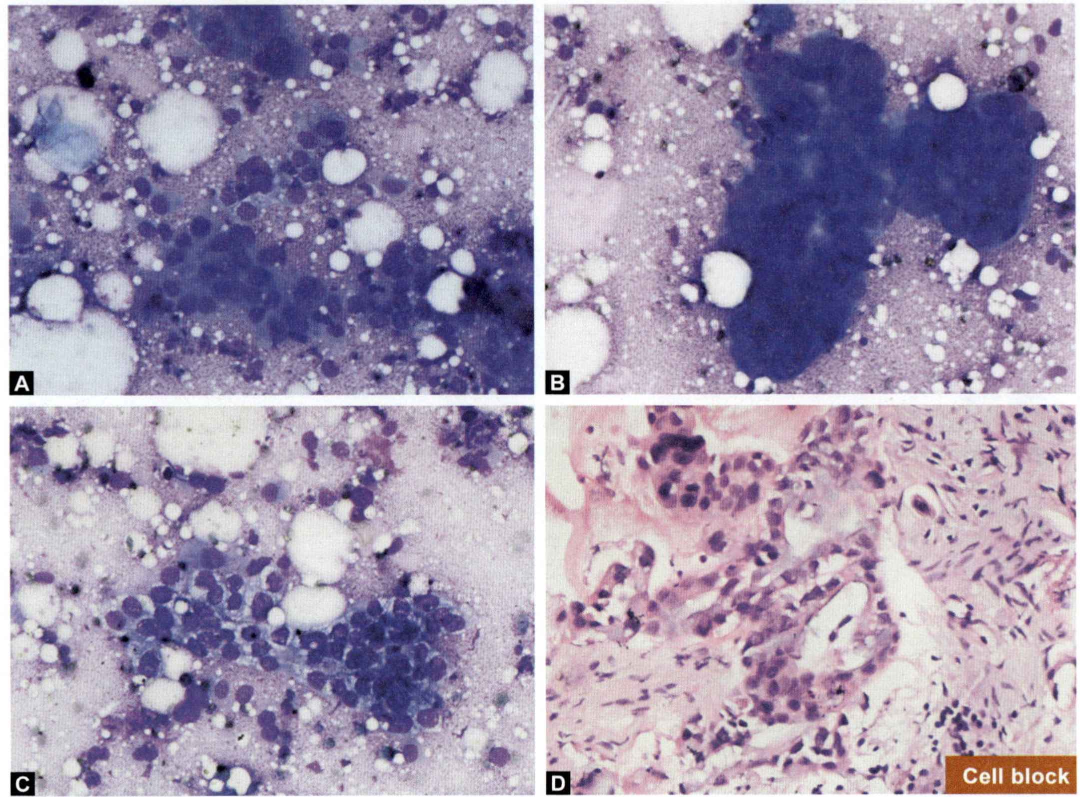

Figures 4.24A to D

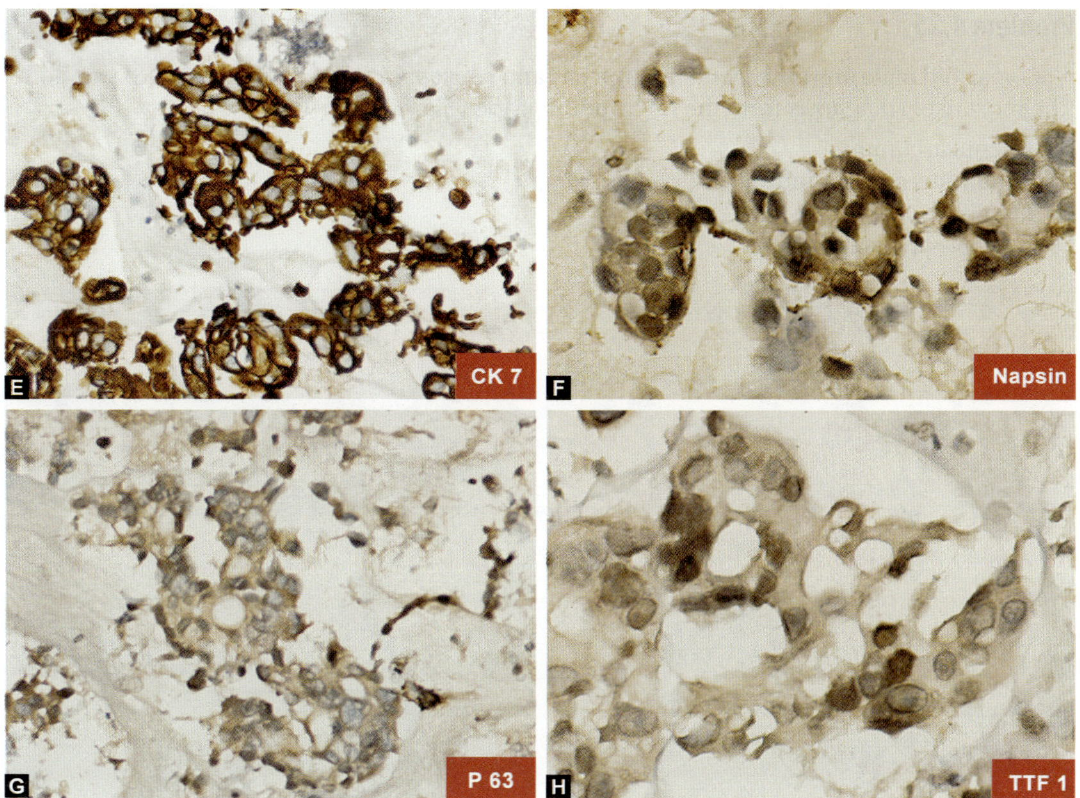

Figures 4.24E to H

Problem 4.25

A 56-year-old male presented with hemoptysis and chest pain for 15 days. EUS TBNA was done. The most likely diagnosis from the microphotograph (Fig. 4.25) is:

A. Adenocarcinoma
B. Squamous cell carcinoma
C. Small cell carcinoma
D. Reactive atypia

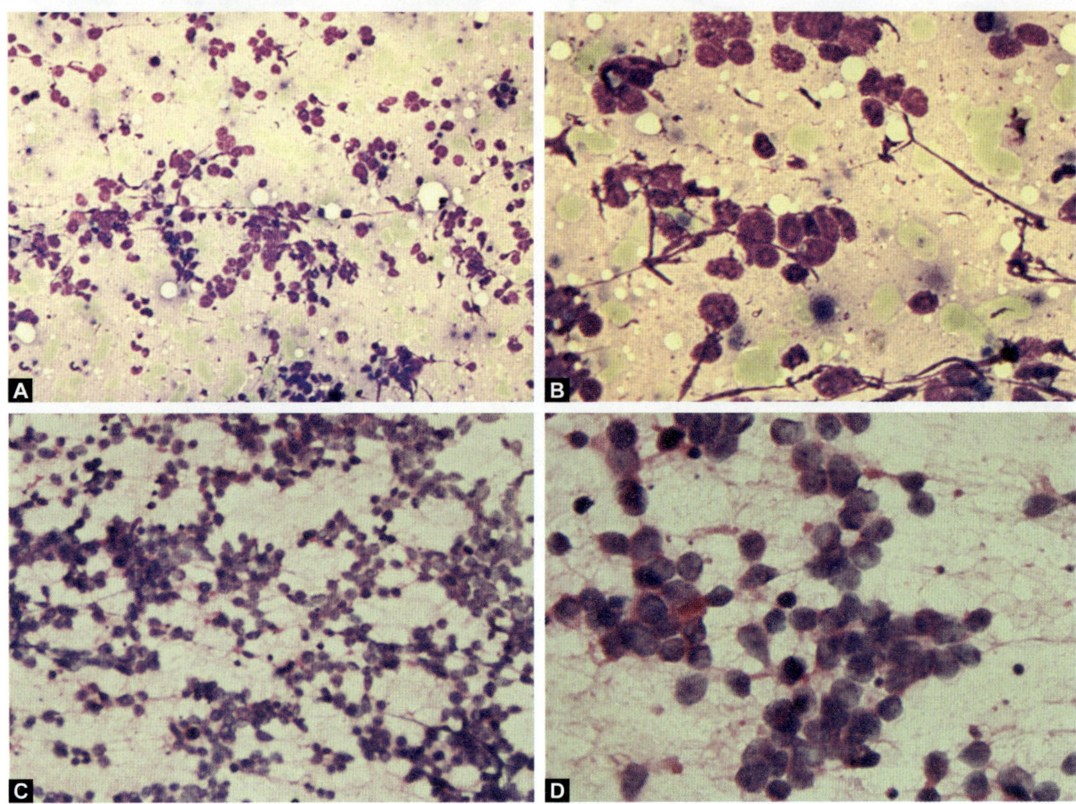

Figures 4.25A to D

Problem 4.26

A 58-year-old male presented with chest pain and hemoptysis for 2 months. He had right sided pleural effusion along with 3 cm diameter mediastinal lymph node. EUS TBNA was done. Detailed immunocytochemistry was also done on cell block material. The most likely diagnosis from Figure 4.26 is:

A. Adenocarcinoma
B. Squamous cell carcinoma
C. Small cell carcinoma
D. Carcinoid

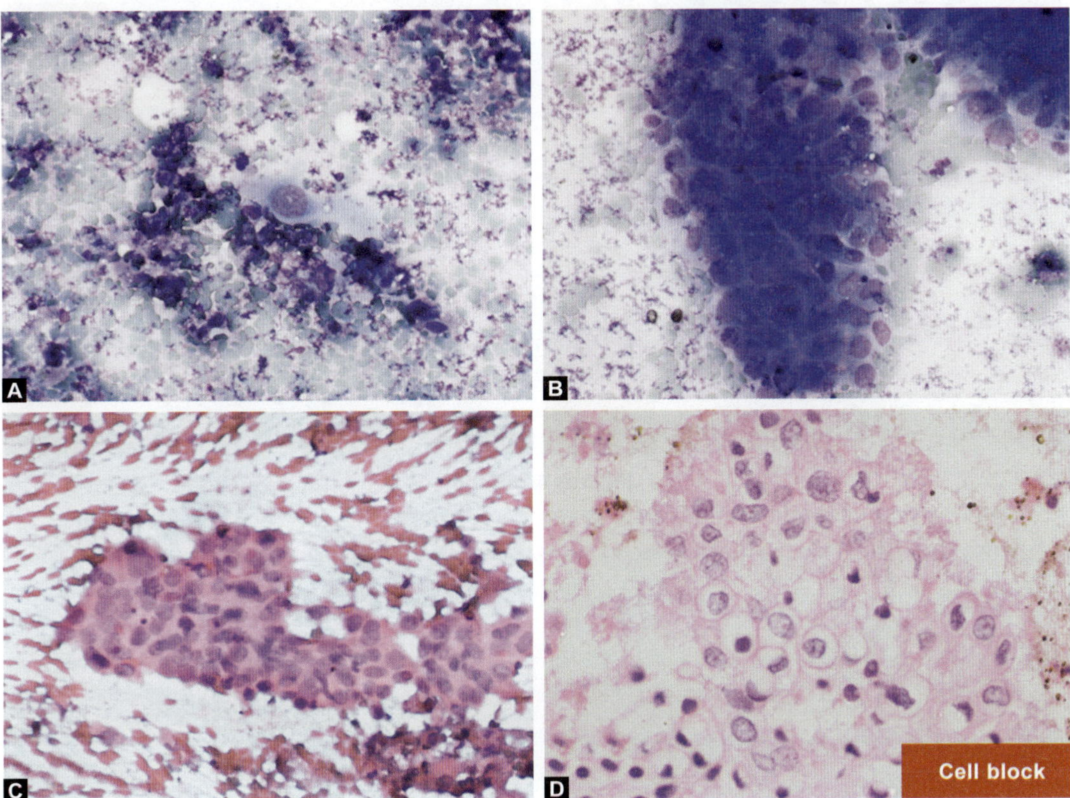

Figures 4.26A to D

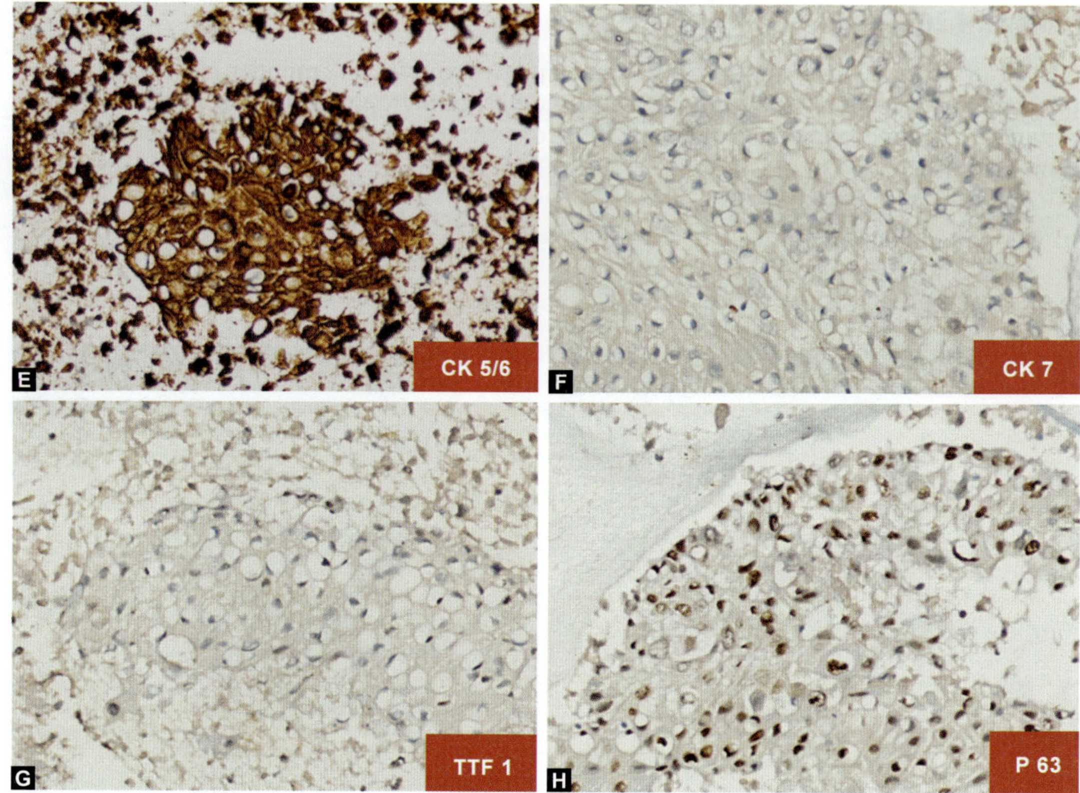

Figures 4.26E to H

Problem 4.27

A 53-year-old male presented with chest pain and fever. He had 3 cm diameter peribronchial space occupying lesion in the left upper lobe of lung. EUS TBNA was done. The most likely diagnosis from Figure 4.27 is:

A. Adenocarcinoma
B. Small cell carcinoma
C. Carcinoid
D. Bronchial hyperplasia

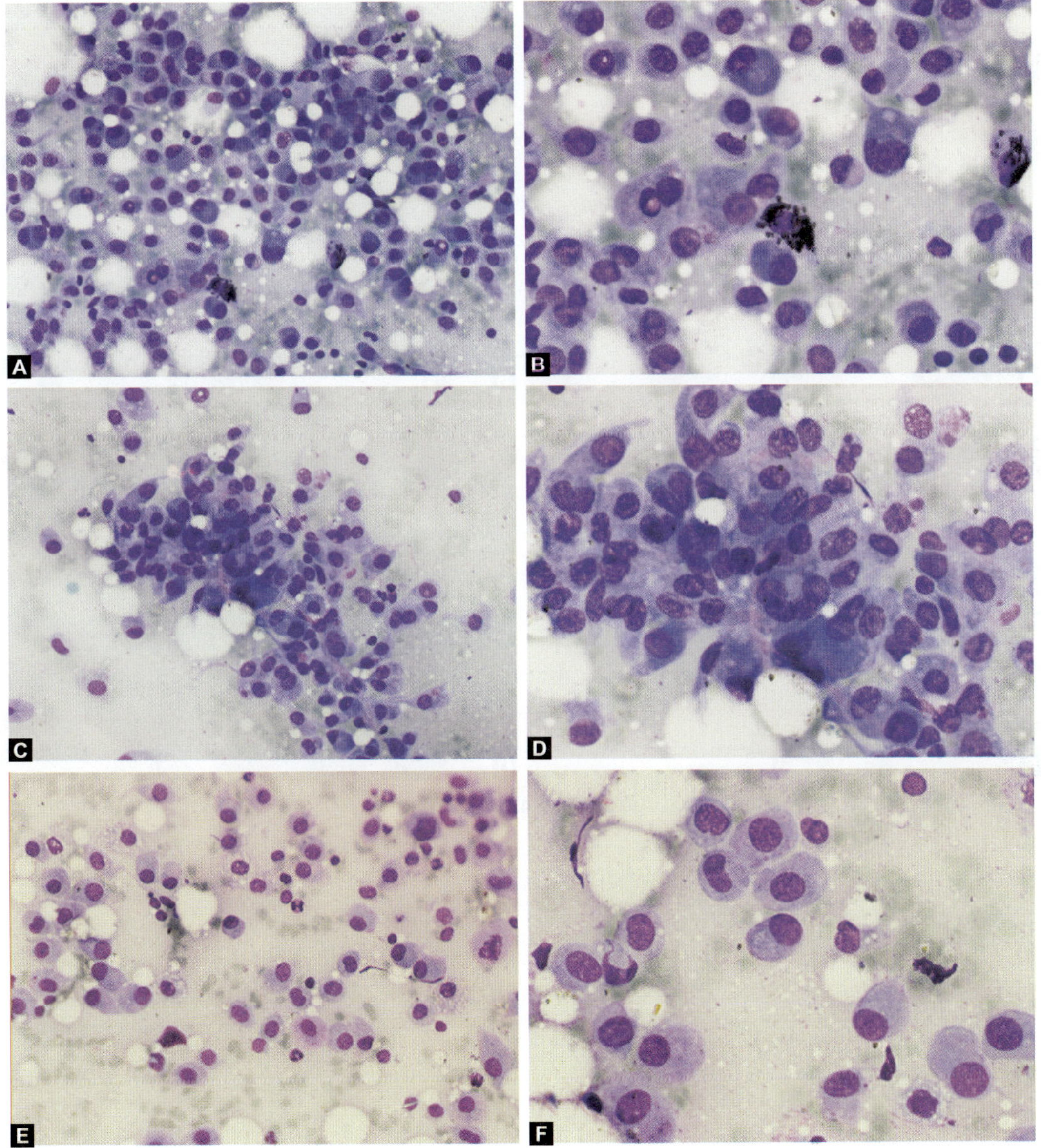

Figures 4.27A to F

Problem 4.28

A 45-year-old female had chest pain for 2 months. The patient showed 2.5 cm diameter space occupying lesion in the right middle lobe of lung. EUS TBNA was done. The most likely diagnosis from Figure 4.28 is:

A. Carcinoid
B. Peripheral neuroectodermal tumor
C. Small cell carcinoma
D. Adenocarcinoma

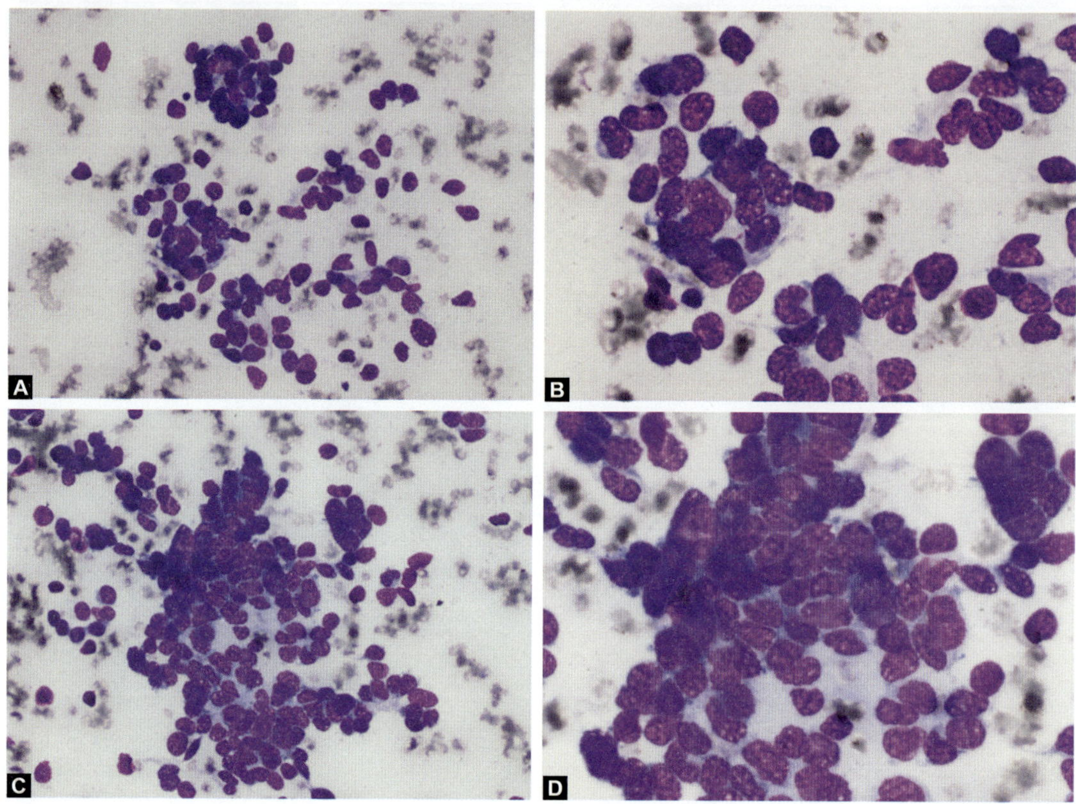

Figures 4.28A to D

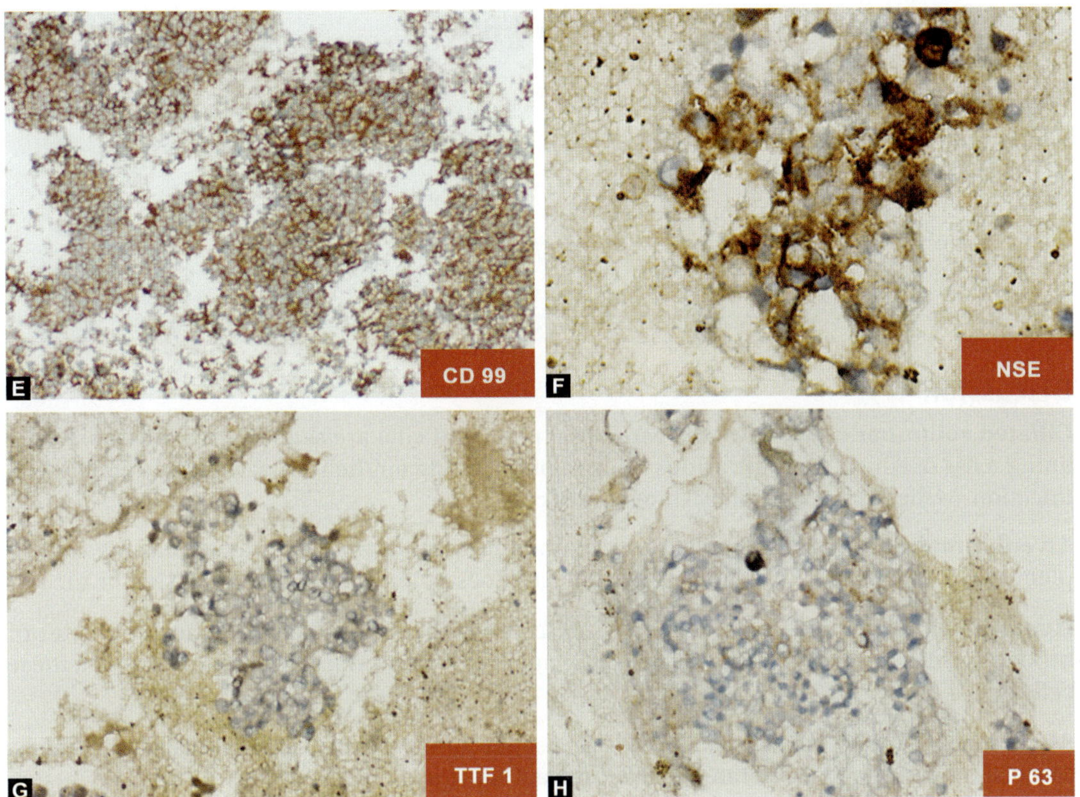

Figures 4.28E to H

ANSWERS

Problem 4.1

A	3. Ciliated columnar cells
B	4. Alveolar macrophages
C	1. Hyperplastic bronchial cells
D	5. Atypical cells
E	2. Curschmann's spiral

Discussion

Ciliated columnar cells: The cells with basal plate and cilia are seen in the respiratory tract lining. Loss of cilia is seen in various injuries of the lining epithelium particularly due to viral infection. Malignant cells usually show loss of cilia.

Alveolar macrophages: The cells show abundant cytoplasm and intracytoplasmic carbon pigments. The presence of alveolar macrophages does not bear any additional significance.

Hyperplastic bronchial cells: Hyperplastic bronchial cells show tight cohesive cluster of bronchial cells. The cells usually do not show nuclear overlapping. Cilia are preserved in these cells.

Atypical cells: The atypical bronchial cells show tight cohesive three-dimensional group of cells. The cells show mild to moderate nuclear atypia. Cilia are lost in these cells.

Curschmann's spiral: This thread like elongated spiral structure is usually seen in chronic bronchitis. These are the mucus threads.

Problem 4.2

D. Adenocarcinoma

Discussion

Description: The cytology smears show multiple clusters of tumor cells along with benign respiratory epithelial cells and alveolar macrophages. The tumor cells are columnar in shape with round enlarged nuclei having prominent nucleoli.

Diagnostic features in the index case: In the index case the following features favor the diagnosis of adenocarcinoma:

- Multiple cohesive three-dimensional cluster of cells
- Round cells with moderately vacuolated cytoplasm
- Enlarged nuclei with moderate nuclear pleomorphism
- Fine chromatin
- Prominent nucleoli.

Differential diagnosis: Reactive bronchial cells usually show well preserved cilia. The cells are also in monolayer sheets rather than three-dimensional clusters. The cells of small cell carcinoma do not show much cytoplasm and the nuclei are round with condensed chromatin. The squamous cell carcinoma is a remote possibility in this case as the cells do not show any squamoid differentiation.

Problem 4.3

C. Aspergillosis

Discussion

Description: The cytology smears show abundant clusters and discrete long filamentous organism with acute angle branching (aspergillus). In addition, the smears also show many atypical squamous cells with enlarged mildly hyperchromatic nuclei (Figs 4.3B to D).

Diagnostic features: In the index case there are clusters of fungal profiles of aspergillus. The fungal hyphae are slender septed with acute angle branching. In addition, the epithelial cells show nuclear atypia. The presence of nuclear atypia in case of aspergillus infection is known. These atypical cells may be mistaken as malignant cells. Therefore, one should be careful in such situation.

Further Reading

1. Crapanzano JP, Zakowski MF. Diagnostic dilemmas in pulmonary cytology. Cancer. 2001;93(6):364-75.

Problem 4.4

B. Hyperplastic bronchial cells

Discussion

Description: The cytology smear shows multiple clusters of bronchial epithelial cells. The cells show enlarged hyperchromatic nuclei. However, the peripheral cells of the clusters show preserved cilia (Figs 4.4E and F).

Diagnostic features of the index case: In the index case the following features suggest the diagnosis of hyperplastic bronchial cells:

- Monolayer cells
- Well preserved cilia
- No significant nuclear pleomorphism.

Differential diagnosis: Adenocarcinoma usually shows three-dimensional clusters with moderate nuclear pleomorphism and loss of cilia.

Problem 4.5

A. Granulomatous inflammation

Discussion

Description: The cytology smears show multiple epithelioid cell granulomas in a background of reactive lymphoid cell.

Diagnostic features: In the index case the diagnosis of granulomatous inflammation is quite straight forward as there are multiple epithelioid cell granulomas in the smear. The common causes of granulomatous inflammation in the mediastinal lymph node are tuberculosis, sarcoidosis and other chronic inflammatory reactions predominantly due to fungal infection. Ziehl-Neelsen (ZN) stain should always be done in such case along with bacterial culture.

Further Reading

1. Chatterjee D, Dey P. Tuberculosis revisited: cytological perspective. Diagn Cytopathol. 2014;42(11):993-1001.

Problem 4.6

D. Pneumocystis infection

Discussion

Description: The cytology smears show abundant respiratory lining epithelial cells, lymphocytes and foamy to bubbly alveolar cast material. Careful observation shows round shaped structures.

Diagnostic features: The diagnostic feature in the index case include foamy to bubbly alveolar cast material. Grocott's methenamine silver stain of the material confirmed the diagnosis.

Description: Pneumocystis infection occurs in the acquired immunodeficiency syndrome (AIDS) or in the immunocompromised patients. The index case was immunocompromised as the patient was a renal transplant case. May Grunwald Giemsa or Papanicolaou's staining helps to detect intracystic sporozoites. However, this should be confirmed by methenamine silver stain or PCP antibody staining.

Further Reading

1. Dahiya S, Mathur SR, Iyer VK, Kapila K, Verma K. Diagnosis of Pneumocystis pneumonia by bronchoalveolar lavage cytology: experience at a Tertiary Care Centre in India. Indian J Chest Dis Allied Sci. 2005;47(4):259-65.

Problem 4.7

D. Adenocarcinoma

Discussion

Description: The cytology smears show predominantly discrete and clusters of malignant cells. The cells show moderate amount of cytoplasm and enlarged nuclei with prominent nucleoli.

Diagnostic points: The following features indicate the diagnosis of adenocarcinoma:

- Discrete cells with round nuclei
- High N/C ratio

- Moderate vacuolated cytoplasm
- Large prominent nucleoli.

Differential diagnosis: The reactive bronchial cells are usually ciliated. N/C ratio is too high in the cell population of this index case. The presence of moderate cytoplasm and prominent nucleoli exclude the possibility of small cell carcinoma.

Problem 4.8

B. Squamous cell carcinoma

Discussion

Description: The cytology smears show many discrete tumor cells with polyhedral appearance and orangeophilic cytoplasm. The cells have enlarged hyperchromatic and pleomorphic nuclei. Occasional fibre cells are also noted.

Diagnostic features in the index case: In this index case the following features favor the diagnosis of squamous cell carcinoma:

- Discrete malignant cells with squamoid differentiation
- The cells are polyhedral with round hyperchromatic nuclei
- Cytoplasmic orangeophilia
- Fiber-like cells.

Differential diagnosis: The cells with reactive atypia do not show such severe nuclear atypia such as hyperchromatic nuclei and nuclear pleomorphism. The possibilities of small cell carcinoma and adenocarcinoma are remote in this index case.

Problem 4.9

C. Adenocarcinoma

Discussion

Description: The cytology smears show many tight cohesive clusters and papillae like arrangement of tumor cells along with abundant benign squamous cells. The individual cells are round with scanty cytoplasm and fine granular chromatin.

Diagnostic features: The diagnostic features in the index case are:

- Cohesive clusters and papillary arrangement of tumor cells
- Round cells with moderate cytoplasm
- Fine granular chromatin.

Differential diagnosis: Reactive bronchial cells may often show papillary clusters. However the nuclear hyperchromasia and pleomorphism are not significant. The cells show preserved cilia.

The cells of squamous cell carcinoma are polyhedral in shape and nuclei are hyperchromatic. Fiber and tadpole cells are also seen.

The cells of small cell carcinoma are small in size and they have condensed chromatin. Cell within the cell appearance and vertebroid like arrangement are also seen.

Problem 4.10

B. Squamous cell carcinoma

Discussion

Description: The cytology smears show many discrete oval to polyhedral cells with orangeophilic cytoplasm having enlarged hyperchromatic nuclei. The nuclear margin is irregular and chromatin is coarse. Occasional fiber cells are also seen.

Diagnostic features: The diagnostic features in the index case include:
- Polyhedral cells
- Hyperchromatic and coarse granular cytoplasm
- Fiber cells.

Problem 4.11

D. Small cell carcinoma

Discussion

Description: The cytology smears show many small clusters of tumor cells. The cells are also arranged in small rows. The individual cells are round with scanty hyperchromatic nuclei.

Diagnostic features: The diagnostic features in the index case include:
- Small clusters and rows of cells
- Small round cells with hyperchromatic condensed nuclei.

Problem 4.12

B. Squamous cell carcinoma

Discussion

Description: The cytology smears show multiple clusters and discrete malignant cells. The cells are oval to polyhedral with moderate amount of cytoplasm and centrally placed hyperchromatic moderately pleomorphic nuclei. Occasional fiber like cells and multinucleated tumor giant cells are also seen.

Diagnostic features in the index case:
- Oval to polyhedral cells
- Moderate amount of cytoplasm and hyperchromatic central nuclei
- Fiber-like cells.

Problem 4.13

C. Small cell carcinoma

Discussion

Description: The cytology smears show multiple tight cohesive clusters and occasional discrete round cells. The individual cells have thin rim of cytoplasm having round monomorphic hyperchromatic nuclei. These cells are positive for chromogranin and CD 56. Tumor cells are negative for CK 5/6, CK7, TTF-1 and p63.

Diagnostic features:
- Small round cells
- Chromogranin and CD 56 positivity.

Problem 4.14

A. Adenocarcinoma

Discussion

Description: The cytology smears show abundant discrete cells with moderate amount of vacuolated cytoplasm. The individual cells show enlarged nuclei with fine chromatin.

The tumor cells are positive for CK 7, TTF-1 and napsin. The cells are negative for CK 5/6.

Diagnostic features: In the index case following features favor the diagnosis of adenocarcinoma:
- The cells with abundant vacuolated cytoplasm
- Nuclear enlargement and pleomorphism
- Positive for TTF-1, napsin and CK 5/6.

Further Reading

1. Travis WD, Brambilla E, Nicholson AG, Yatabe Y, Austin JH, Beasley MB, et al. WHO Panel. The 2015 World Health Organization classification of lung tumors: Impact of Genetic, Clinical and Radiologic Advances Since the 2004 Classification. J Thorac Oncol. 2015;10(9):1243-60.

Problem 4.15

D. Aspergillosis

Discussion

Description: The cytology smear shows alveolar macrophages, respiratory epithelial cells and many fungal organisms. The fungi show long slender hyphae with acute angle branching.

Diagnostic features in the index case: In this index case the features suggestive of aspergillosis are:
- The presence of fungal structure
- The fungi are slender, septed with acute angle branching.

Disease description: The aspergillus infection is usually seen in the immunocompromised patient. In the index case the patient is a leukemic patient under chemotherapy and this may be the underlying factor. The presence of fungi eliminate all other possibilities.

Problem 4.16

D. Alveolar proteinosis

Discussion

Description: The cytology smears show acellular bluish (in May Grunwald Giemsa stain) homogenous material and alveolar macrophages. Occasional blob like material is also noted.

Diagnostic features:

- Homogenous blob like material.

This material is PAS positive.

Disease description: Pulmonary alveolar proteinosis (PAP) is a rare lung disease characterized by abnormal collection of proteinaceous material rich in pulmonary surfactant material in the lung alveoli.[1,2] The animal model shows that deficiency of granulocyte-macrophage colony stimulating factor (GM-CSF) produce similar type of disease in mice.

The patient usually complaints of slow onset dyspnea, dry cough and loss of weight. The chest radiograph shows nonspecific features or at times "bat-wing configuration" which is suggestive of PAP. None of the clinical and radiological features are helpful in diagnosis. Bronchoalveolar lavage is the important diagnostic modality in PAP. The lavage yields milky white material. The smear shows abundant alveolar macrophages filled with proteinaceous material that are PAS positive. The smear also shows many fat globules in Papanicolaou's stain. The presence of more than 18 such fat globules is strongly suggestive of PAP.[3]

References

1. Khan A, Agarwal R. Pulmonary alveolar proteinosis. Respir Care. 2011;56(7):1016-28.
2. Borie R, Danel C, Debray MP, Taille C, Dombret MC, Aubier M, et al. Pulmonary alveolar proteinosis. Eur Respir Rev. 2011;20(120):98-107.
3. Chou CW, Lin FC, Tung SM, Liou RD, Chang SC. Diagnosis of pulmonary alveolar proteinosis: usefulness of papanicolaou-stained smears of bronchoalveolar lavage fluid. Arch Intern Med. 2001;161(4):562-6.

Problem 4.17

D. Hodgkin's lymphoma

Discussion

Description: The histology sections (Figs 4.17A to F) show reactive lymphoid cells along with large mono and binuclear cells with large prominent nucleoli. The lymph node shows fibrosis. The large atypical cells are negative for CD 45. However, these cells are positive for CD 15 and CD30.

The cytology smears (Figs 4.17G to L) show abundant squamous cells and many discrete large atypical cells. These cells have enlarged nuclei with prominent nucleoli. Occasional bi-nucleated cells are also seen.

Diagnostic features: The histology and immunocytochemistry of the lymph node indicate the diagnosis of Hodgkin's lymphoma. The large atypical cells in BAL are difficult to categorize. However, considering the overall history and histopathology of the lymph node, the most likely diagnosis is infiltration by Hodgkin's lymphoma in the lung.

Disease description: In overall 15 to 40% cases of Hodgkin's lymphoma lung may be involved.[1] This is usually caused by the extension of lymphoid follicles to the lung parenchyma or by extension of peribronchiolar lymphoid cells to adjacent lung alveoli.[2] Occasionally sputum cytology or bronchial brush cytology may reveal the Reed-Sternberg cells or large atypical cells.[3] The identification of such cells is helpful to avoid any extensive or aggressive surgery.

References

1. Fuentes Pradera J, Arriola Arellano E, Cisneros JM, Quiroga E. Adenopatías mediastínicas y masa pulmonar cavitada como forma de presentación de la enfermedad de Hodgkin. Medicina Clínica. 2001;10:398-9.
2. Bakan ND, Camsari G, Gur A, et al. A 21-year-old male with productive cough, hemoptysis, chest pain, and weight loss. Respiration. 2007;74:706-9.
3. Kasuganti D, Cimbaluk D, Gattuso P. Reed-Sternberg cells in bronchial brushings from a patient with Hodgkin's disease. Diagn Cytopathol. 2006;34:850-1.

Problem 4.18

B. Pneumocystis infection

Discussion

Description: The cytology smears show occasional respiratory lining epithelial cells, alveolar macrophages and foamy to bubbly alveolar cast material.

Diagnostic features: The diagnostic feature in the index case include foamy to bubbly alveolar cast material. Grocott's methenamine silver stain of the material confirms the diagnosis.

Description: Pneumocystis infection occurs in the acquired immunodeficiency syndrome (AIDS) or in the immunocompromised patients. The index case was immunocompromised as the patient was a renal transplant case. May Grunwald Giemsa (MGG) or Papanicolaou's staining helps to suggest intracystic sporozoites. The findings in MGG and Papanicolaou's staining should be confirmed by methenamine silver stain or PCP antibody staining.

Further Reading

1. Dahiya S, Mathur SR, Iyer VK, Kapila K, Verma K. Diagnosis of pneumocystis pneumonia by bronchoalveolar lavage cytology: Experience at a tertiary care center in India. Indian J Chest Dis Allied Sci. 2005;47(4):259-65.

Problem 4.19

B. Primary non-small cell carcinoma of lung

Discussion

Description: The cytology smears show multiple cohesive clusters of malignant cells. The cells show mild to moderate amount of vacuolated cytoplasm with moderate nuclear pleomorphism.

Diagnostic features: The index case shows following diagnostic features:

- Clusters of cells
- Moderate amount of vacuolated cytoplasm
- Moderately pleomorphic nuclei.

Problem 4.20

A. Mucormycosis

Discussion

Description: The cytology smears show abundant polymorphs, foamy macrophages and fungal organisms. The fungi show broad hyphae with wide angle branching.

Diagnostic features: The fungi with broad hyphae and wide angle branching indicate the diagnosis of mucormycosis infection.

Disease description: Pulmonary mucormycosis is a life threatening infection (40 to 60% mortality). It typically occurs in immunocompromised patient such as transplant recipients and rarely in diabetic patients. The incidence of pulmonary mucormycosis is approximately 25 % of all cases of mucormycosis infections.[1] The pulmonary mucormycosis may occur due to inhalation of the fungal spore. The patient usually presents with nonspecific symptoms such as fever and cough. Bronchoalveolar lavage is an important technique to detect this infection. The demonstration of fungal hyphae is possible by MGG stain. However, special stains such as Gomori's methenamine silver shows better demonstration of the fungi.[2] The fungal hyphae are broad, non-or pauci-septate, and ribbon like. The hyphae has branching angle more than 90.

References

1. Hamillos G, Samonis G, kontoyiannis DP. Pulmonary mucormycosis. Semin Respir Crit Care Med. 2011;32(6):693-702.
2. Lass-Flörl C. Zygomycosis: conventional laboratory diagnosis. Clin Microbiol Infect. 2009;15(Suppl 5):60-5.

Problem 4.21

D. Langerhans cell histiocytosis

Discussion

Description: The cytology smears show many multinucleated giant cells, mononuclear large cells with prominent longitudinal groove and benign alveolar macrophages. The mononuclear cells are strongly positive for S 100 and CD 1a.

Diagnostic features:

- Large mononuclear cells with longitudinal nuclear groove
- Multinucleated giant cells
- Strong S 100 and CD 1a positivity
- Clinical history of multiple solid cystic lung lesion.

Disease description: Pulmonary Langerhans cell histiocytosis (PLCH) is an uncommon cystic disease of lung and is often associated with systemic involvement. PLCH may occur in isolated condition in lung only. The incidence of Langerhans cell histiocytosis (LCH) is 3 to 5 cases per million children per year.[1] The exact incidence of PLCH is unknown. The prevalence of PLCH in case of interstitial lung disease is only 3 to 5%.[2] The patient usually presents with cough and dyspnea. Chest radiograph usually shows bilateral multiple small nodule about 1 cm in diameter with frequent cystic changes. BAL is an important test to detect PLCH in children particularly in the index case. The cytology smears show large number of Langerhans cell (LC). These cells are mononuclear or multinuclear with enlarged nuclei and prominent longitudinal groove. LC are positive for S 100 and CD 1a. CD 1a is a relatively specific marker for LC and the presence of more than 5% CD1a-positive cells is considered as diagnostic.[3]

References

1. Carstensen H, Ornvold K. The epidemiology of Langerhans cell histiocytosis in Denmark, 1975-89. Med Pediatr Oncol. 1993;21:387-8.
2. Thomeer M, Demedts M, Vandeurzen K. Registration of interstitial lung diseases by 20 centres of respiratory medicine in Flanders. Acta Clin Belg. 2001;56:163-72.
3. Tazi A. Adult pulmonary Langerhans' cell histiocytosis. Eur Respir J. 2006;27:1272-85.

Problem 4.22

C. Nocardial infection

Discussion

Description: The cytology smears show abundant polymorphs, multinucleated giant cells and long slender filamentous branching organisms. These organisms are positive in modified ZN stain (Fig. 4.22F).

Diagnostic features: The diagnostic features in the index case include:
- Long slender filamentous branching organisms
- Positive in modified Ziehl-Neelsen (ZN) stain.

Disease description: Nocardia are in the group of Actinomycetales. They are aerobic gram-positive bacteria. Nocardia infection usually occurs in immunocompromised patients. The organism evokes inflammatory reaction and occasional epithelioid cell granulomas may be seen. This infection may be missed as pulmonary tuberculosis.[1] The organisms are stained lightly in routine May Grunwald Giemsa and Papanicolaou's stain. Therefore special stains such as modified ZN stain is needed to identify nocardia.[2] Culture is needed for final confirmation of the organism.

References

1. Chopra V, Ahir GC, Chand G, Jain PK. Pulmonary nocardiosis mimicking pulmonary tuberculosis. Ind J Tub. 2001;48: 211-3.
2. Sharma S, Singh P, Garg R, Dey P. Pulmonary nocardiosis in fine-needle aspiration cytology smears: Report of four cases. Diagn Cytopathol. 2016;44(4):347-50.

Problem 4.23

B. Histoplasma infection

Discussion

Description: The cytology smears show respiratory lining epithelial cells, alveolar macrophages and many histiocytes loaded with small safety-pin shaped organisms.

Disease: Histoplasma infection usually occurs in immune compromised patients particularly in HIV infected patients. The lung is infected by inhalation of conidia of histoplasma. The most of the pulmonary histoplasma infection resolve spontaneously. Only the patient with immunodeficiency develops acute or chronic pulmonary disease. The sputum or bronchoalveolar lavage is effective in the detection of histoplasma infection. The organisms are small and intracellular and may be missed in routine screening. Special stains such as methenamine silver helps to detect histoplasma infection.

Further Reading

1. Blumenfeld W, Gan GL. Diagnosis of histoplasmosis in broncholavage fluid by intracytoplasmic localization of silver-positive yeast. Acta Cytol. 1991;35:710-2.

Problem 4.24

A. Adenocarcinoma

Discussion

Description: The cytology smears show discrete and many clusters of tumor cells. The cells are round to oval with moderate amount of vacuolated cytoplasm. The nuclei are enlarged with moderate pleomorphism and fine nuclear chromatin. The cell block preparation (Fig. 4.24D) shows tumor cell with gland like arrangement.

Diagnostic features: The diagnostic features in the index case include:

- ❑ Cells with moderate amount of vacuolated cytoplasm
- ❑ Gland like arrangement in cell block
- ❑ TTF-1, napsin and CK 7 positivity along with negative p 63.

Differential diagnosis: Adenocarcinoma should be distinguished from squamous cell carcinoma because of specific tyrosine kinase inhibitor therapy in adenocarcinoma. The cases of adenocarcinoma show typical positivity of TTF-1, napsisn and CK 7; whereas, squamous cell carcinomas are positive for p63, p40 and CK 5/6.

Further Reading

1. Travis WD, Brambilla E, Nicholson AG, Yatabe Y, Austin JH, Beasley MB, et al. WHO Panel. The 2015 World Health Organization classification of lung tumors: Impact of Genetic, Clinical and Radiologic Advances Since the 2004 Classification. J Thorac Oncol. 2015;10(9):1243-60.

Problem 4.25

C. Small cell carcinoma

Discussion

Cytology description: The smears show abundant discrete and loose clusters of small round cells. The cells show hyperchromatic nuclei with inconspicuous nucleoli. Frequent nuclear threading and moulding are also present.

Diagnostic features in the index case: In this index case the following features suggest the diagnosis of small cell carcinoma:

- Loose clusters and discrete cells
- Small cells with scanty cytoplasm
- Nuclear threading
- Round hyperchromatic nuclei
- Nuclear molding.

Differential diagnosis: The cells do not show any squamoid differentiation and the possibility of squamous cell carcinoma is not entertained. The possibility of adenocarcinoma is not considered as there is no glandular arrangement and the individual cells do not show any features of adenocarcinoma.

Problem 4.26

B. Squamous cell carcinoma

Discussion

Description: The cytology smears show oval to polyhedral cells with moderate amount of cytoplasm and central moderately pleomorphic hyperchromatic nuclei. Cell block section (Fig. 4.26D) shows sheets of polyhedral cells. The tumor cells are positive for CK 5/6 and p 63; and negative for CK 7 and TTF-1.

Diagnostic features: In the index case following features help in diagnosis:

- Polyhedral cells with central hyperchromatic nuclei
- Positive CK 5/6 and p 63
- Negative CK 7 and TTF-1.

Problem 4.27

C. Carcinoid

Discussion

Description: The cytology smear shows abundant discrete cells with moderate amount of cytoplasm and central to eccentric nuclei. The nuclei are monomorphic with inconspicuous nucleoli and fine stippled chromatin.

Diagnostic feature in the index case:

- Discrete cells
- Monomorphic nuclei
- Abundant cytoplasm
- Stippled chromatin.

Problem 4.28

B. Peripheral neuroectodermal tumor

Discussion

Description: The cytology smears show discrete and loose clusters of tumor cells with occasional vague rosette like structures (Fig. 4.28C). The cells show enlarged nuclei with salt and pepper like chromatin. The tumor cells are positive for CD 99 and NSE, and are negative for TTF-1 and p63.

Diagnostic features:

- ❑ Round cells with salt and pepper like chromatin
- ❑ Positivity for CD 99 and NSE.

Differential diagnosis: The tumor will come in the differential diagnosis of malignant tumor with predominant round cell morphology such as carcinoid, small cell carcinoma, etc. However, the typical CD 99 (Mic 2) and NSE positivity indicate the diagnosis of peripheral neuroectodermal tumor.

Further Reading

1. Gachechiladze M, Skarda J, Ibrahim M, Tichý T, Flodr P, Latálová P, et al. Primitive neuroectodermal tumor (PNET) of the lung in an adult woman. World J Surg Oncol. 2014;12:374. doi: 10.1186/1477-7819-12-374.
2. Kahn AG, Avagnina A, Nazar J, Elsner B. Primitive neuroectodermal tumor of the lung. Arch Pathol Lab Med. 2001;125(3):397-9.

CHAPTER 5

Gastrointestinal Tract

Introduction

Esophagus

Squamous Cell Carcinoma

Squamous cell carcinoma is the most common malignancy of the esophagus and includes more than 85% of esophageal malignancies. The cytological features show the morphology of squamous cell carcinoma similar to other areas of the body.

Adenocarcinoma

Adenocarcinoma of the esophagus arises from the lower part of the esophagus. Cytology smears show the glandular arrangement of cells, loose clusters and discrete cells. Individual cells show a moderate amount of vacuolated cytoplasm with moderately pleomorphic nuclei with prominent nucleoli.

Small Cell Carcinoma

Small cell carcinoma (SCC) is an uncommon tumor and accounts for less than 5% of esophageal carcinoma. The cells are small, slightly larger than lymphocytes. Nuclei are hyperchromatic with inconspicuous nucleoli.

Stomach

Adenocarcinoma

There are two types of gastric adenocarcinoma, intestinal and diffuse gastric type. The characteristic features of intestinal type of gastric adenocarcinoma include:

- Single cells, loose cohesive clusters of cells
- Columnar or cuboidal cells with marked nuclear pleomorphism
- Enlarged hyperchromatic nuclei with high N/C ratio and prominent nucleoli
- Irregular nuclear margin.

Diffuse or gastric type shows:
- Predominantly single cells and occasional loose clusters
- The cells are round with abundant vacuolated cytoplasm
- Nuclei are crescent and pushed to the periphery giving signet ring appearance
- Hyperchromatic nucleus
- Prominent large nucleoli
- Pool of mucin may also be seen.

Gastrointestinal Stromal Tumor

Gastrointestinal stromal tumor (GIST) is most commonly seen in the stomach followed by the small intestine and rectum. The behavior of the tumor depends on its size and mitotic activity.
- *Very low risk tumor*: Less than 2 cm in size and mitotic figures less than 5/50 high power filed (HPF)
- *Low risk tumor*: 2–5 cm in size and mitotic figures less than 5/50 HPF
- *Intermediate risk*: 5–10 cm in size and mitotic figures less than 5/50 HPF
- *High risk*: Any size and mitotic figures more than 10/50 HPF.

CD 117 (C-Kit) positivity is essential for diagnosis of GIST. In addition, the cells are also positive for CD34 (membranous pattern of positivity) and focally positive for smooth muscle actin.

The salient cytological features of GIST include:
- Clusters, fascicles and also dissociated cells
- Palisading nuclei
- Spindle cells with
  - Scanty to moderate cytoplasm
  - Mild to moderate nuclear pleomorphism with coarse granular chromatin and indistinct nucleoli
- Frequent mitosis, necrosis are noted in malignant GIST.

QUESTIONS

Problem 5.1

A 57-year-old female had dysphagia for 2 months. She had anorexia, pallor and loss of weight. Endoscopic guided esophageal brush cytology was done. The most likely diagnosis from Figure 5.1 is:

A. Squamous cell carcinoma
B. Adenocarcinoma
C. No malignant cells
D. Viral infection

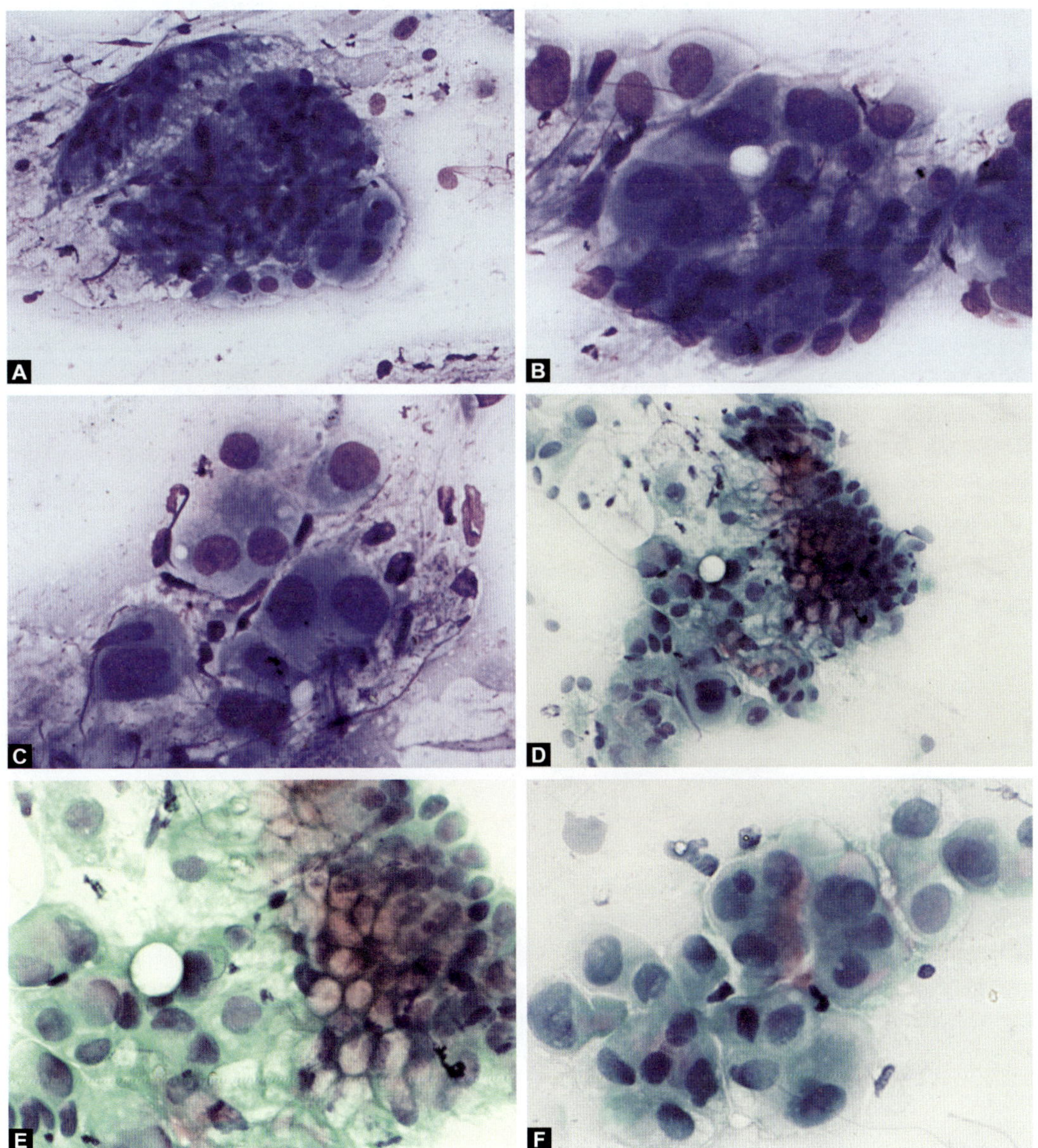

Figures 5.1A to F

Problem 5.2

A 65-year-old female presented with anorexia, dyspepsia and loss of weight. The endoscopic brush cytology of ulcer of the stomach was done. The most likely diagnosis from Figure 5.2 is:

A. Squamous cell carcinoma
B. Adenocarcinoma
C. Dysplasia
D. No malignancy: reactive atypia

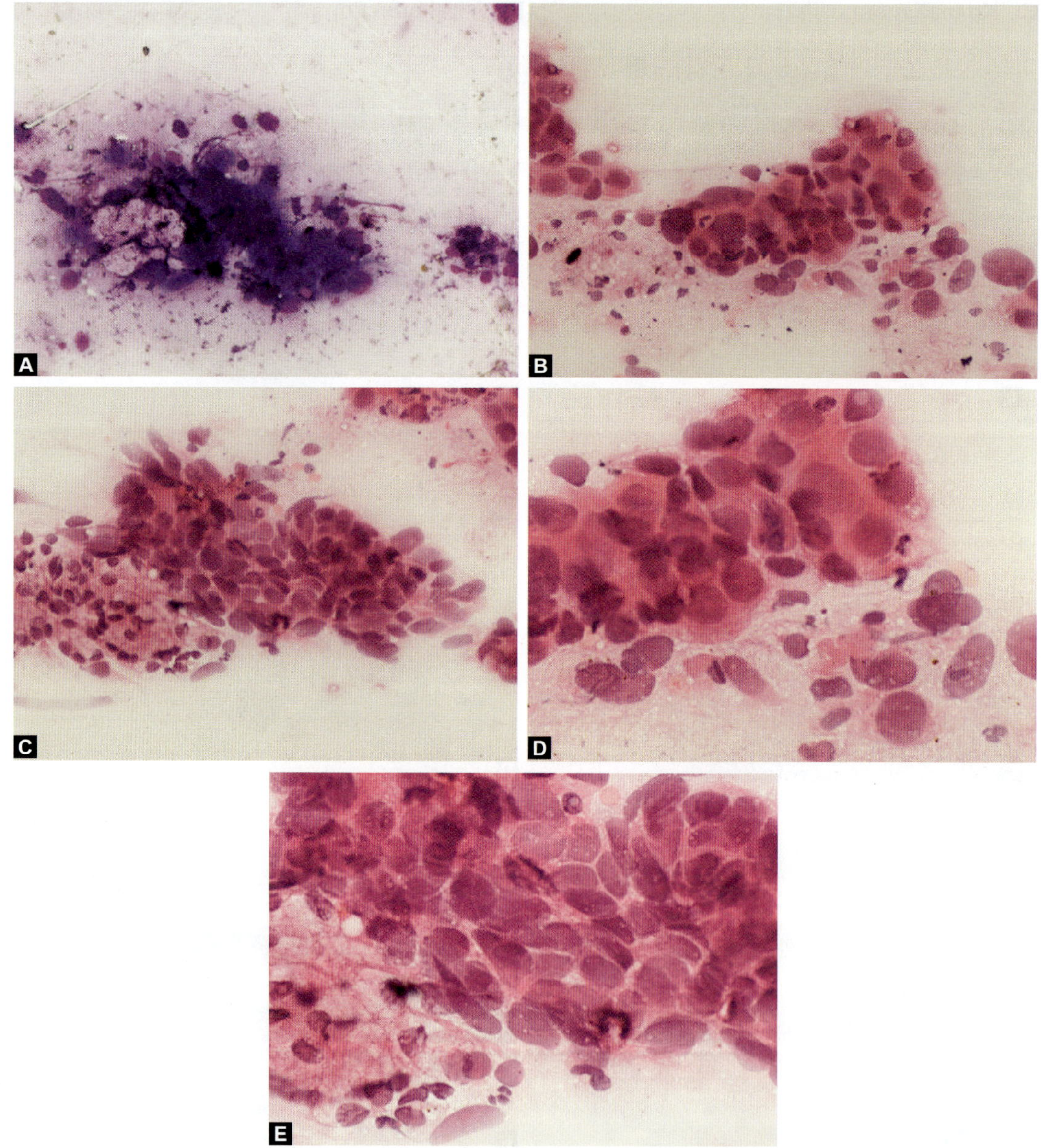

Figures 5.2A to E

Problem 5.3

A 43-year-old male patient had renal transplant about three year back. The patient now complaints of dysphagia, pain throat and white patch in esophagus. The esophageal brush cytology was done. What is the diagnosis from the microphotograph of the smear (Fig. 5.3)?

A. Dysplasia
B. Squamous cell carcinoma
C. Adenocarcinoma
D. Candidal esophagitis

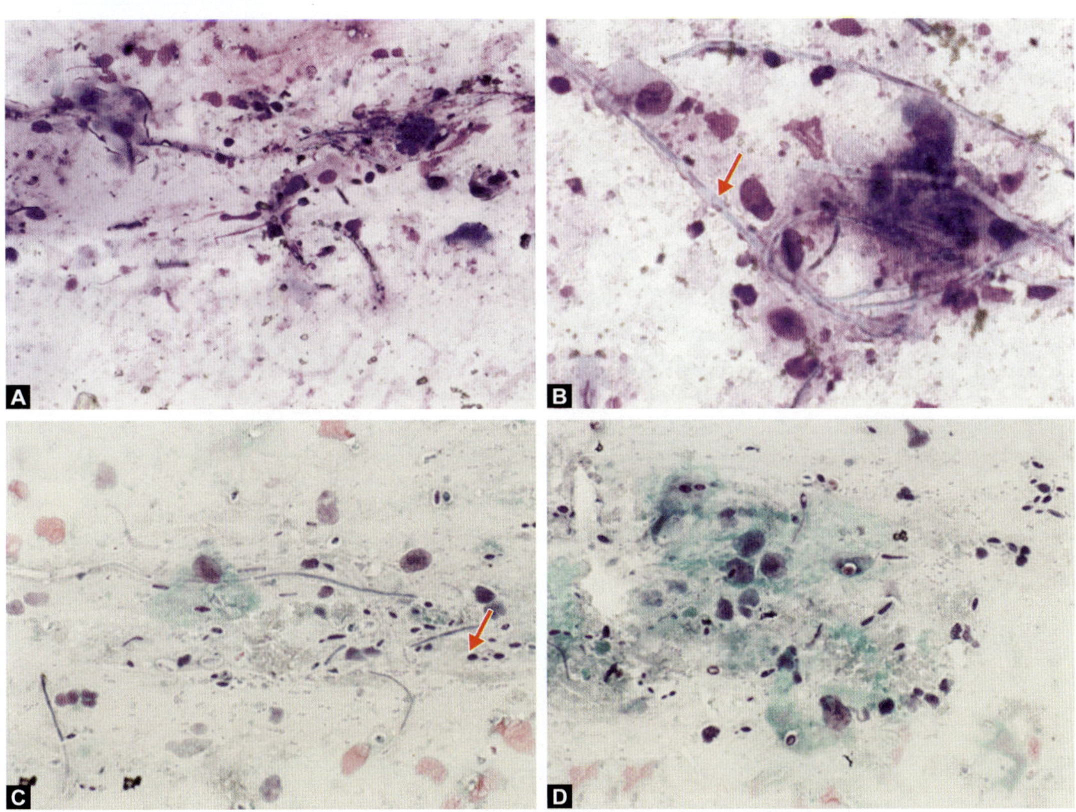

Figures 5.3A to D

Problem 5.4

A 43-year-old male has obstructive jaundice for 2 months. He had small elevated mucosal lesion in the common bile duct. Endoscopic USG guided FNAC was done. The most likely diagnosis from Figure 5.4 is:

A. Adenocarcinoma
B. Chronic inflammation
C. Non-Hodgkin lymphoma
D. Mild dysplasia

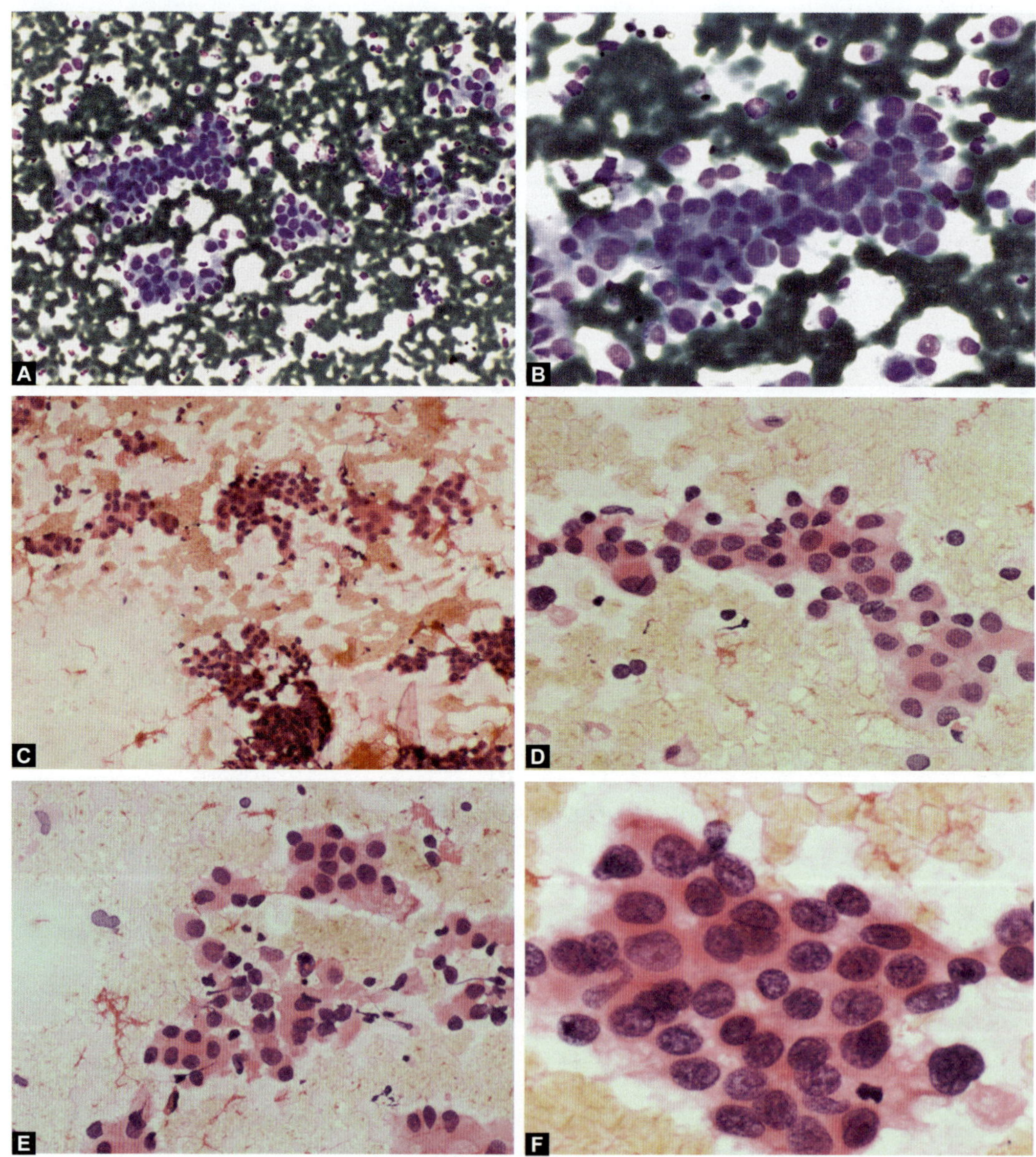

Figures 5.4A to F

Problem 5.5

A 61-year-old female was diagnosed as oral squamous cell carcinoma 3 year back. She also developed rectal adenocarcinoma two year back and was treated surgically. Now the patient has anorexia, intermittent diarrhea and loss of weight. Her upper GI endoscopy showed 2 cm diameter mucosal thickening in duodenum. Endoscopic USG FNAC was done. The most likely diagnosis from Figure 5.5 is:

A. Adenocarcinoma
B. Squamous cell carcinoma
C. Normal mucosa
D. Dysplastic epithelium

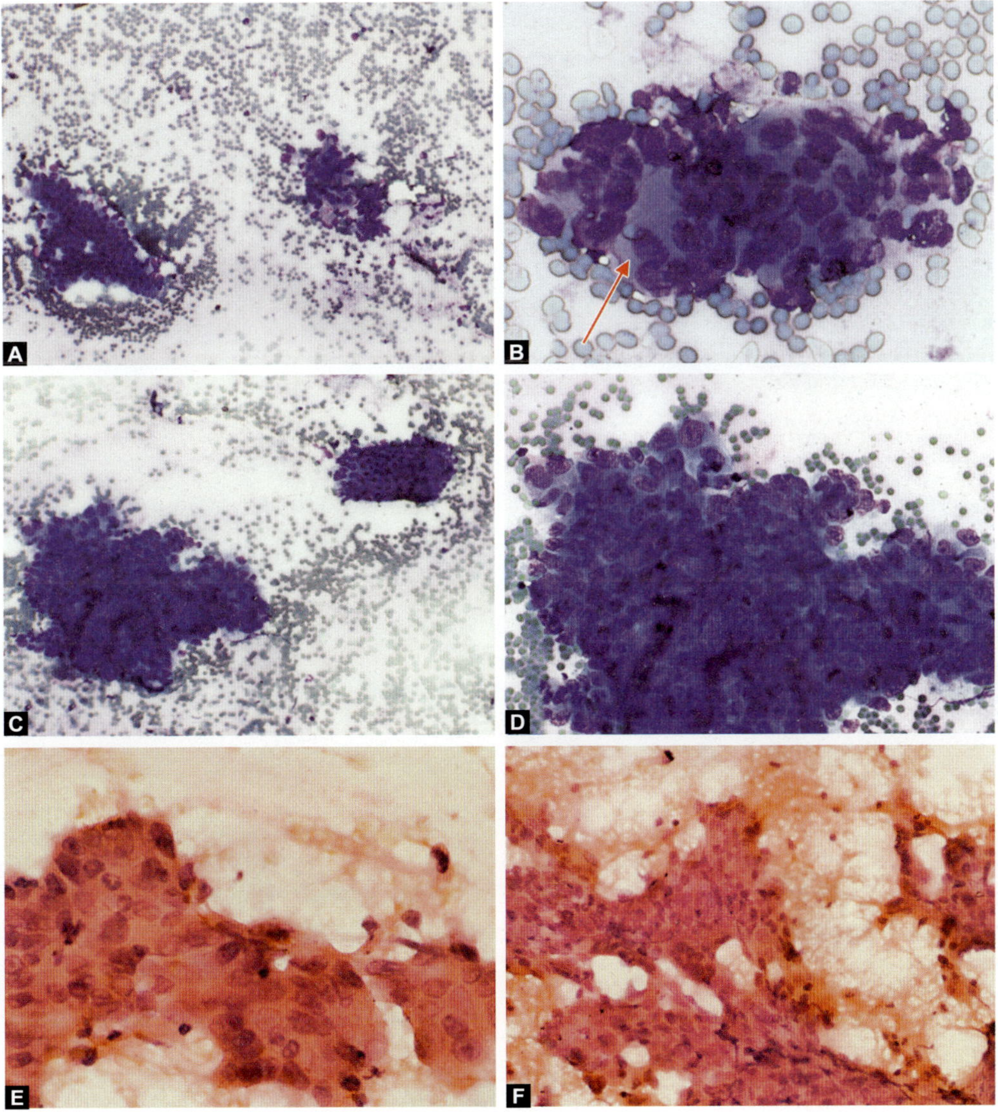

Figures 5.5A to F

Problem 5.6

A 53-year-old man was operated for cholecystectomy one year back. The histopathology report is not available. The patient developed stricture and obstructive jaundice. Endoscopic retrograde cholangiopancreatography (ERCP) was done and the brush cytology smear was taken from the stricture of common bile duct. The possible diagnosis from Figure 5.6 is:

A. Normal bile ductular cells
B. Adenocarcinoma
C. Chronic inflammation
D. Stone induced atypia

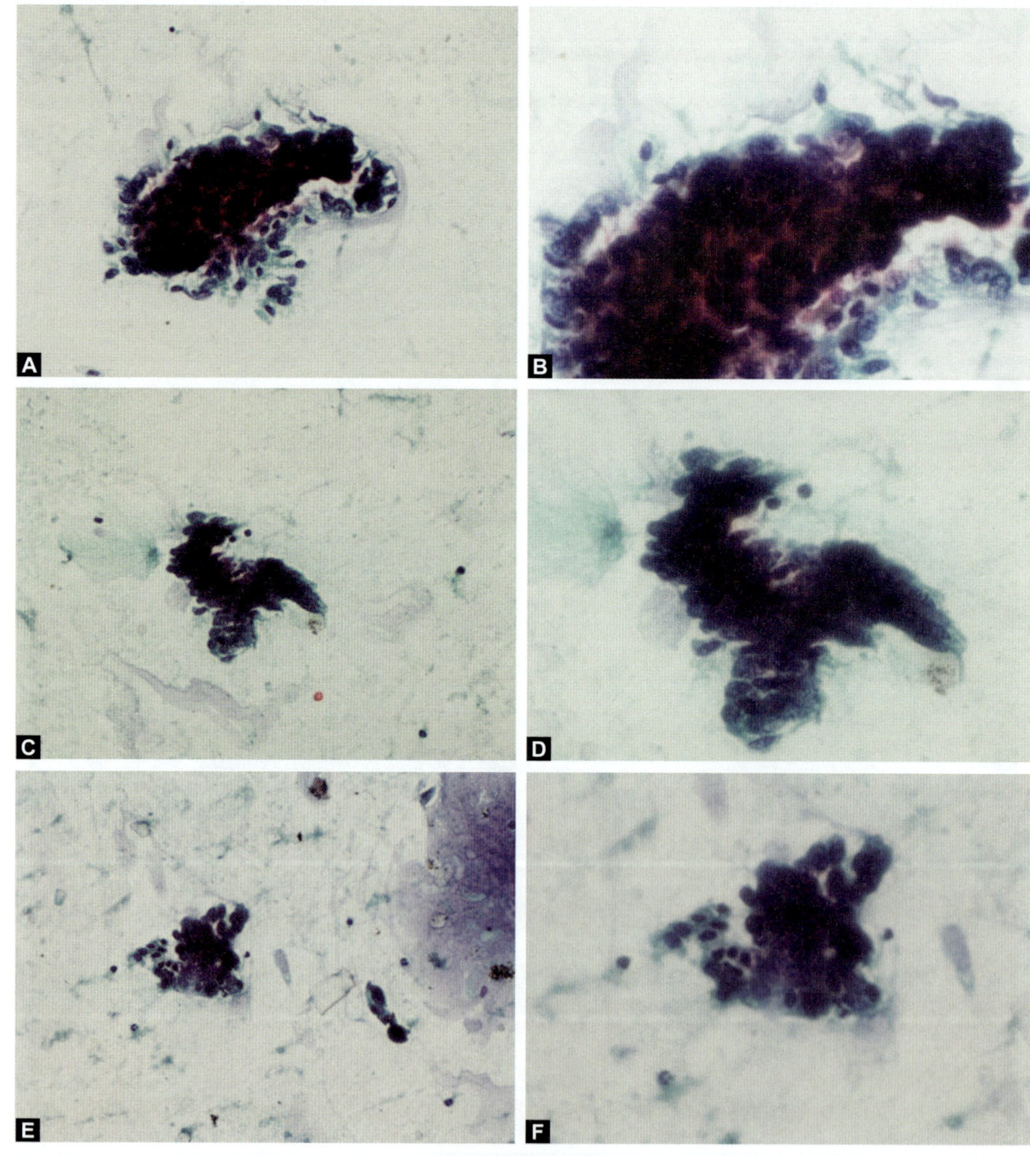

Figures 5.6A to F

Problem 5.7

A 37-year-old female presented with dysphagia. The endoscopy showed esophageal stricture. Endoscopic brush cytology of the esophagus was done. What is your diagnosis from Figure 5.7?

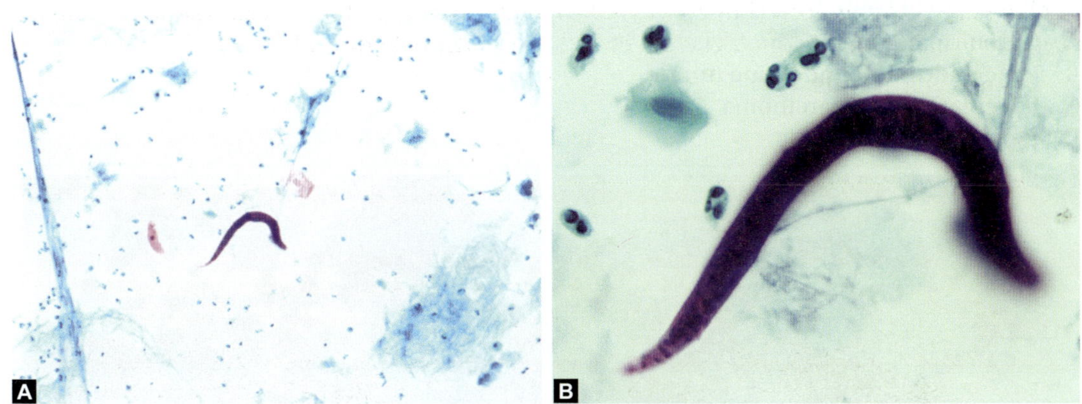

Figures 5.7A and B

Problem 5.8

A 59-year-old male presented with jaundice and loss of weight for 3 months. USG of the abdomen showed 2.5 cm diameter space occupying lesion in the head of pancreas. Endoscopic ultrasound guided FNAC of head of pancreas was done. The most likely diagnosis from Figure 5.8 is:

A. Chronic pancreatitis
B. Pancreatic neuroendocrine tumor
C. Pancreatic adenocarcinoma
D. Benign pancreatic cells

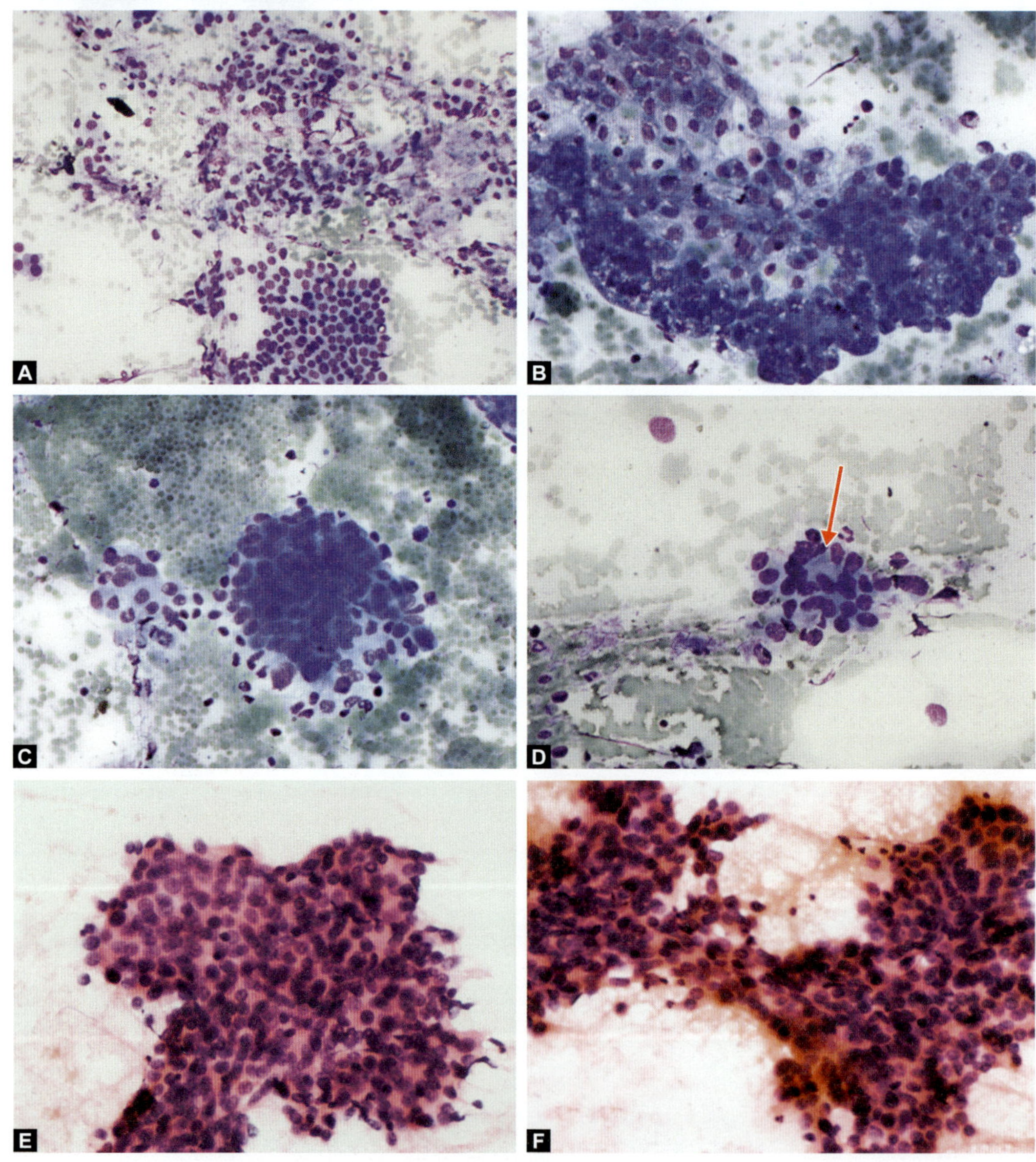

Figures 5.8A to F

Problem 5.9

A 35-year-old male presented with anorexia and loss of weight. Endoscopy of the upper gastrointestinal tract shows thickened mucosa in the body of stomach. Endoscopic guided FNAC was done from the swelling in the stomach. The most likely diagnosis from Figure 5.9 is:

A. Gastrointestinal stromal tumor
B. Leiomyoma
C. Adenocarcinoma
D. Normal smooth muscle

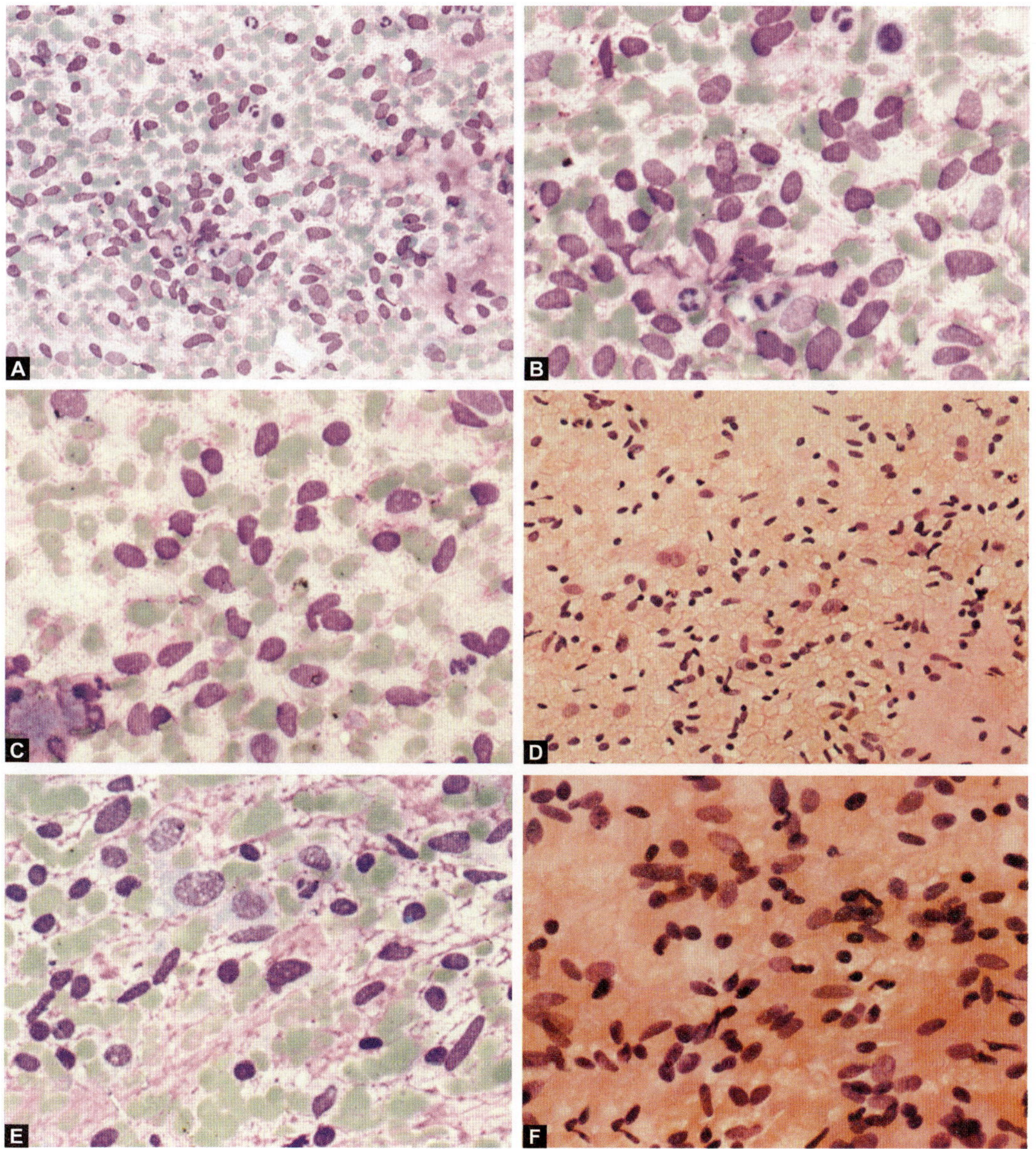

Figures 5.9A to F

Problem 5.10

A 31-year-old female complaints with anorexia and loss of weight for 2 months. Endoscopic USG guided FNAC was done from the thickened wall of the stomach. The most likely diagnosis form Figure 5.10 is:

A. Adenocarcinoma
B. Gastritis
C. Non-Hodgkin lymphoma
D. Reactive lymphoid hyperplasia

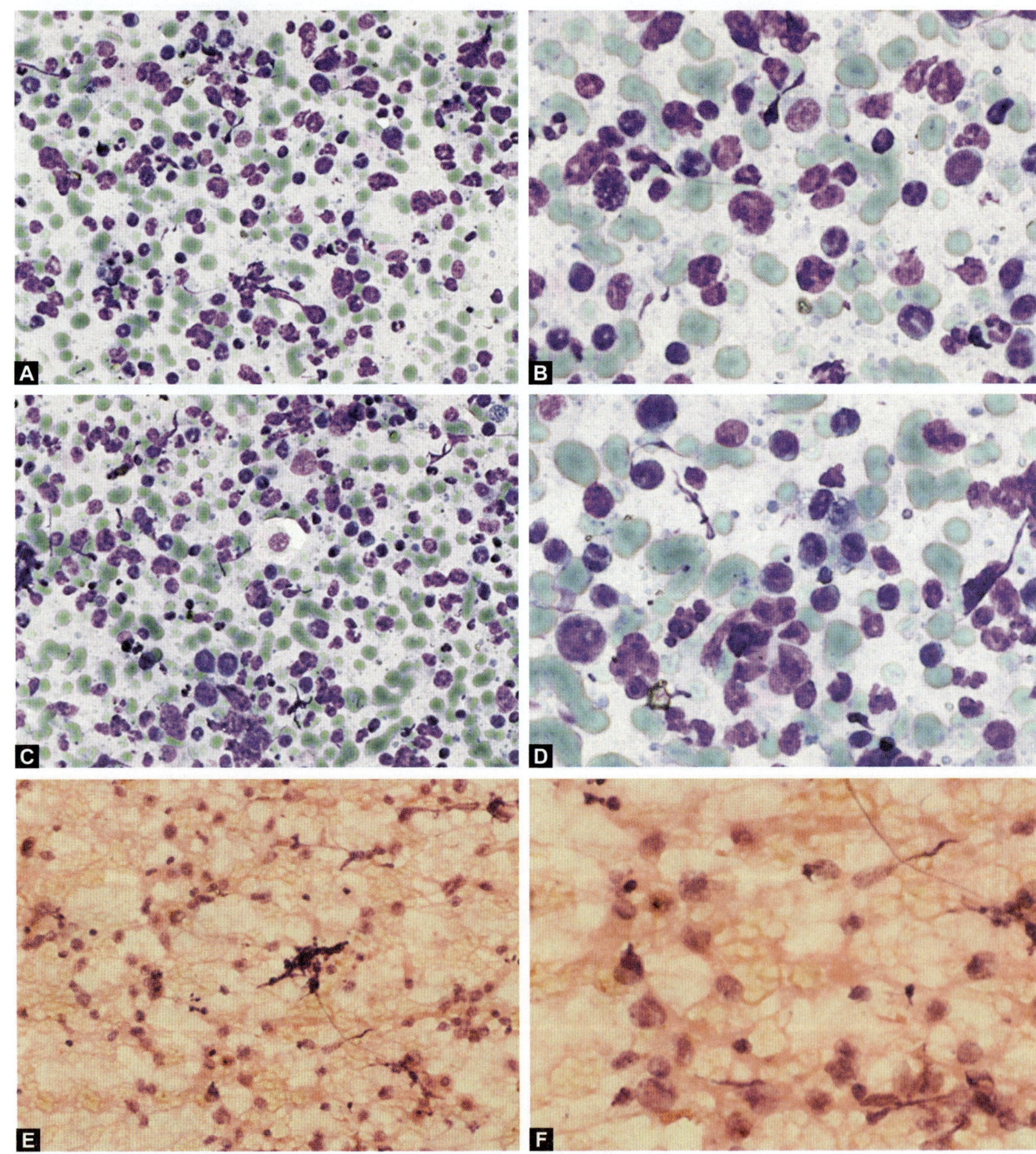

Figures 5.10A to F

Problem 5.11

A 48-year-old male presented with dysphagia, anorexia and significant weight loss. Esophageal brush cytology smear was done. The most likely diagnosis from the microphotographs (Fig. 5.11) of the cytology smear is:

A. Adenocarcinoma
B. Candidal infection
C. Reactive atypical cells
D. Squamous cell carcinoma

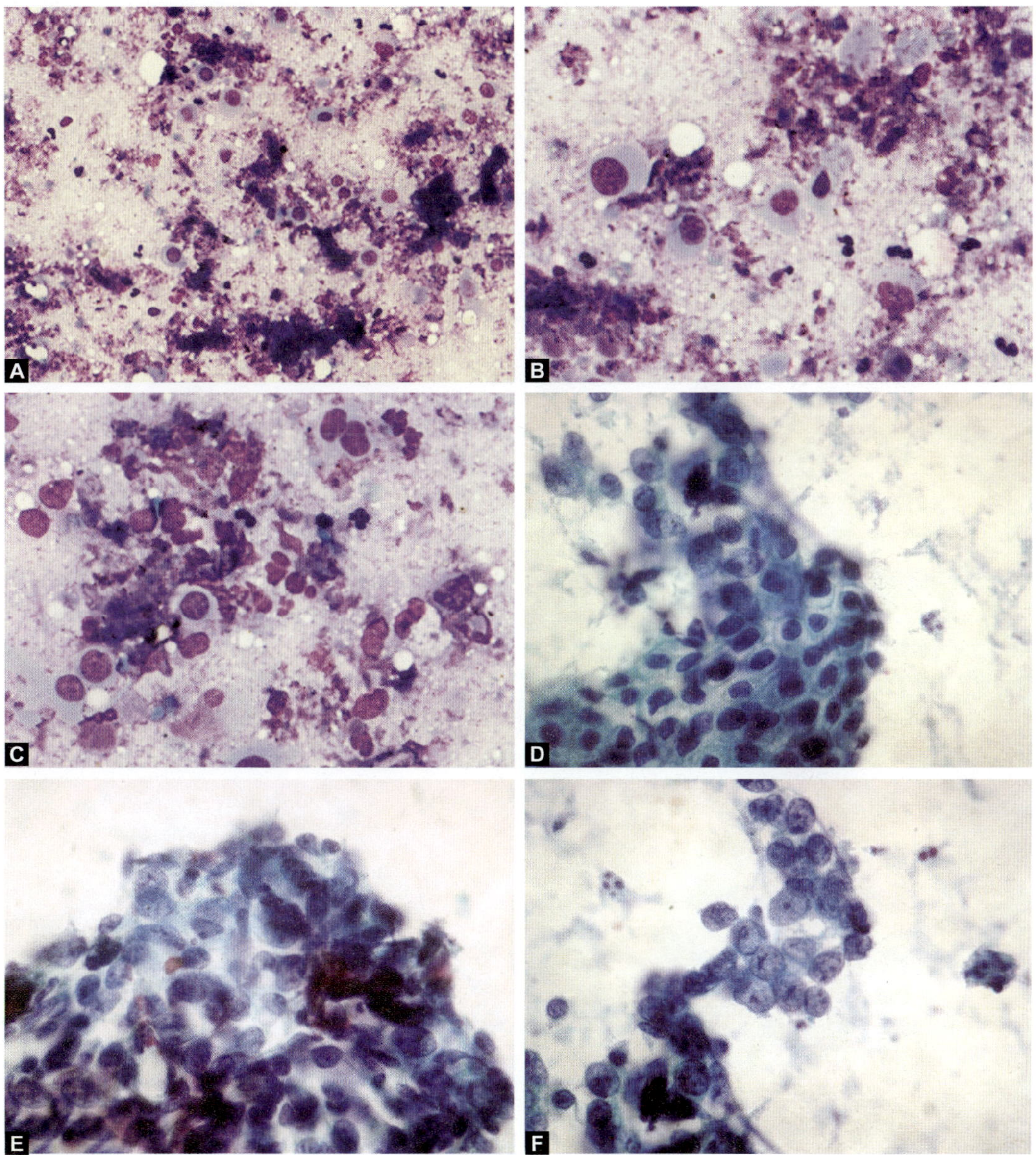

Figures 5.11A to F

Problem 5.12

A 49-year-old female had obstructive jaundice and marked loss of weight. Endoscopic USG guided FNAC of pancreas was done from 3 cm diameter mass of the head of pancreas. The most likely diagnosis from Figure 5.12 is:

A. Adenocarcinoma of pancreas
B. Chronic inflammation
C. Non-Hodgkin lymphoma
D. Pancreatic neuroendocrine tumor

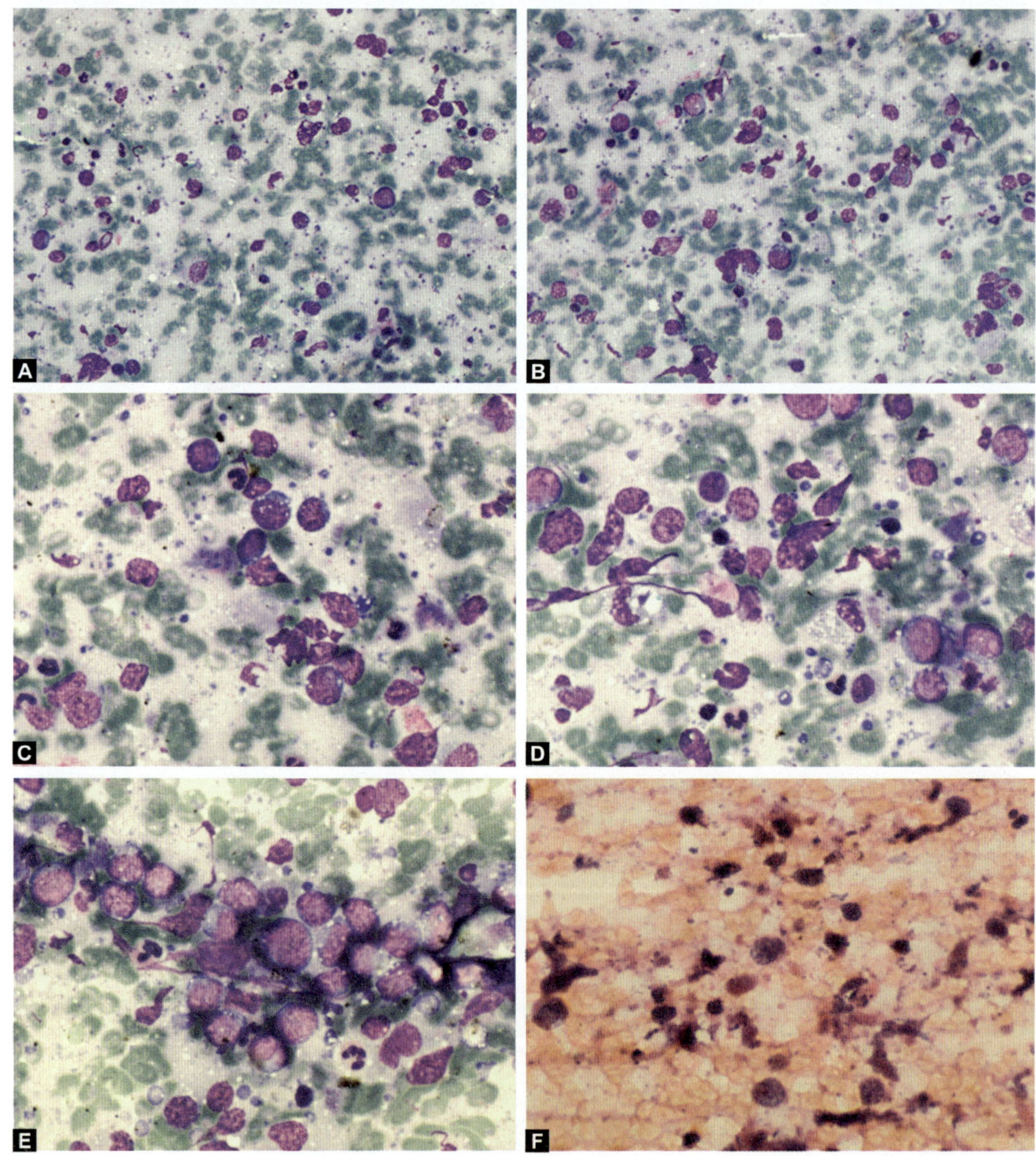

Figures 5.12A to F

Problem 5.13

A 39-year-old female presented with heartburn and dysphagia for 2 months. The upper gastrointestinal endoscopy shows a stricture in the lower esophagus. Esophageal brush cytology was done. The most likely diagnosis from Figure 5.13 is:

A. Squamous cell carcinoma
B. Candidal infection
C. Barrett's esophagus
D. Adenocarcinoma

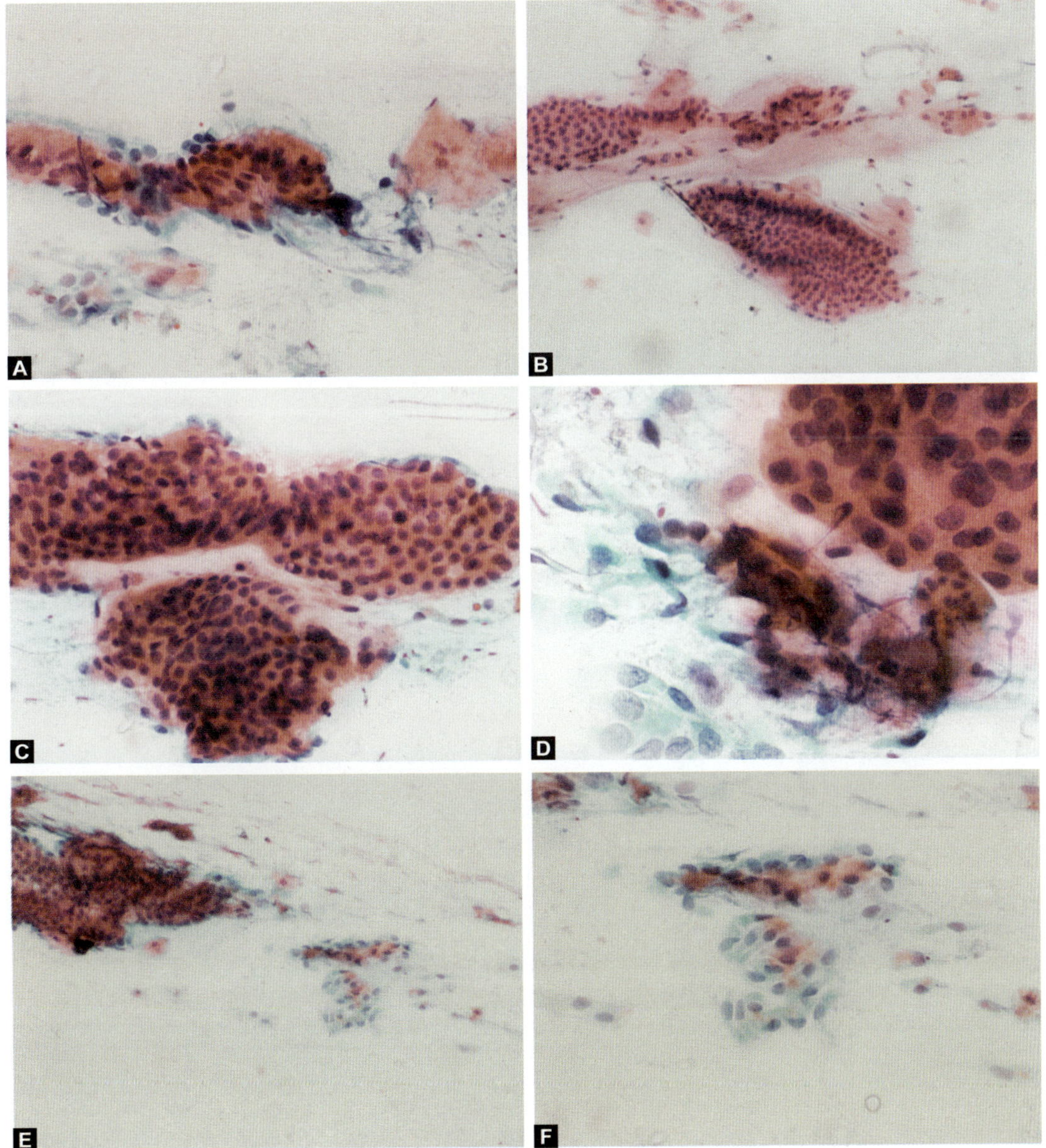

Figures 5.13A to F

Problem 5.14

A 49-year-old male presented with acidity and indigestion for 3 months. Endoscopy showed a large 5 cm diameter ulcerated elevated lesion in the antral region of the stomach. Endoscopic ultrasound guided FNAC was done. The most likely diagnosis from Figure 5.14 is:

A. Adenocarcinoma
B. Chronic gastritis
C. Gastrointestinal stromal tumor
D. Lymphoma

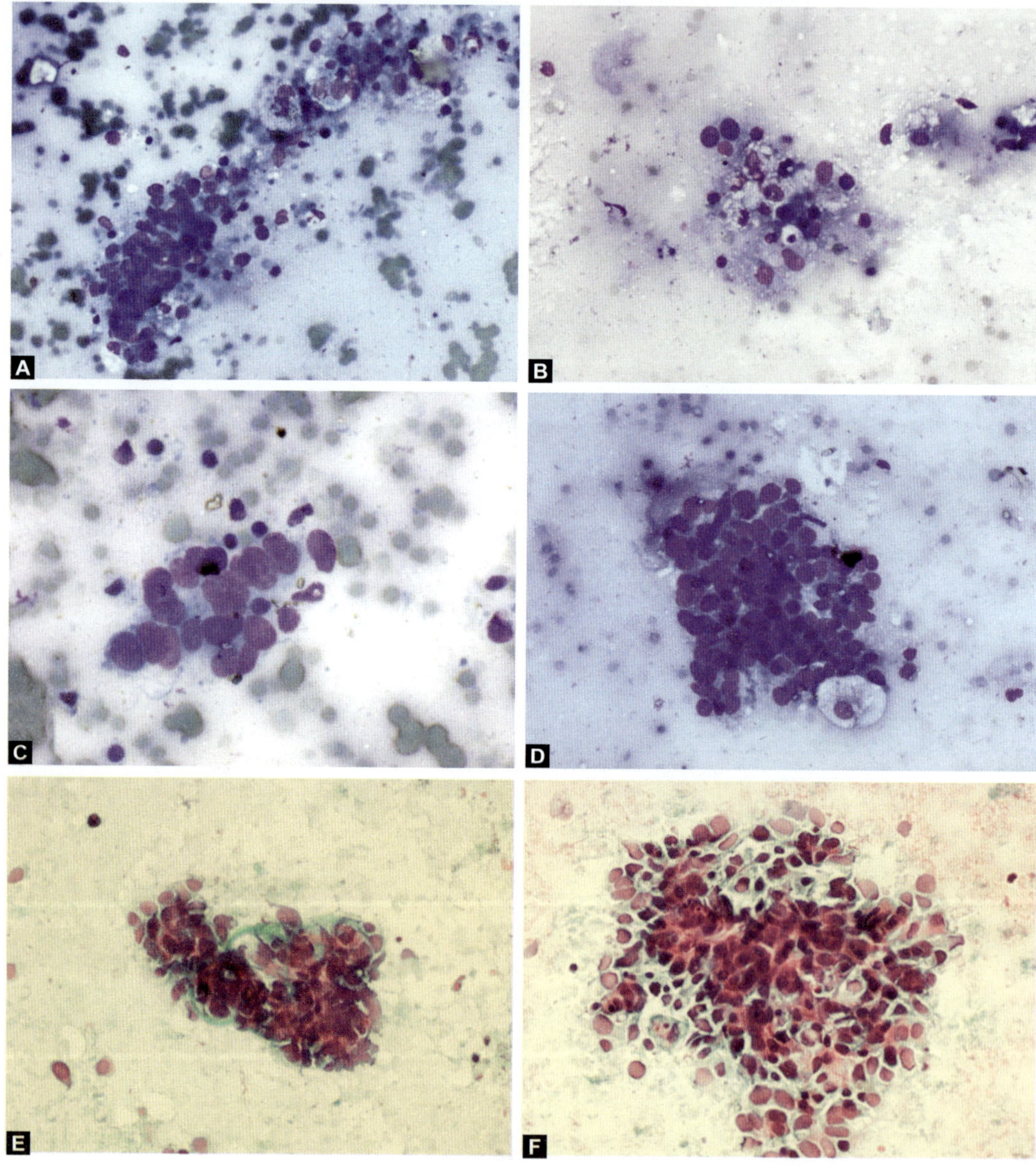

Figures 5.14A to F

Problem 5.15

A 47-year-old female patient presented with obstructive jaundice for 3 months. Endoscopic common bile duct brush cytology was done. Her serum IgG4 level was raised (220 mg/dL). The most likely diagnosis from Figure 5.15 is:

A. IgG4 disease
B. Adenocarcinoma of bile duct
C. Dysplasia of the duct epithelium
D. Within normal limit

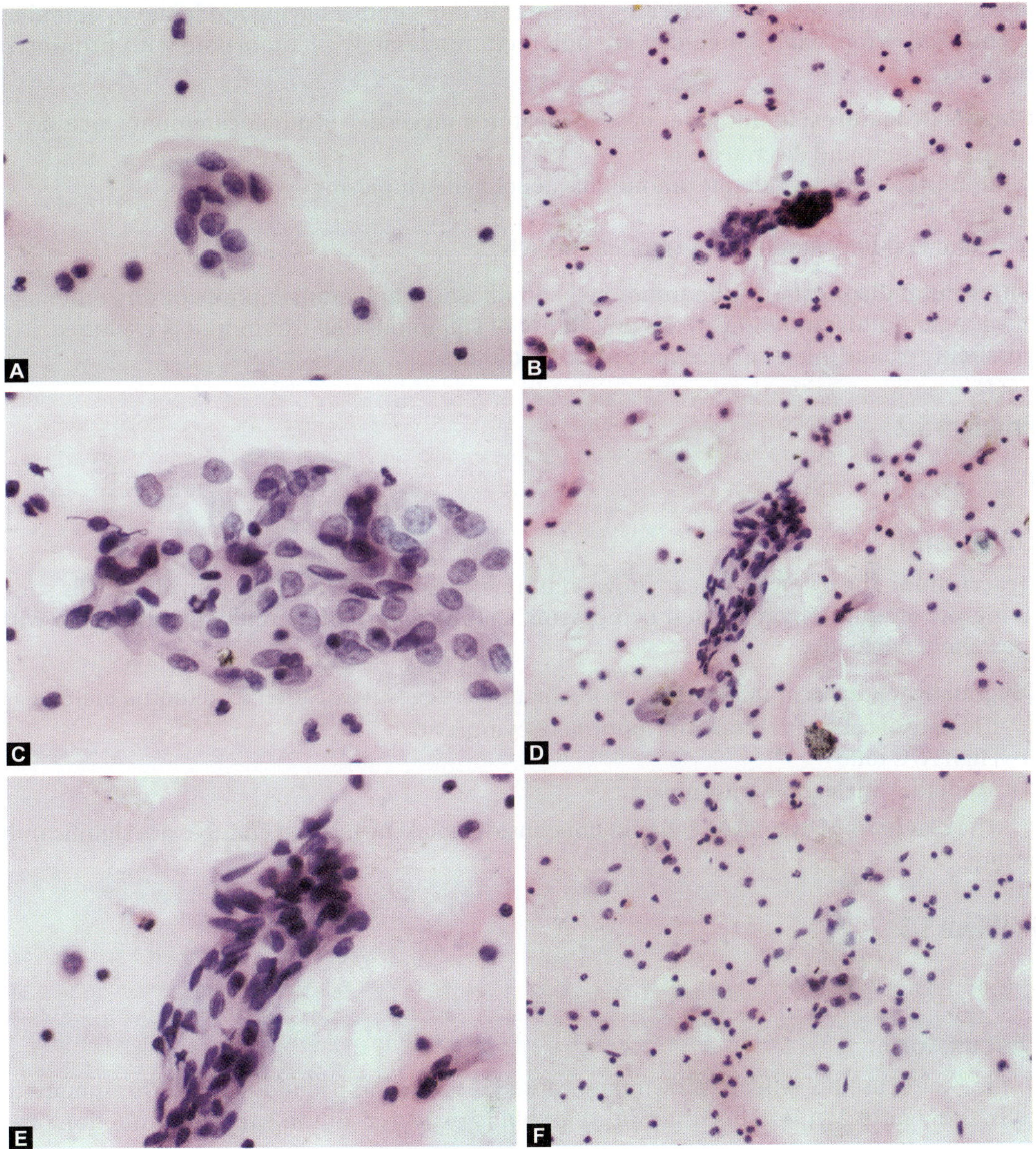

Figures 5.15A to F

ANSWERS

Problem 5.1

B. Adenocarcinoma

Discussion

Description: The cytology smear shows multiple clusters of tumor cells admixed with benign looking columnar cells. The cells are columnar in appearance having moderate to abundant vacuolated cytoplasm. The nuclei are enlarged, moderately pleomorphic with prominent nucleoli.

Diagnostic features: In this index case the features suggestive of adenocarcinoma include:
- The clusters and discrete cells
- Round cells with moderate amount of vacuolated cytoplasm
- High N/C ratio
- Columnar appearance of the malignant cells.

Differential diagnosis: As the tumor cells do not show any orangeophilia or any other sign of squamoid differentiation so the possibility of squamous cell carcinoma is excluded. The possibility of viral infection is remote in this index case.

Problem 5.2

B. Adenocarcinoma

Discussion

Description: The cytology smear shows multiple clusters of cells. The individual cells show moderate amount of vacuolated cytoplasm. The nuclei are moderately pleomorphic with prominent nucleoli.

Diagnostic features: In this index case the features suggestive of adenocarcinoma include:
- Clusters of cells with moderate amount of cytoplasm
- High N/C ratio
- Moderate nuclear pleomorphism.

The diagnosis of dysplastic cells are not considered as the nuclear changes are significantly severe.

Problem 5.3

D. Candidal esophagitis

Discussion

In the index case the smears show many candidal spore and pseudohyphae. The squamous cells show mild atypia. This amount of atypia is expected in the candidal infection. The candidal esophagitis usually occurs in immunocompromised patients. The index patient is renal transplant recipient and therefore immunocompromised. The possibility of malignancy is remote in this case.

Problem 5.4

A. Adenocarcinoma

Discussion

Description: The smears show occasional benign looking ductal cells and multiple loose clusters of tumor cells. The cells show mild to moderate nuclear enlargement and pleomorphism. The nuclear chromatin is irregularly clumped.

Diagnostic features:

- Clusters and discrete cells with moderate cytoplasm
- Moderate nuclear pleomorphism
- Prominent nucleoli.

Differential diagnosis: The nuclear atypia is significant and therefore possibility of dysplasia is excluded. The possibility of non-Hodgkin lymphoma is not considered as the tumor cells are in clusters and contain moderate amount of cytoplasm. Lymphoglandular bodies are also absent in the background.

Problem 5.5

A. Adenocarcinoma

Discussion

Description: The cytology smears show abundant clusters of cells with occasional gland like arrangement (Fig. 5.5B). The individual cells show moderate nuclear atypia and pleomorphism.

Diagnostic features: In the index case the following features suggest the diagnosis of adenocarcinoma:

- Tumor cells admixed with benign epithelial cells
- The tumor cells are in clusters with occasional gland like arrangement
- The cells show moderate nuclear atypia and pleomorphism.

Differential diagnosis: The diagnosis of metastatic squamous cell carcinoma is a very remote possibility as this event is rare and the cells also do not show any squamoid differentiation. It is difficult to differentiate the primary versus metastatic adenocarcinoma on the basis of cytomorphology. However, the clinical history and the small mucosal elevation in the duodenum favor a third primary malignancy.

Problem 5.6

B. Adenocarcinoma

Discussion

Description: The cytology smears show multiple three-dimensional clusters of tumor cells. The individual cells are columnar in shape with moderate nuclear enlargement, pleomorphism and prominent nucleoli.

Diagnostic features:

- The three-dimensional tight cohesive clusters
- The nuclear overlapping, enlargement, and pleomorphism and prominent nucleoli.

Differential diagnosis: The atypia is significant and therefore the possibility of reactive atypia or dysplasia is not considered.

Problem 5.7

Esophageal filariasis

Discussion

In the index case the diagnosis of filariasis is straight forward. Filariasis causing esophageal stricture and dysphagia is rare. However in endemic country like India filaria may present as various odd manifestations.

Further Reading

1. Rawal A, Hayer J, Dey P. Microfilariae in esophageal stricture. Acta Cytol. 2002;46(1):70-1.

Problem 5.8

C. Pancreatic adenocarcinoma

Discussion

Description: The cytology smears show many three-dimensional cohesive clusters of tumor cells along with glands. The individual cells show moderate nuclear enlargement and pleomorphism.

Diagnostic features:

- Discrete and also three-dimensional cohesive clusters of tumor cells
- Gland like arrangement
- Cells show significant nuclear enlargement and pleomorphism.

Differential diagnosis: The possibility of chronic pancreatitis is not considered due to significant nuclear overlapping, high N/C ratio and nuclear pleomorphism. The smears of pancreatic neuroendocrine tumor usually show discrete cells with moderate cytoplasm. The cytoplasm often contains reddish granules. The nuclear chromatin is salt and pepper like.

Problem 5.9

A. Gastrointestinal stromal tumor

Discussion

Description: The cytology smears show abundant discrete and occasional loose clusters of spindle cells with scanty to moderate cytoplasm having oval to spindle shaped nuclei. The nuclei show mild pleomorphism with inconspicuous nucleoli. Background shows acellular pinkish material.

Diagnostic features:
- Loose clusters and also dissociated cells
- Spindle cells with scanty to moderate cytoplasm
- Spindle shaped nuclei with mild pleomorphism
- Coarse granular chromatin, indistinct nucleoli.

CD 117 (C-Kit) positivity is essential for diagnosis of GIST. The tumor cells are also positive for CD34 (membranous pattern of positivity) and focally positive for smooth muscle actin.

Differential diagnosis: The leiomyoma and normal smooth muscle does not yield much cellularity in the visceral organ.

Further Reading

1. Boggino HE, Fernandez MP, Logroño R. Cytomorphology of gastrointestinal stromal tumor: diagnostic role of aspiration cytology, core biopsy, and immunochemistry. Diagn Cytopathol. 2000;23(3):156-60.
2. Miettinen M, Lasota J. Gastrointestinal stromal tumors—definition, clinical, histological, immunohistochemical, and molecular genetic features and differential diagnosis. Virchows Arch. 2001;438(1):1-12.

Problem 5.10

C. Non-Hodgkin lymphoma

Discussion

Description: The cytology smears show discrete monomorphic lymphoid cell population with background lymphoglandular bodies. The individual cells are round with deep blue cytoplasm. The nuclei are mildly pleomorphic with irregular margin. Nucleoli are prominent.

Diagnostic features: The features suggestive of non-Hodgkin lymphoma include:
- Discrete monomorphic cell population
- Background lymphoglandular bodies
- The cells are round with deep blue cytoplasm.

Differential diagnosis: The cases of chronic gastritis and reactive lymphoid cells usually show reactive polymorphic population of lymphoid cells. The possibility of adenocarcinoma is not entertained as the cells are predominantly discrete compared to clusters of cells and occasional gland like arrangement in adenocarcinoma. Poorly differentiated adenocarcinoma may come in the differential diagnosis of non-Hodgkin lymphoma. Immunocytochemistry is helpful (CD 45 positive cells in lymphoma and CK positive cells in adenocarcinoma) in such condition.

Problem 5.11

D. Squamous cell carcinoma

Discussion

Description: The cytology smears show multiple clusters and discrete tumor cells. The cells are polyhedral with centrally placed moderately pleomorphic hyperchromatic nuclei.

Diagnostic features: The features suggestive of squamous cell carcinoma include:
- Clusters and discrete tumor cells
- The cells are oval to polyhedral
- Moderate amount of cytoplasm
- The nuclei of the cells are enlarged and pleomorphic.

Differential diagnosis: The tumor cells show significant nuclear abnormalities and therefore the diagnosis of reactive atypia is excluded. Candida may be associated with carcinoma due to superficial ulceration and infection. In this index case there was no evidence of any candida in the smear.

Problem 5.12

C. Non-Hodgkin lymphoma

Discussion

Description: The cytology smears show abundant discrete monomorphic population of lymphoid cells with background lymphoglandular bodies. The tumor cells are round with deep blue vacuolated cytoplasm. The nuclei are round with fine chromatin and occasional nucleoli.

Diagnostic features: The features suggestive of non-Hodgkin lymphoma (NHL) in this case are:
- Discrete monomorphic lymphoid cell population
- Abundant lymphoglandular bodies
- Round to oval cells with deep blue cytoplasm.

Differential diagnosis: Chronic inflammation usually shows polymorphic population of cells. Pancreatic neuroendocrine tumor shows discrete cells with occasional rosette like structure. The cells may show reddish granules and salt and pepper chromatin. NHL of pancreas is very uncommon and it comprises only 0.5% of all pancreatic tumors. The common types of NHL of pancreas are low-grade MALTOMA, follicular lymphoma and diffuse large B-cell lymphoma. Detailed immunophenotyping is mandatory in such cases for exact subtyping of NHL.

Problem 5.13

C. Barrett's esophagus

Discussion

Description: The cytology smears show multiple clusters and discrete benign looking squamous epithelial cells admixed with benign columnar cells.

Diagnostic features in this index case:
- The endoscopic finding of abnormality of lower esophagus
- Admixture of columnar cells along with squamous epithelial cells.

Disease description: Barrett's esophagus is characterized by replacement of squamous epithelium of lower esophagus by metaplastic glandular epithelial cells. Near about 10% gastroesophageal reflux disease is associated with Barrett's esophagus. The most of the adenocarcinoma of lower esophagus is developed from Barrett's esophagus. So this is a premalignant condition. It is important to identify any dysplasia of the intestinal metaplastic cell of Barrett's esophagus.

Problem 5.14

A. Adenocarcinoma

Discussion

Description: The cytology smears show multiple clusters of epithelial cells with occasional gland like arrangement (Fig. 5.14C). The tumor cells have moderate amount of vacuolated cytoplasm. The nuclei are enlarged with moderate pleomorphism.

Diagnostic features:
- Gland like structure
- Cells with vacuolated cytoplasm and moderate nuclear pleomorphism.

Differential diagnosis: Chronic gastritis may show nuclear atypia. However the index case shows significant atypical cells. The cells of gastrointestinal stromal tumor are usually oval to spindle shaped and often show background pinkish stromal material. Lymphoma shows predominantly discrete cells with background lymphoglandular bodies.

Problem 5.15

A. IgG4 disease

Discussion

Description: The cytology smears show many loose clusters of bile duct epithelial cells admixed with lymphocytes and plasma cells. The individual epithelial cells show mild nuclear atypia.

Diagnostic features:
- High IgG4 serum level
- Epithelial cells admixed with lymphocytes and plasma cells.

Disease description: IgG4-related disease is characterized by tumor like tissue, consistent infiltration by IgG4 positive plasma cells, storiform fibrosis and an elevated serum IgG4 concentration. This is mostly detected as incidentally. The disease may involve different organ and may be mistaken as malignancy due to nuclear atypia.

Further Reading

1. Stone JH, Khosroshahi A, Deshpande V, et al. Recommendations for the nomenclature of IgG4-related disease and its individual organ system manifestations. Arthritis Rheum. 2012;64:3061-7.
2. Stone JH, Zen Y, Deshpande V. IgG4-related disease. N Engl J Med. 2012;366:539-51.

CHAPTER 6

Urine Cytology

Introduction

Urothelial cell carcinoma (UCC) is the most common cancers of urinary bladder and constitutes more than 90% of all malignant bladder tumors. The patient commonly presents with hematuria, dysuria, and increased urinary frequency. WHO and the International Society of Urological Pathology classification of UCC [1] are highlighted in the Box 6.1.

Box 6.1: Classification of urothelial neoplasm

- **Hyperplasia**
  - Flat hyperplasia
  - Papillary hyperplasia
- **Flat lesions with atypia**
  - Reactive atypia
  - Dysplasia
  - Carcinoma in situ
- **Papillary neoplasia**
  - Papilloma
  - Inverted papilloma
  - Papillary neoplasia of low malignant potential
  - Papillary carcinoma, low grade
  - Papillary carcinoma, high grade

Invasive neoplasm

The main aim of urinary cytology is to detect the flat lesions which may be missed in cystoscopy and high-grade UCC. The consensus approach to the reporting of urinary cytology is accepted in Paris, 2013 in the 18th International Congress of Cytology. Box 6.2 highlights the Paris System for Reporting Urinary Cytology.

Box 6.2: Paris system for reporting urinary cytology

- Unsatisfactory
- Negative for high-grade UCC
- Atypical urothelial cells
- Suspicious for high-grade urothelial cell carcinoma
- High-grade urothelial cell carcinoma
- Low-grade urothelial cell carcinoma
- Other malignancies: Primary and secondary

The main goal of the system is to detect more number of high-grade UCC.

Criteria of Diagnosis

- **Atypical urothelial cells**: This is the challenging area and the diagnosis is difficult. The main diagnostic criteria are:
  - The nucleo-cytoplasmic ratio of the urothelial cells is between 0.5 and 0.7. These cells should be non-superficial and non-degenerated urothelial cells.
  - Any one of the three features should be present:
    - Hyperchromasia
    - Irregular clumped chromatin
    - Irregular nuclear membrane.
- **Suspicious for high-grade urothelial cell carcinoma**:
  - The nucleo-cytoplasmic ratio of the urothelial cells should be more than 07.
  - Hyperchromasia
  - Any one of the two features should be present:
    - Irregular clumped chromatin
    - Irregular nuclear membrane.
- **High-grade urothelial cell carcinoma:** The lesions should have following features:
  - High cellularity
  - Dissociated and loose clusters
  - Moderately pleomorphic nuclei
  - Moderate to marked nuclear hyperchromasia
  - High N/C ratio
  - Irregular nuclear membrane
  - Coarse granular chromatin
  - Large prominent nucleoli
  - Atypical mitosis.
- **Low-grade urothelial cell carcinoma:** These lesions are difficult to diagnose in urine cytology. The presence of papillae with fibrovascular core may indicate this lesion, however this finding is very rare. Low-grade urothelial cell carcinoma lesions may show increased cellularity with minimal atypia and the absence of umbrella cells.

Other Malignancies

Squamous Cell Carcinoma

Pure squamous cell carcinoma is rare in the urinary bladder. It is usually related to Schistosoma hematobium infection. The malignant cells show oval to polyhedral cells with hyperchromatic pleomorphic nuclei. The cytoplasm of the cells shows orangeophilia indicating intracellular keratinization.

Adenocarcinoma

The adenocarcinoma is rare in the urinary tract. The tumor resembles mucin secreting gastrointestinal carcinoma. The cells show columnar in appearance with pleomorphic nuclei and prominent nucleoli. Cells with glandular appearance may be noted. Abundant mucin may be present in the background.

Small Cell Carcinoma

This is a rare aggressive tumor of the bladder. Cytomorphology is similar to that of small cell carcinoma of other parts of the body. Smears show discrete small round cells with scanty cytoplasm. Small cell carcinoma in urinary bladder is positive for chromogranin, synaptophysin, CK7 and EMA.

Reference

1. Epstein JI, Amin MB, Reuter VE, Mostofi FK. The World Health Organization/International Society of Urological Pathology consensus classification of urothelial (transitional cell) neoplasms of the urinary bladder. Am J Surg Pathol. 1998;22:1435-48.

QUESTIONS

Problem 6.1

Match the cells in urine from Figure 6.1

A.	1. Urothelial cell
B.	2. Umbrella cells
C.	3. Chemotherapy changes
D.	4. Seminal vesicle cells
E.	5. Atypical urothelial cell

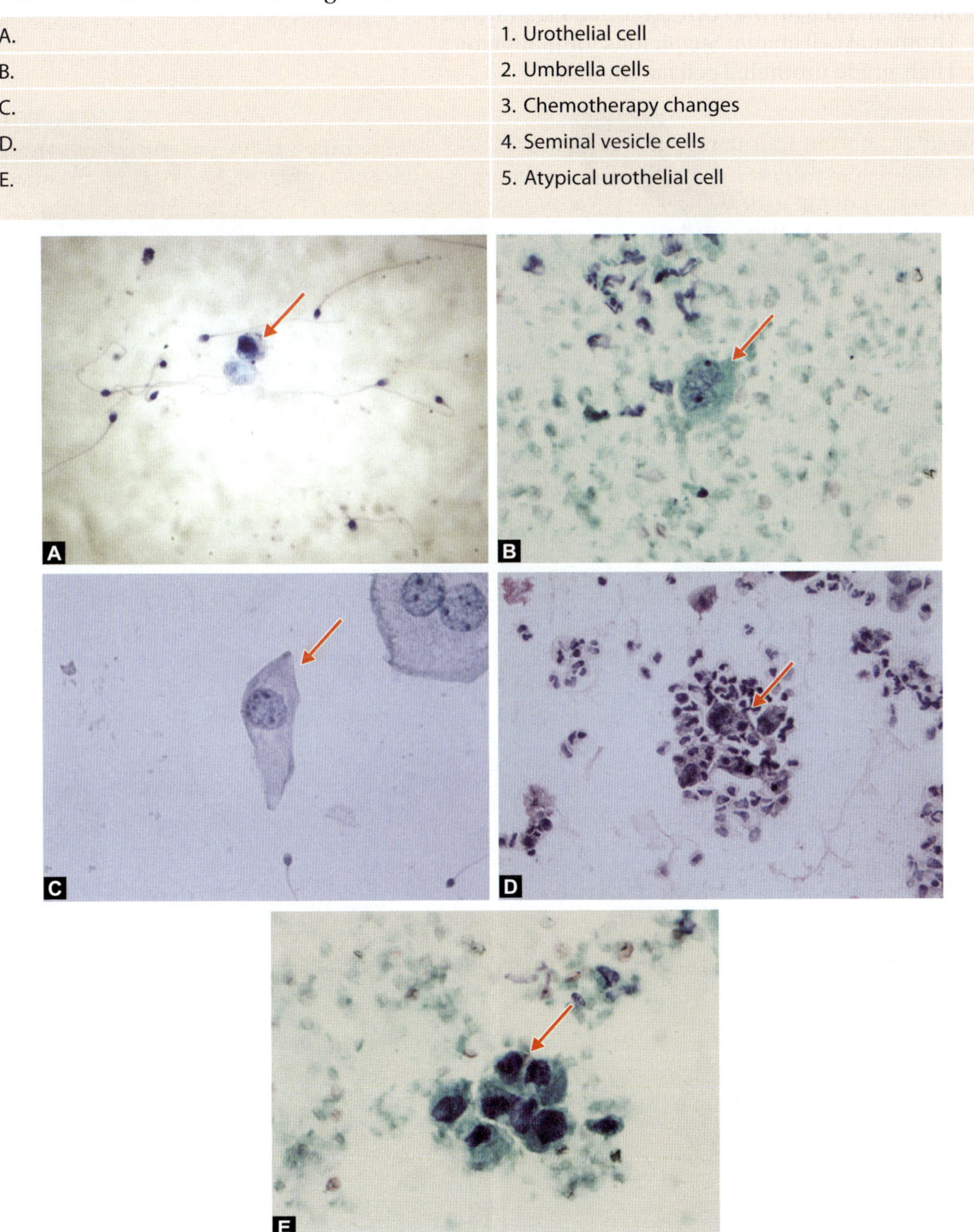

Figures 6.1A to E

Problem 6.2

A 56-year-old male presented with intermittent hematuria for one month. Urine cytology examination was done. The most likely diagnosis from Figure 6.2 is:

A. No evidence of malignancy
B. Urothelial cell atypia: Undetermined significance
C. Urothelial cell atypia: Suspicious for malignancy
D. High-grade urothelial cell carcinoma

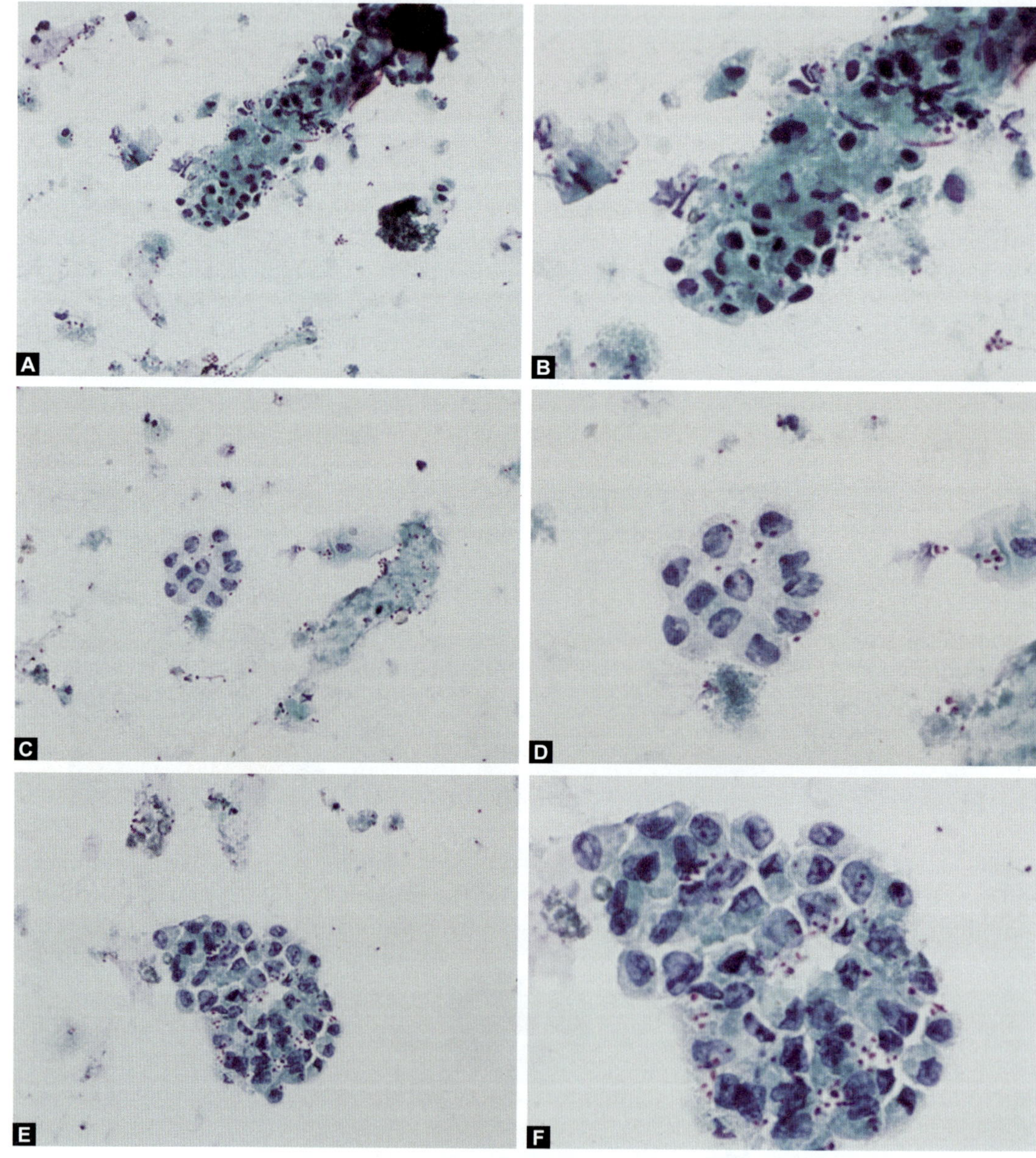

Figures 6.2A to F

Problem 6.3

A 43-year-old male presented with hematuria, dysuria and fever. Urine cytology was done. The most likely diagnosis from Figure 6.3 is:

A. UCC
B. Tuberculosis
C. Reactive atypia
D. Cystitis

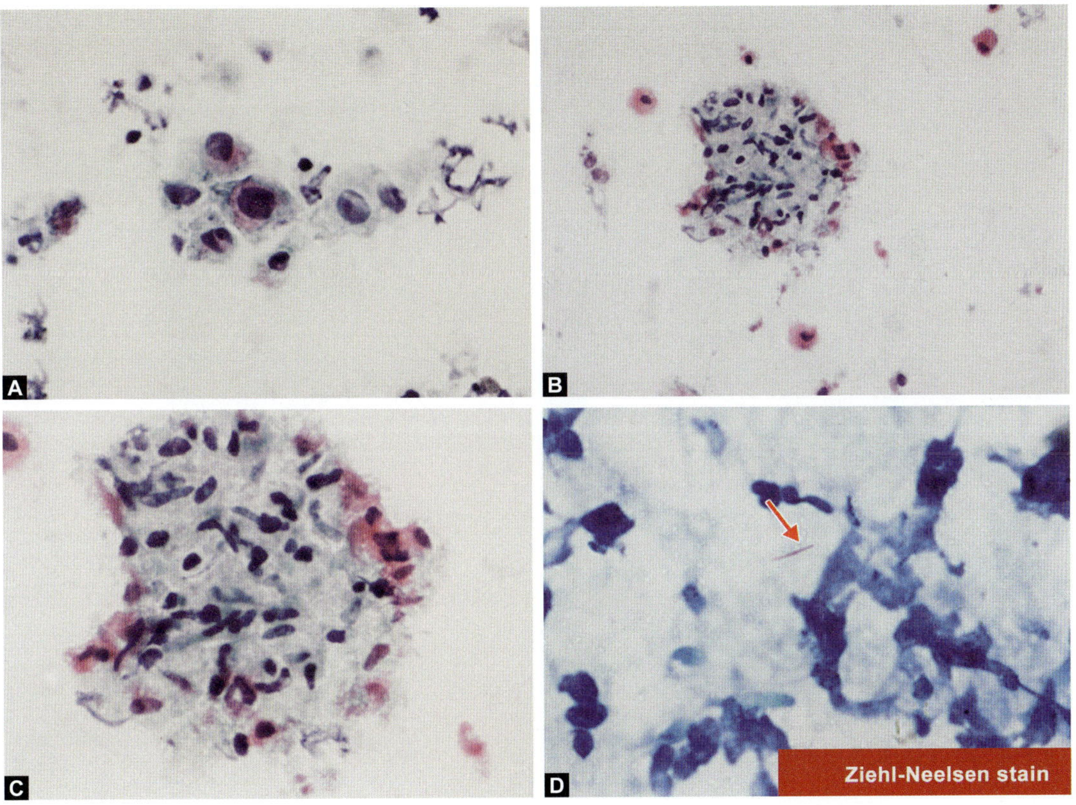

Figures 6.3A to D

Problem 6.4

A 56-year-old male presented with hematuria for 7 days. The routine urine examination was done. What is the most possible diagnosis from the Figure 6.4?

A. No malignant cells seen
B. Reactive atypical cells present
C. Low-grade UCC
D. High-grade UCC

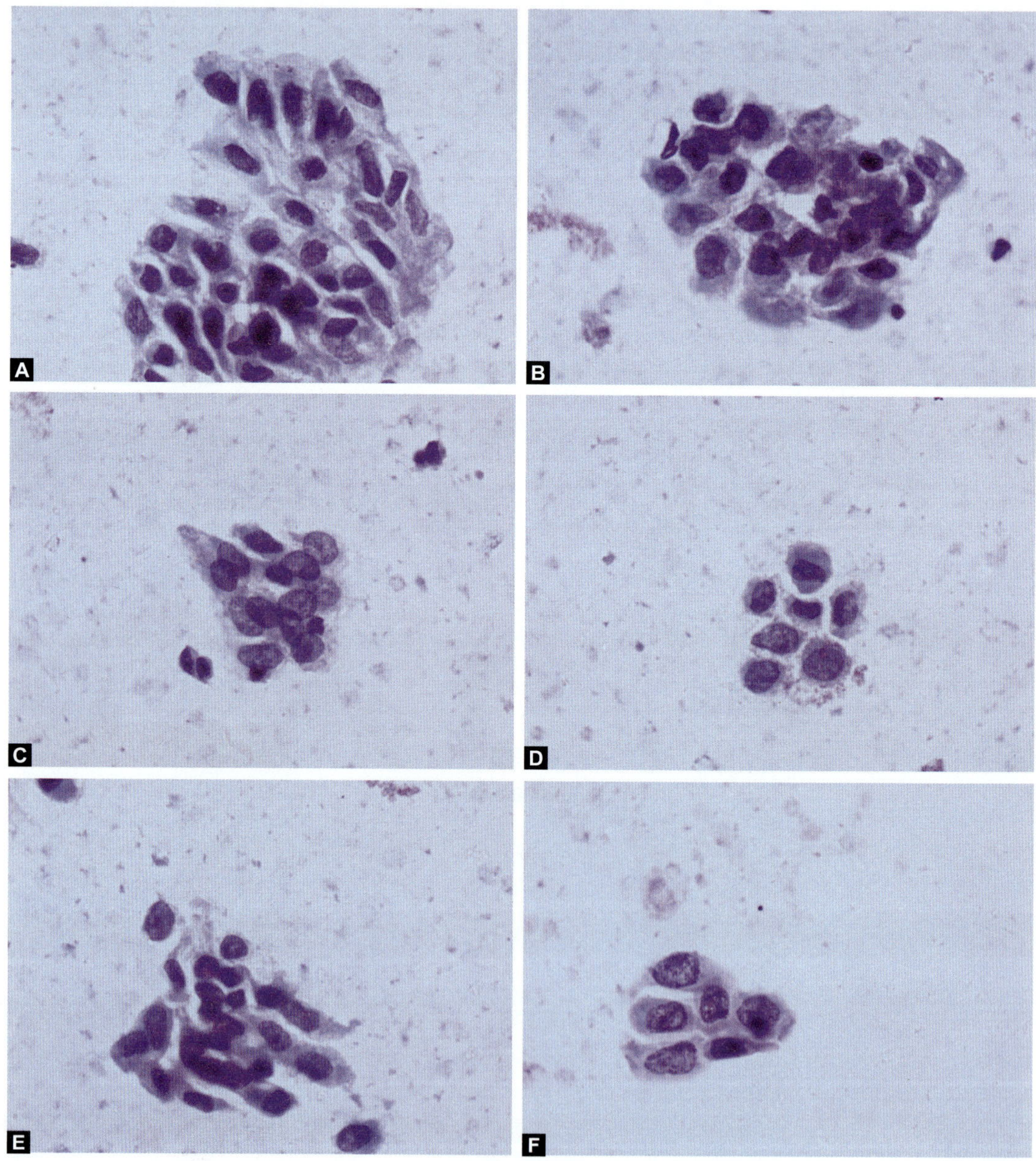

Figures 6.4A to F

Problem 6.5

A 56-year-old male patient had intermittent hematuria for 15 days. He had no other complaints. Urine cytological examination was done. The most likely diagnosis from Figure 6.5 is:

A. Low-grade UCC
B. High-grade UCC
C. Reactive atypia only
D. Suspicious for carcinoma

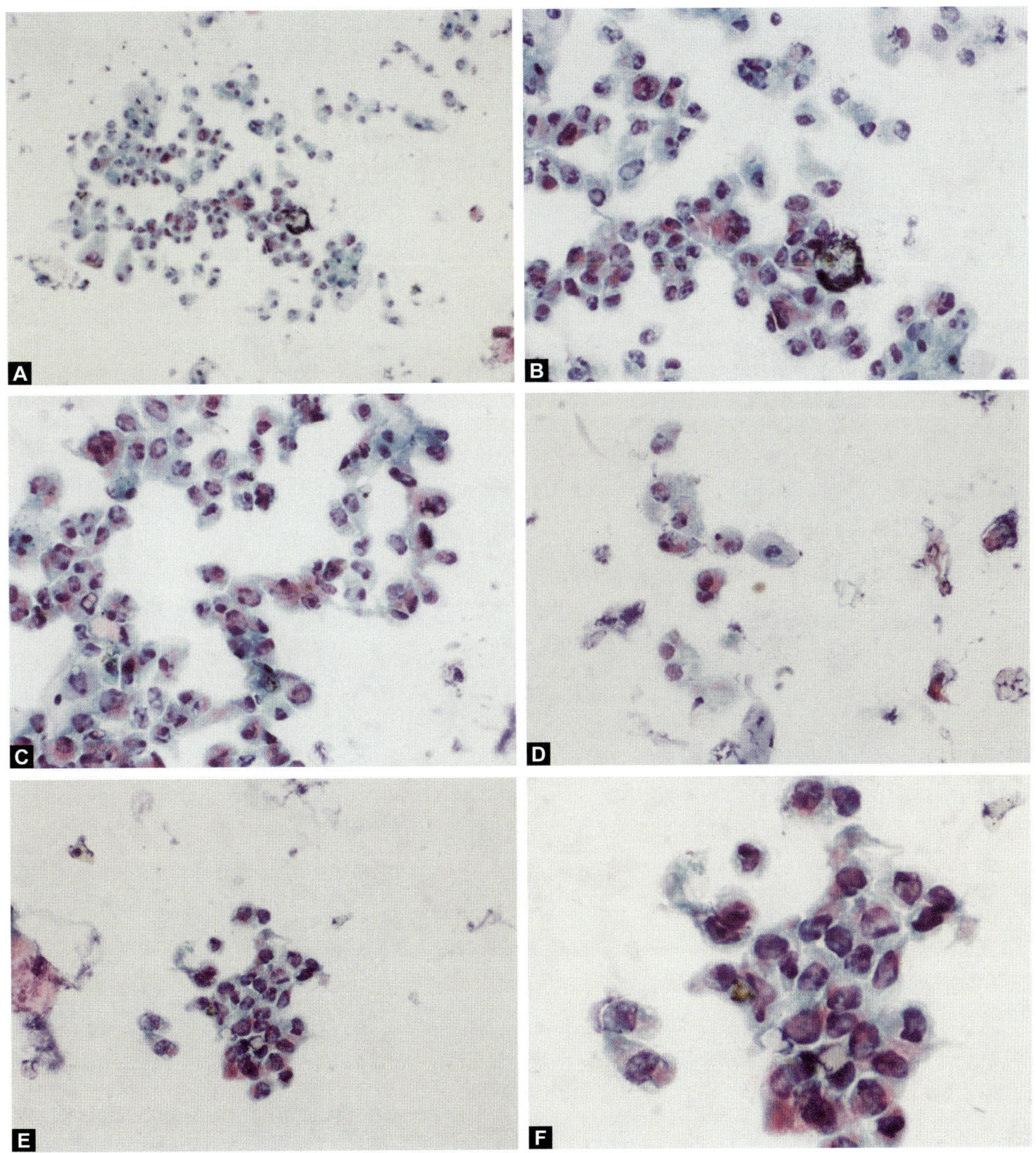

Figures 6.5A to F

Problem 6.6

A 67-year-old man presented with intermittent hematuria for 2 months. The most likely diagnosis from Figure 6.6 is:

A. Reactive atypia
B. Low-grade UCC
C. Inflammation only
D. No evidence of malignancy

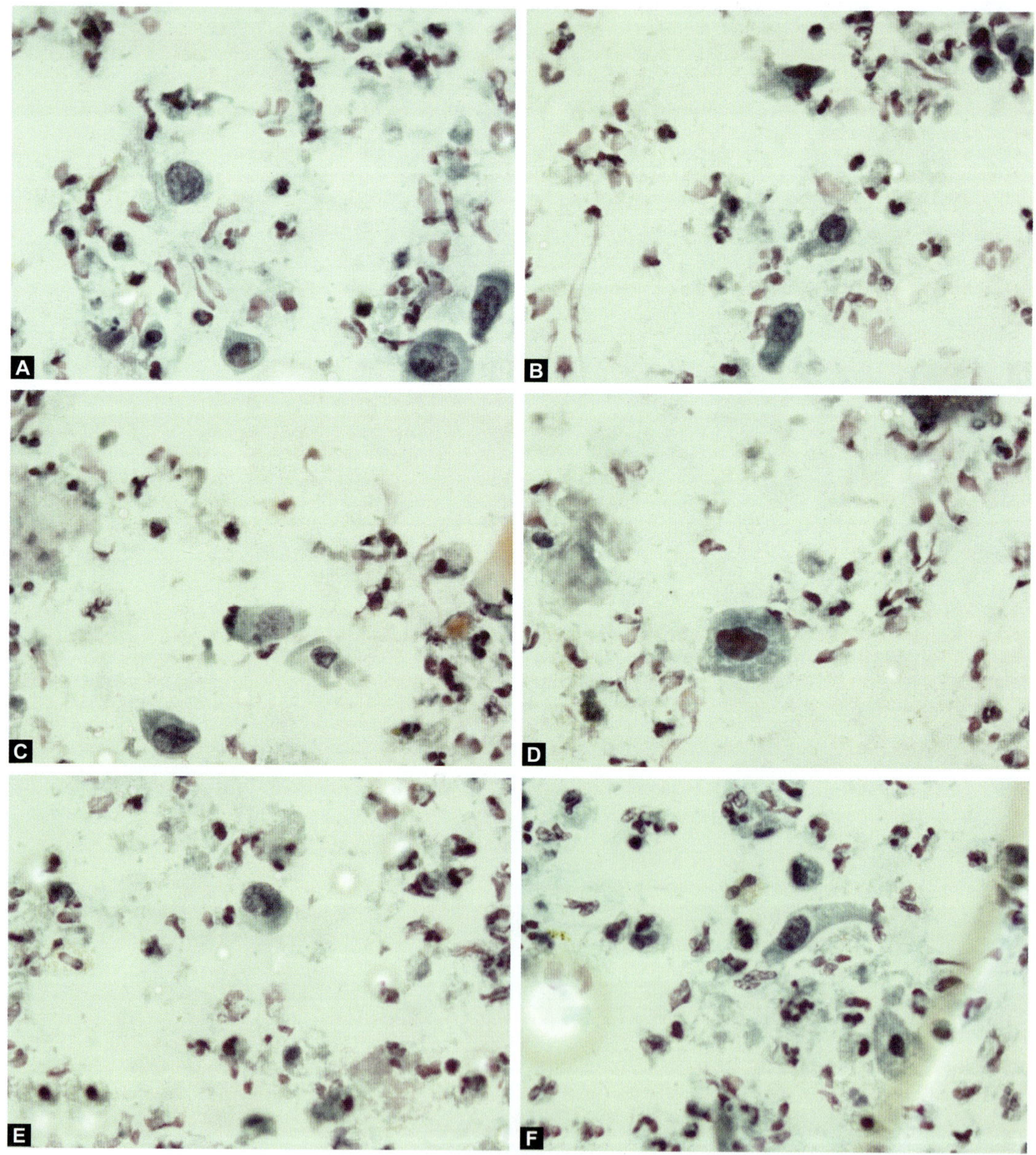

Figures 6.6A to F

Problem 6.7

A 50-year-old female presented with hematuria for one month. The most likely diagnosis from the microphotographs (Fig. 6.7) of urine cytology smear is:

A. Cystitis
B. Low-grade UCC
C. Atypia
D. High-grade UCC

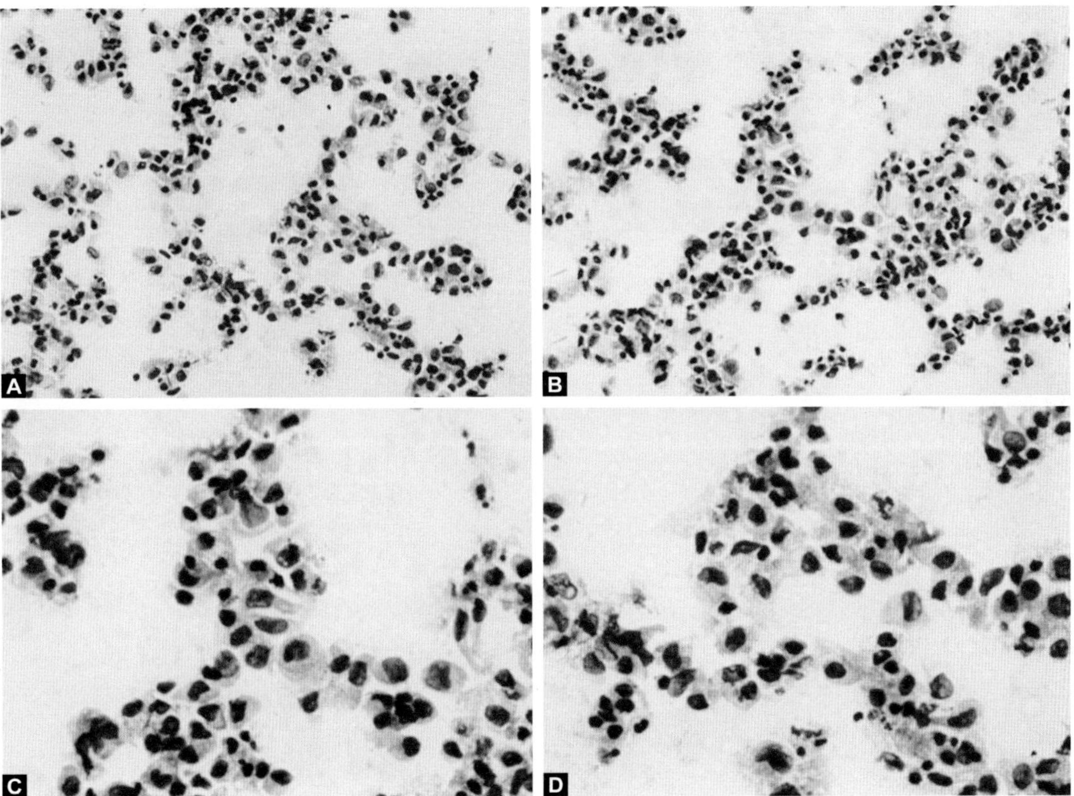

Figures 6.7A to D

Problem 6.8

A 58-year-old male had urothelial carcinoma. The patient is treated by BCG therapy. Urine cytology was done. The most likely diagnosis from Figure 6.8 is:

A. UCC
B. BCG induced changes
C. Cystitis
D. Atypia suggestive of urothelial cell carcinoma

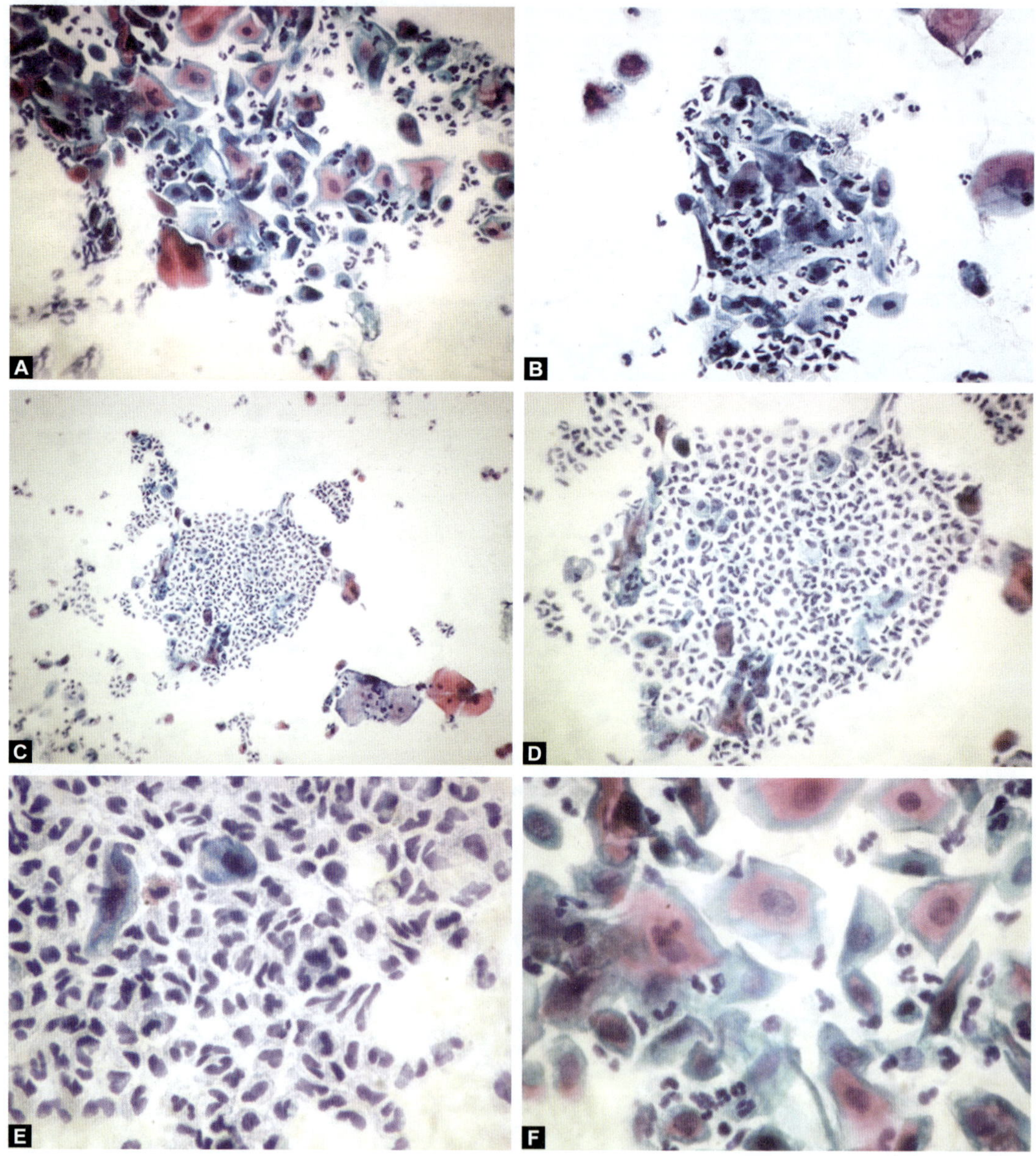

Figures 6.8A to F

Problem 6.9

A 36-year-old male presented with painless hematuria. Routine urine examination was done. What is the most likely diagnosis from these microphotographs (Fig. 6.9)?

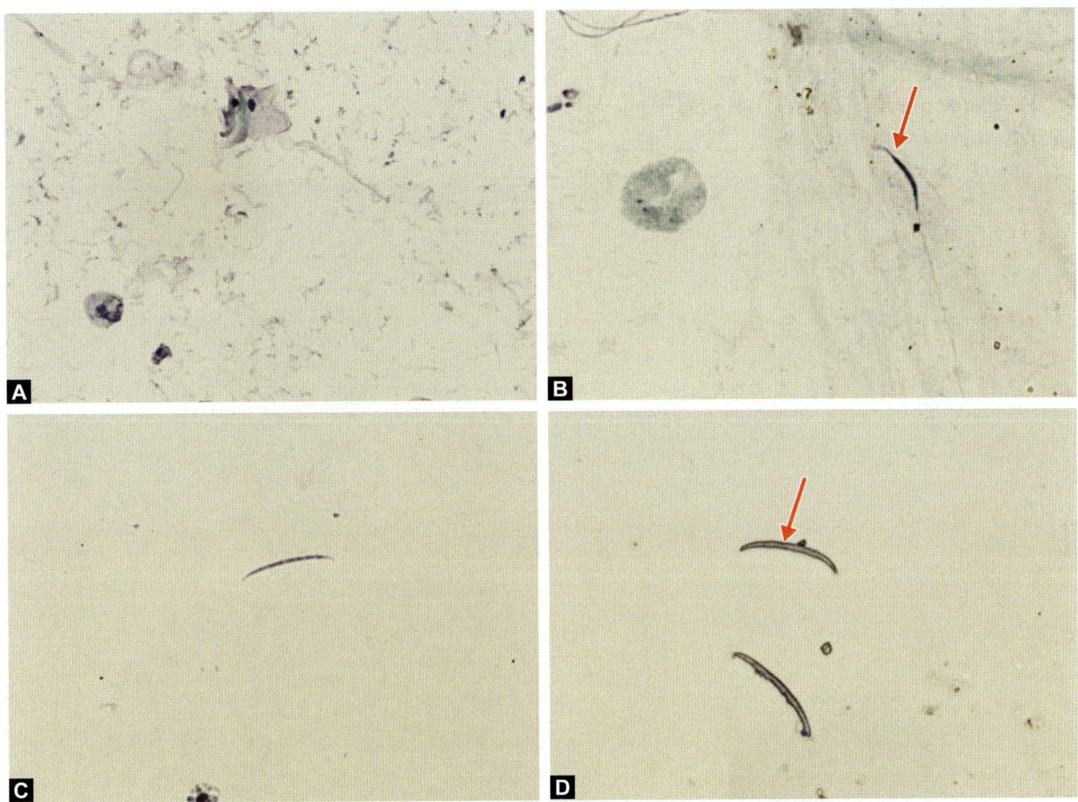

Figures 6.9A to D

Problem 6.10

A 57-year-old male was treated for urothelial cell carcinoma of urinary bladder. The patient received Bacille Calmette-Guerin (BCG) therapy. Routine urine cytology was done. The most likely diagnosis from the Figure 6.10 is:

A. Cystitis
B. BCG induced changes
C. UCC, recurrence
D. Atypia possibly reactive

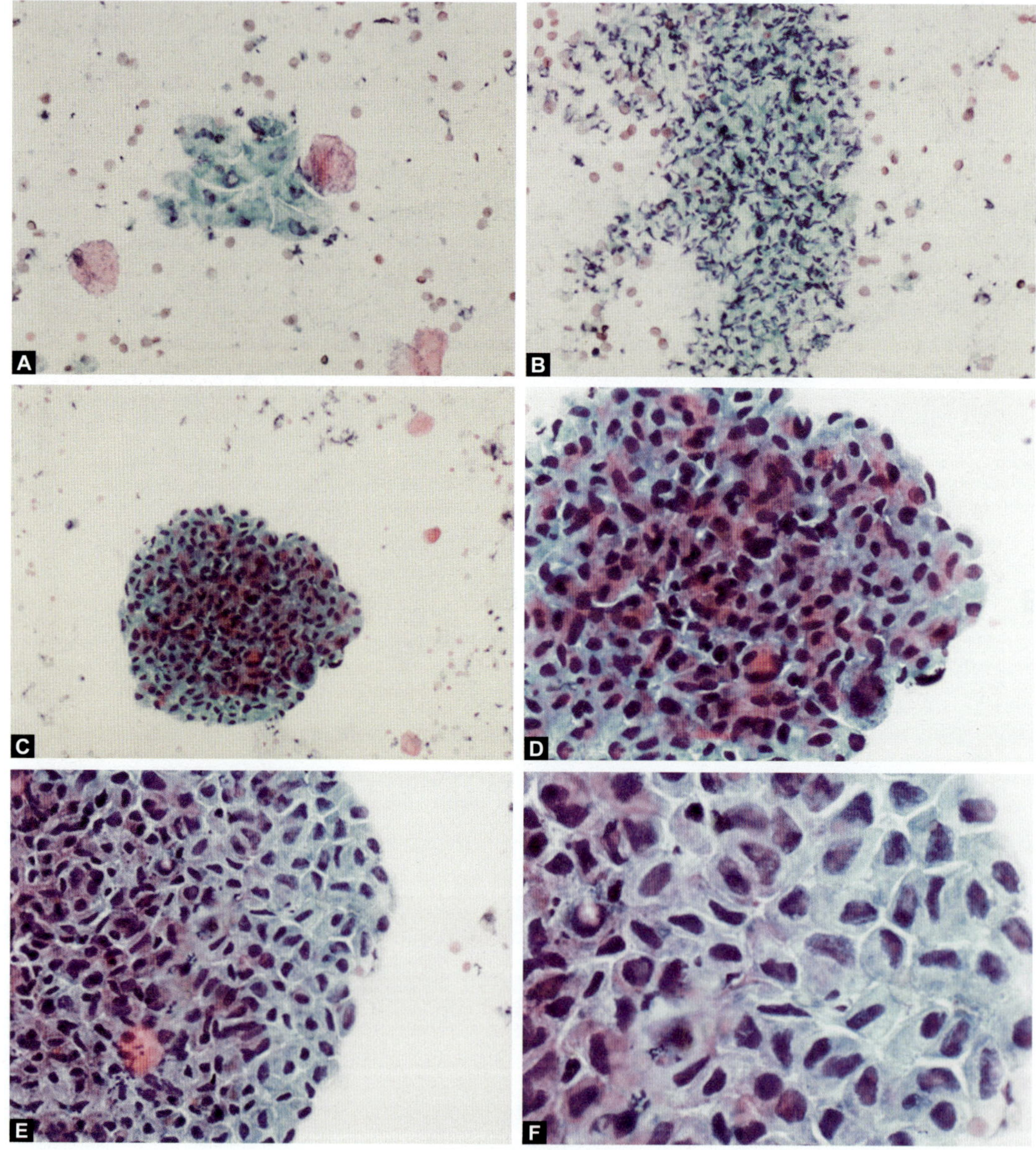

Figures 6.10A to F

Problem 6.11

A 45-year-old male had multiple retroperitoneal and cervical lymph nodes. Fine needle aspiration cytology of the retroperitoneal lymph node was done and microphotographs were shown (Figs 6.11A and B). The patient was treated accordingly. After one year the patient presented with hematuria. The urine cytological examination was done (Figs 6.11C to F). What is the most likely diagnosis?

A. Non-Hodgkin lymphoma
B. Low-grade UCC
C. High-grade UCC
D. Chronic inflammation

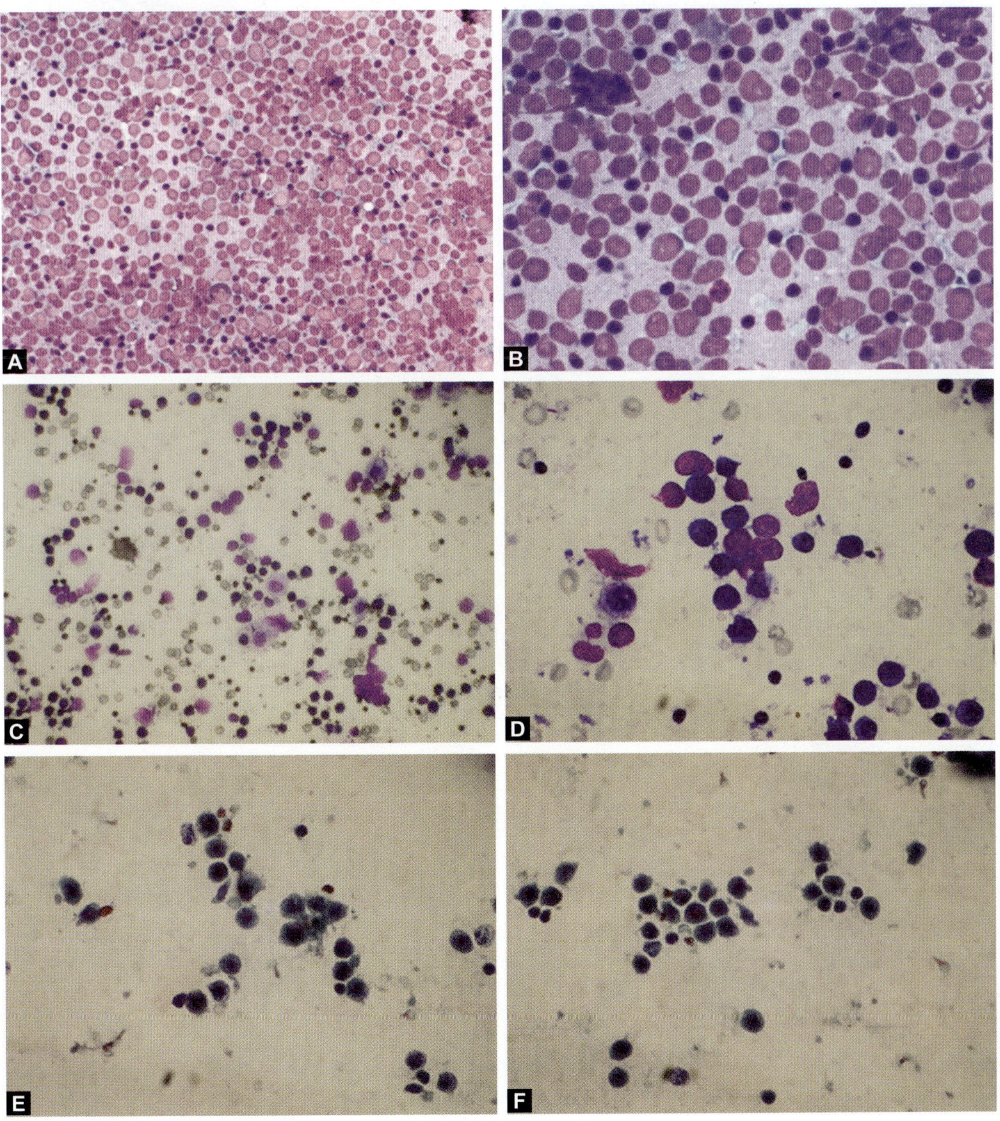

Figures 6.11A to F

Problem 6.12

A 57-year-old male presented with hematuria for 10 days. Urine cytology was done. The most likely diagnosis from Figure 6.12 is:

A. Urothelial cell carcinoma
B. Squamous cell carcinoma
C. Atypical urothelial cells
D. Normal urothelial cells

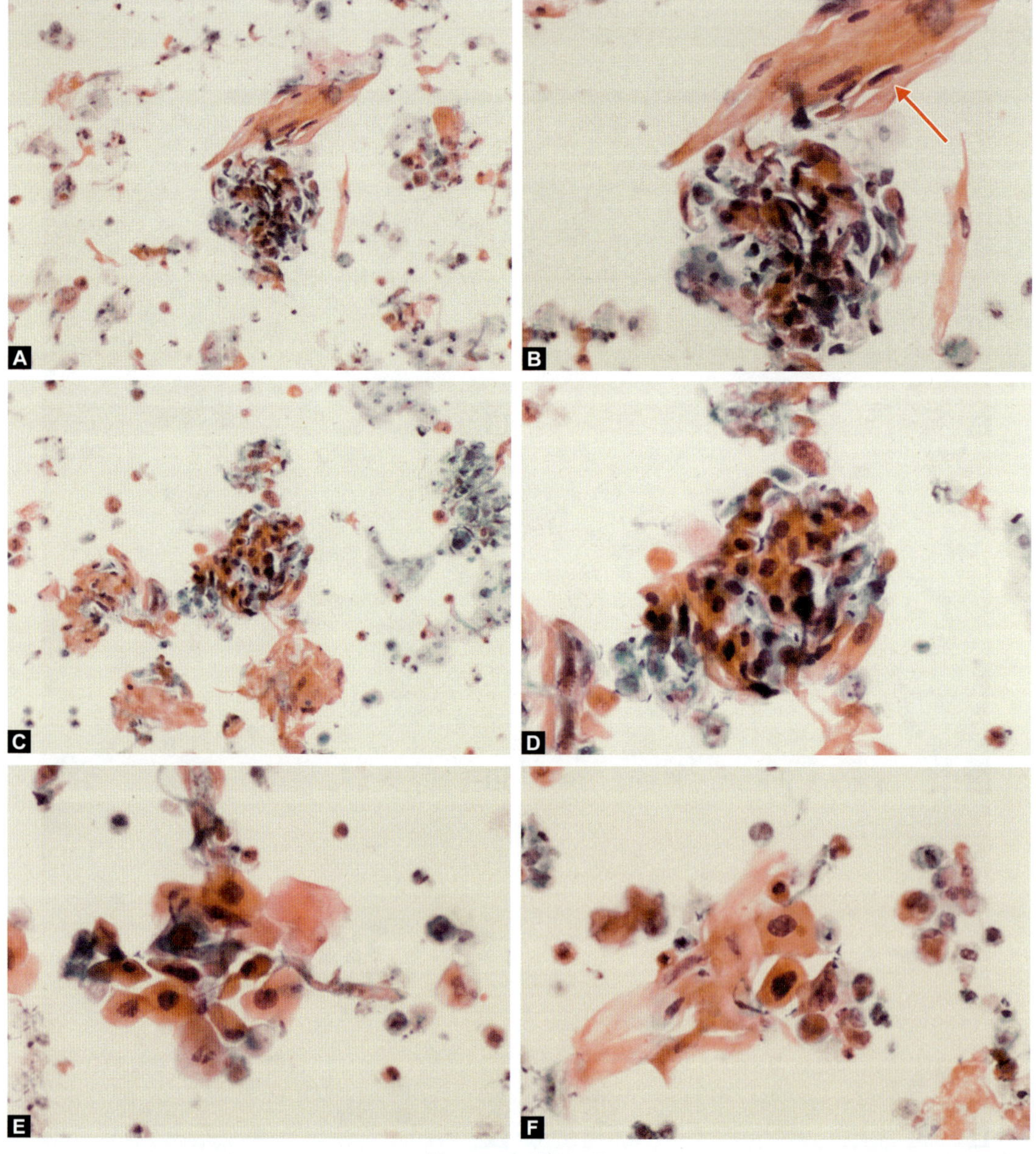

Figures 6.12A to F

Problem 6.13

A 57-year-old male presented with painless hematuria. The most likely diagnosis from the microphotograph (Fig. 6.13) of the urine cytology smear is:

A. Reactive atypia
B. High-grade UCC
C. Low-grade UCC
D. Cystitis

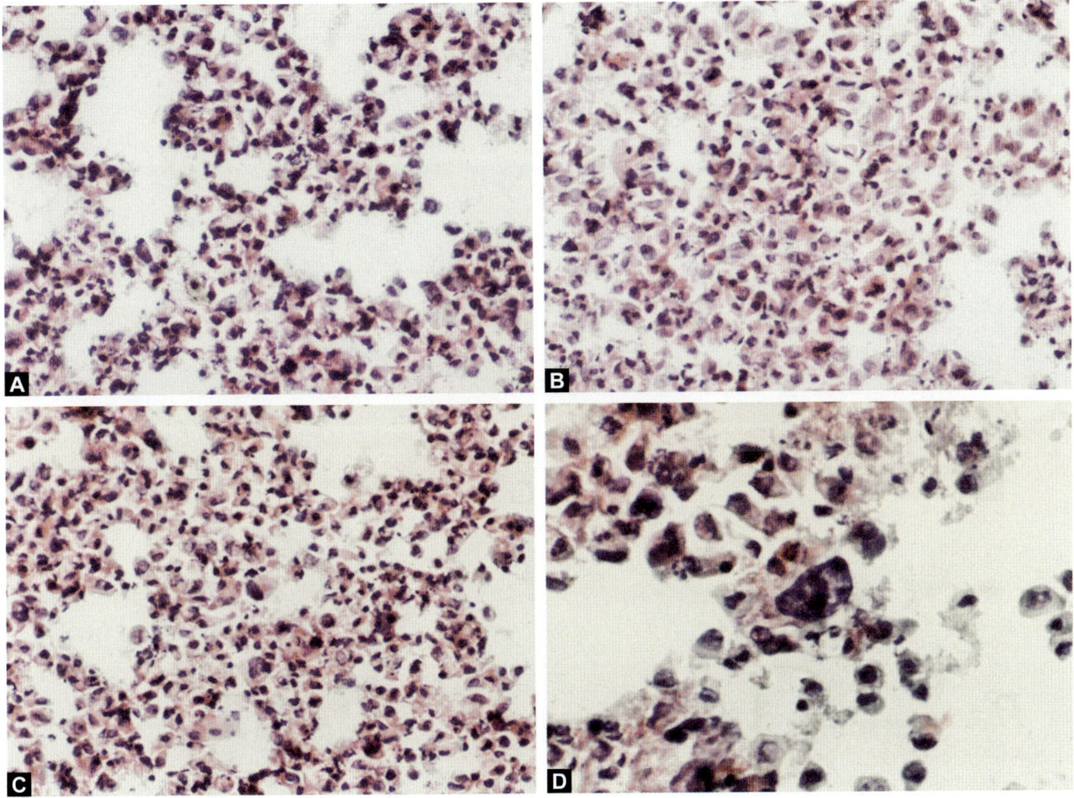

Figures 6.13A to D

Problem 6.14

A 75-year-old man presented with painless, gross hematuria without any history of dysuria or features of lower urinary tract syndrome. The digital rectal examination revealed a mildly enlarged prostate. The serum prostate specific antigen (PSA) level was borderline elevated (5.9 ng/mL). Ultrasonography revealed normal bilateral kidneys and focal thickening in the posterior wall of urinary bladder. Routine urine cytology followed by transurethral resection of bladder swelling was done. The most likely diagnosis from the microphotographs of the urine cytology smear (Figs 6.14A to F) and histology (Figs 6.14G to L) of the bladder is:

A. Urothelial cell carcinoma
B. Urothelial cell carcinoma with villoglandular differentiation
C. Inverted papilloma
D. Cystitis

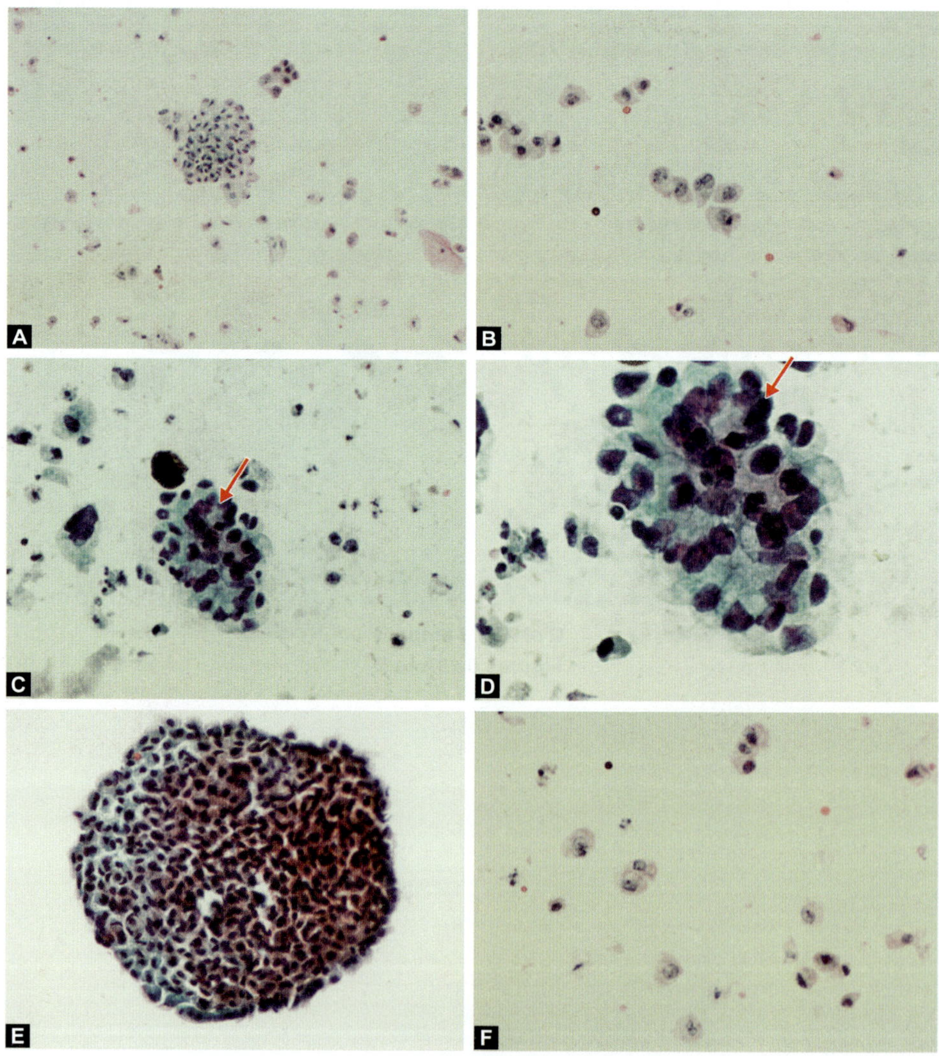

Figures 6.14A to F

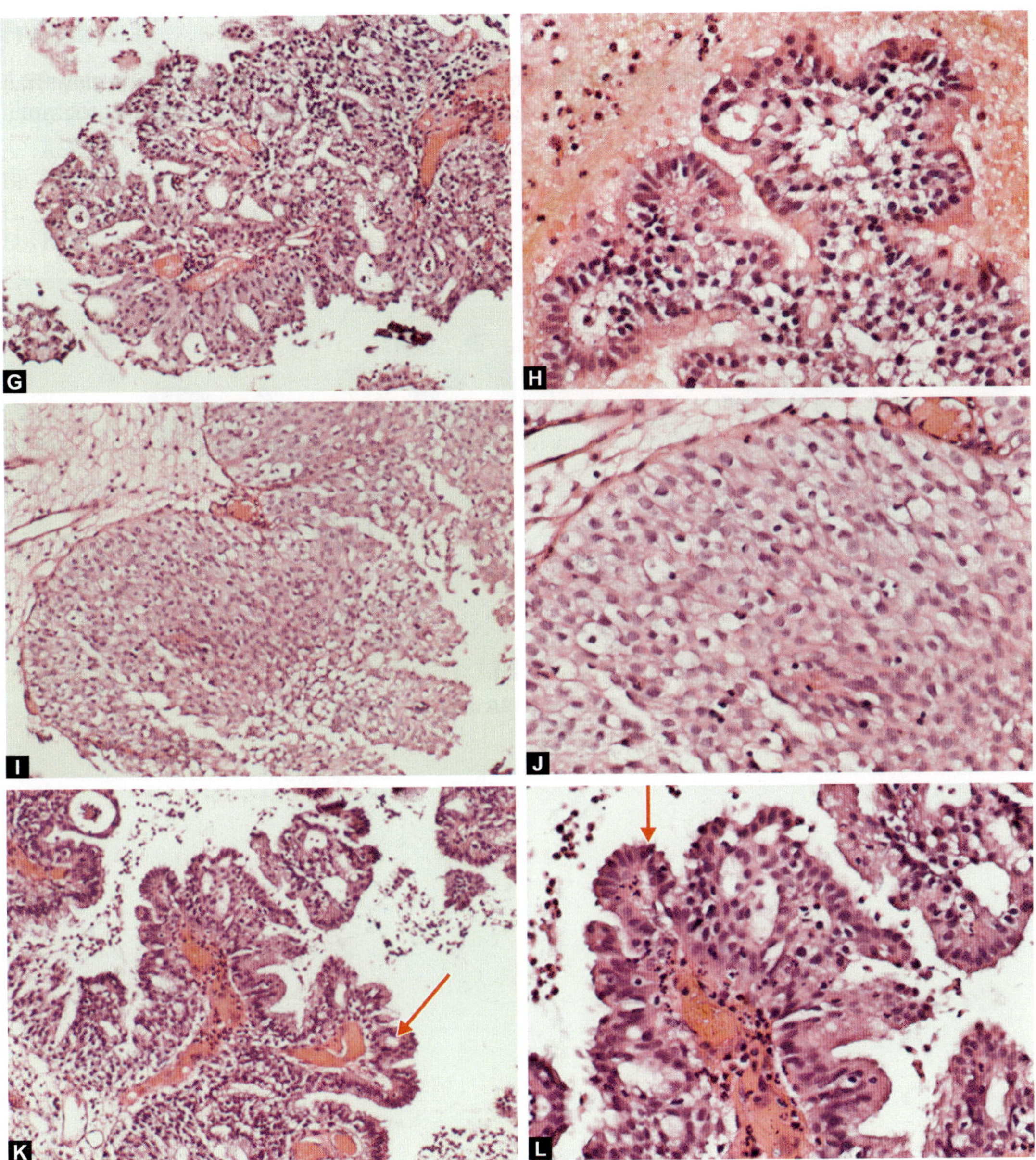

Figures 6.14G to L

Problem 6.15

A 43-year-old male was operated for rectal carcinoma 2 year back. Presently the patient has painless hematuria for 15 days. The urine cytology of the patient was done. The microphotographs of the smear are shown in Figure 6.15. The most likely diagnosis is:

A. Urothelial cell carcinoma
B. Stone in the bladder
C. No malignant cells
D. Metastatic adenocarcinoma

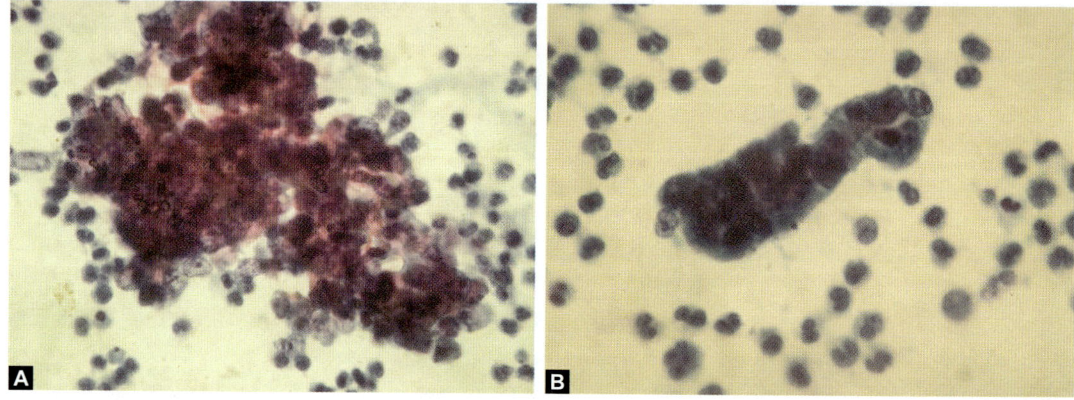

Figures 6.15A and B

Problem 6.16

A 56-year-old male presented with painless hematuria for 2 months. The patient was previously operated for leiomyosarcoma of prostate 3 years back. The urine cytology was done. The most likely diagnosis from Figure 6.16 is?

A. Negative for malignancy
B. High-grade UCC
C. Low-grade UCC
D. Other: Spindle cell tumor possibly sarcoma

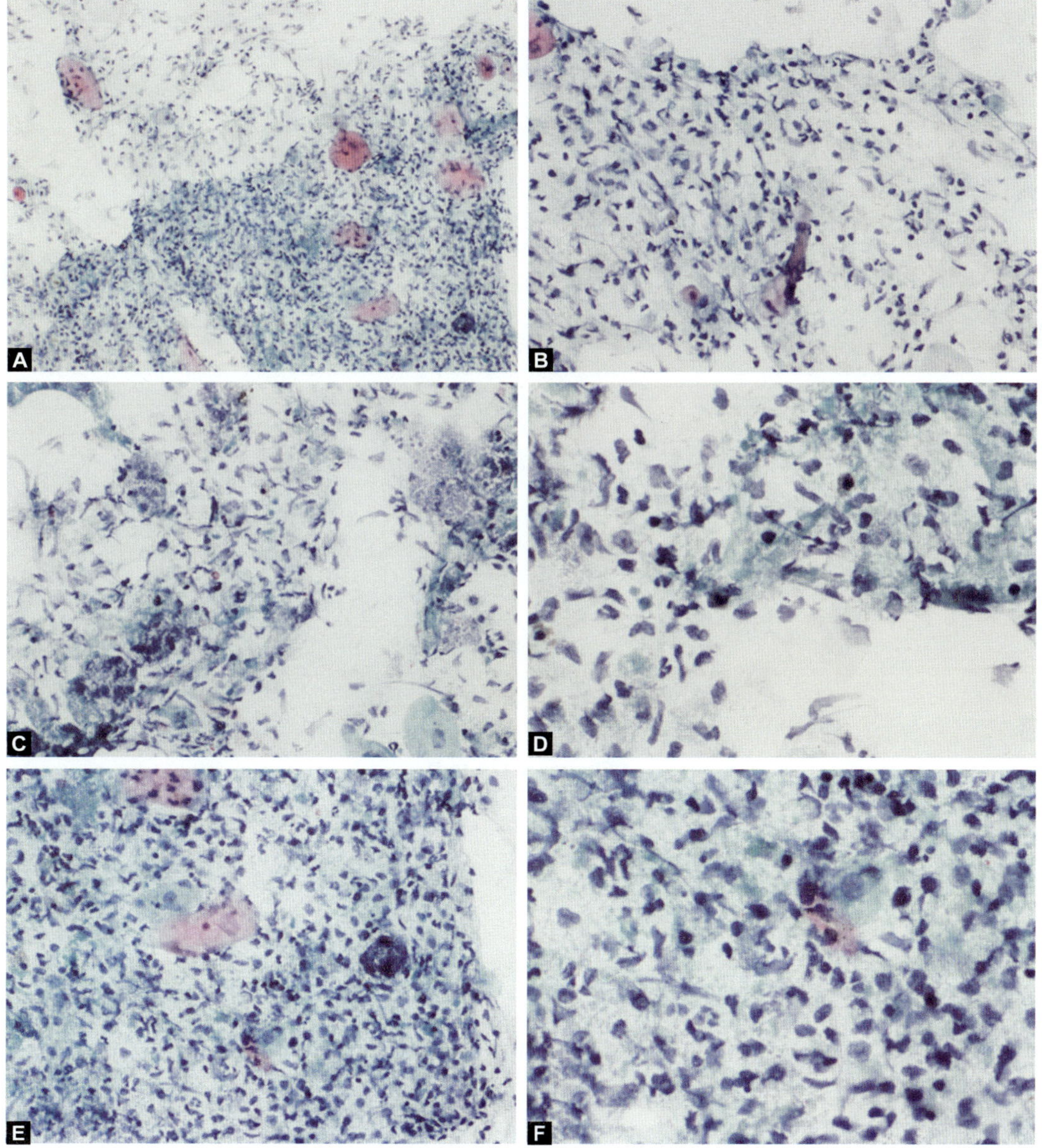

Figures 6.16A to F

Problem 6.17

A 65-year-old male presented with hematuria. Urine cytology examination was done. The most likely diagnosis from Figure 6.17 is:

A. Low-grade UCC
B. Small cell carcinoma
C. Reactive atypia
D. No evidence of any malignancy

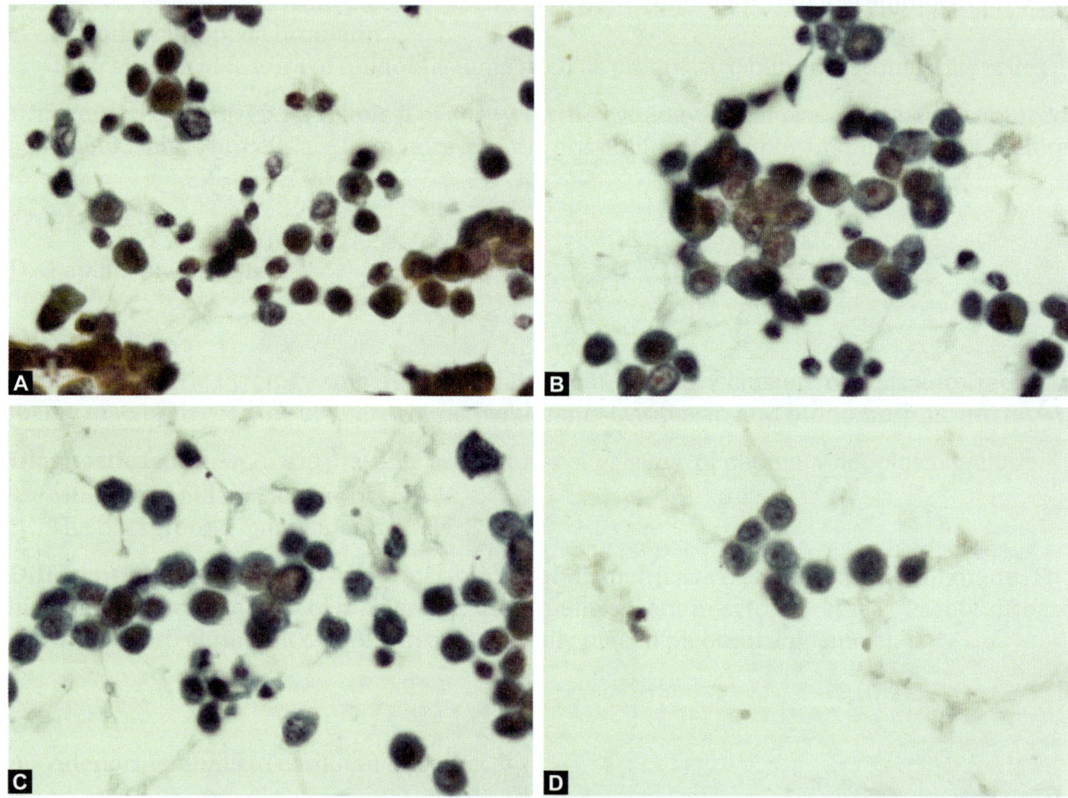

Figures 6.17A to D

Problem 6.18

A 56-year-old male presented with painless hematuria for 15 days. Routine urine cytology was done. The most likely diagnosis from Figure 6.18 is:

A. Atypical urothelial cell
B. Atypical urothelial cell suspicious for high-grade UCC
C. High-grade UCC
D. Adenocarcinoma of urinary bladder

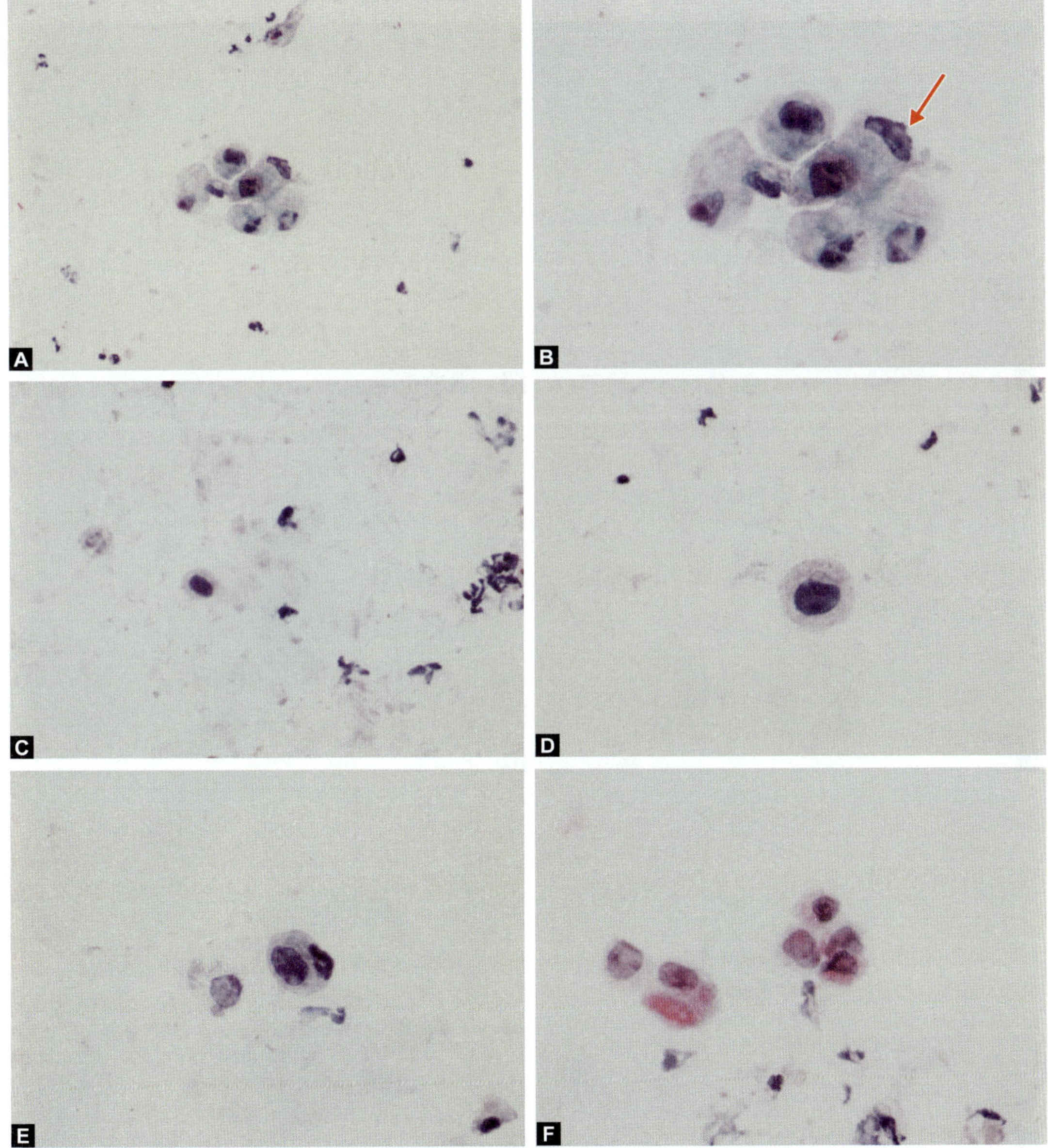

Figures 6.18A to F

Problem 6.19

A 67-year-old male had painless hematuria for 7 days. He had no other complaints. Microphotographs of the urine cytology smear are shown. The most likely diagnosis from Figure 6.19 is:

A. Atypia
B. Atypia suspicious for high-grade UCC
C. No abnormality
D. Low-grade UCC

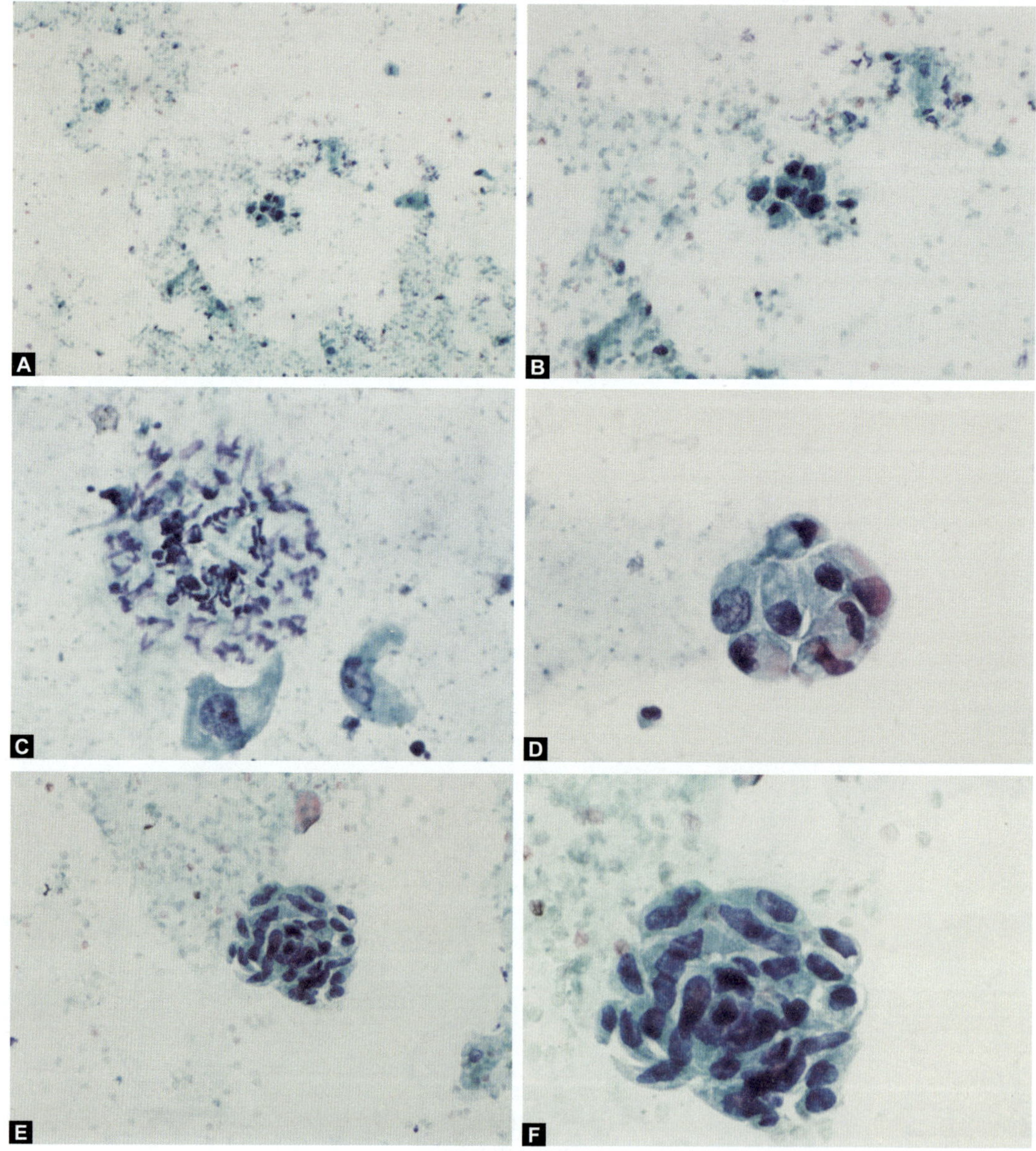

Figures 6.19A to F

Problem 6.20

A 52-year-old female with a known history of squamous cell carcinoma of vulva presented with hematuria. The most likely diagnosis from Figure 6.20 is:

A. Squamous cell carcinoma
B. Urothelial cell carcinoma
C. Atypia
D. Atypia possibly high-grade UCC

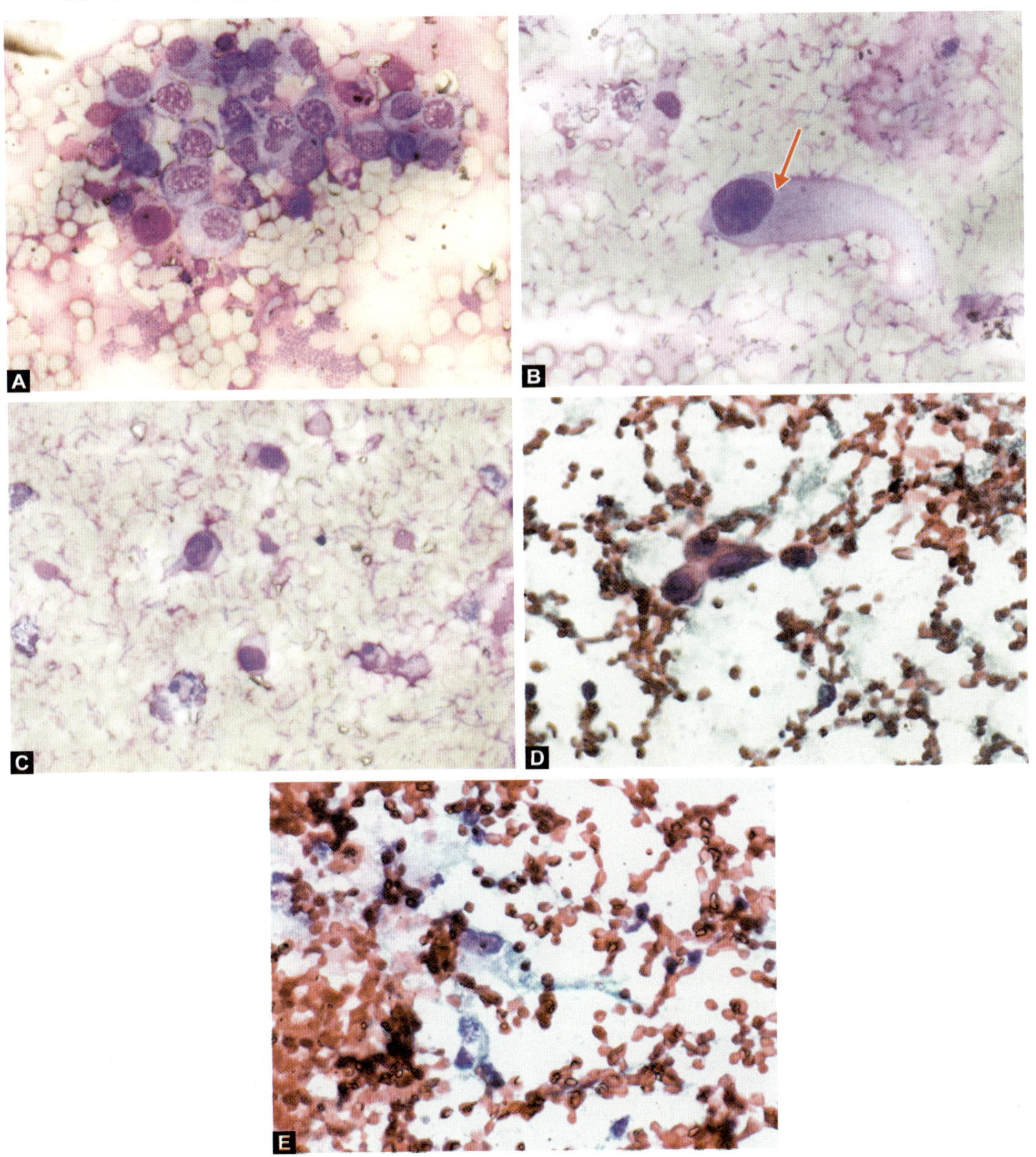

Figures 6.20A to E

Problem 6.21

A 67-year-old male presented with intermittent hematuria for 25 days: Urine cytological examination was done. The most likely diagnosis from Figure 6.21 is:

A. Atypia
B. Atypia: High-grade UCC cannot be excluded
C. High-grade UCC
D. Seminal vesicle cells

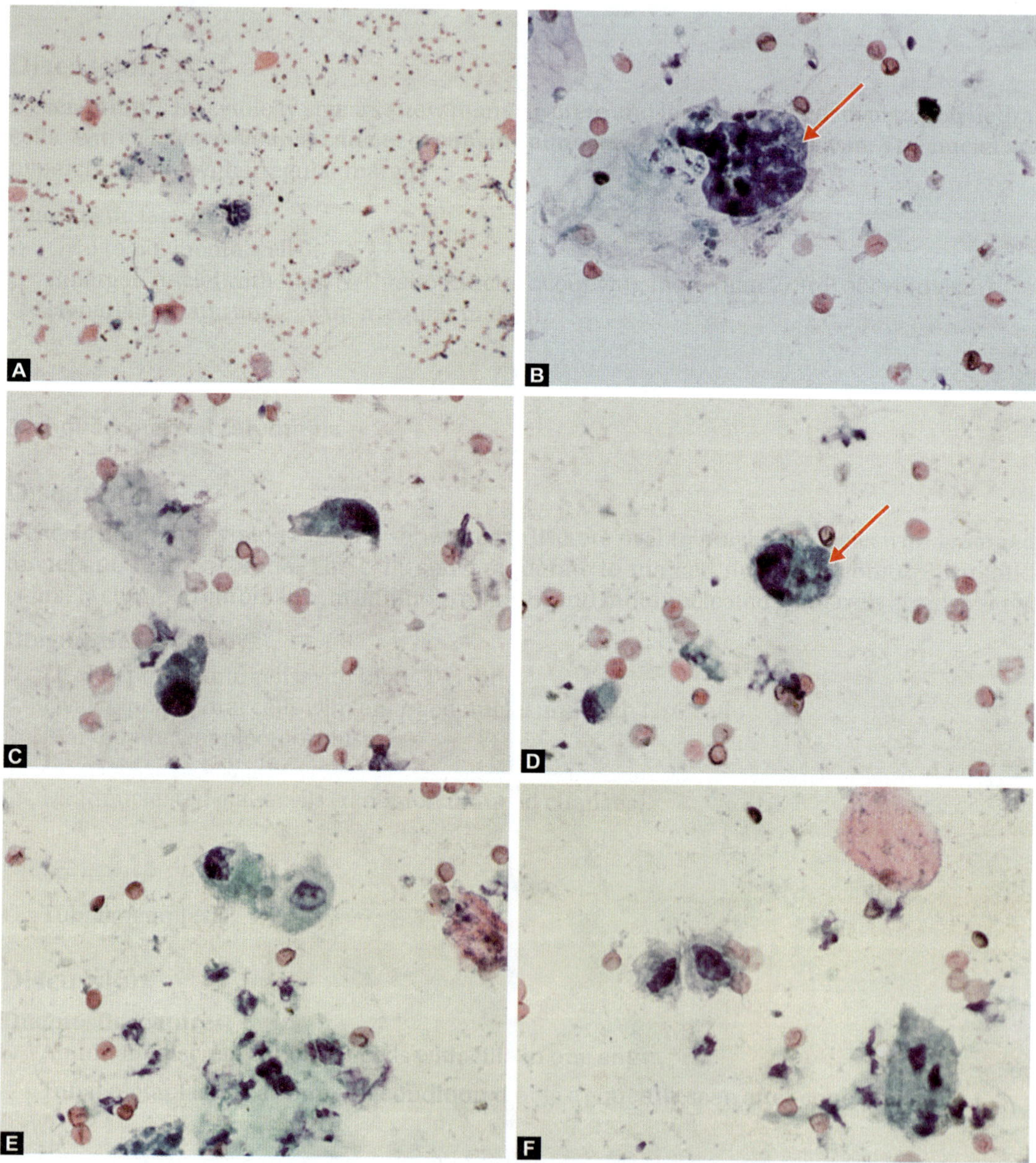

Figures 6.21A to F

Problem 6.22

A 56-year-old female was operated for uterine leiomyosarcoma 20 year back. She had total abdominal hysterectomy with bilateral salpingo-oophorectomy followed by radiation therapy. Now the patient presented with hematuria for 15 days. Urine cytological examination was done. The most likely diagnosis from Figure 6.22 is:

A. Negative for malignancy
B. Low-grade UCC
C. High-grade UCC
D. Consistent with sarcoma

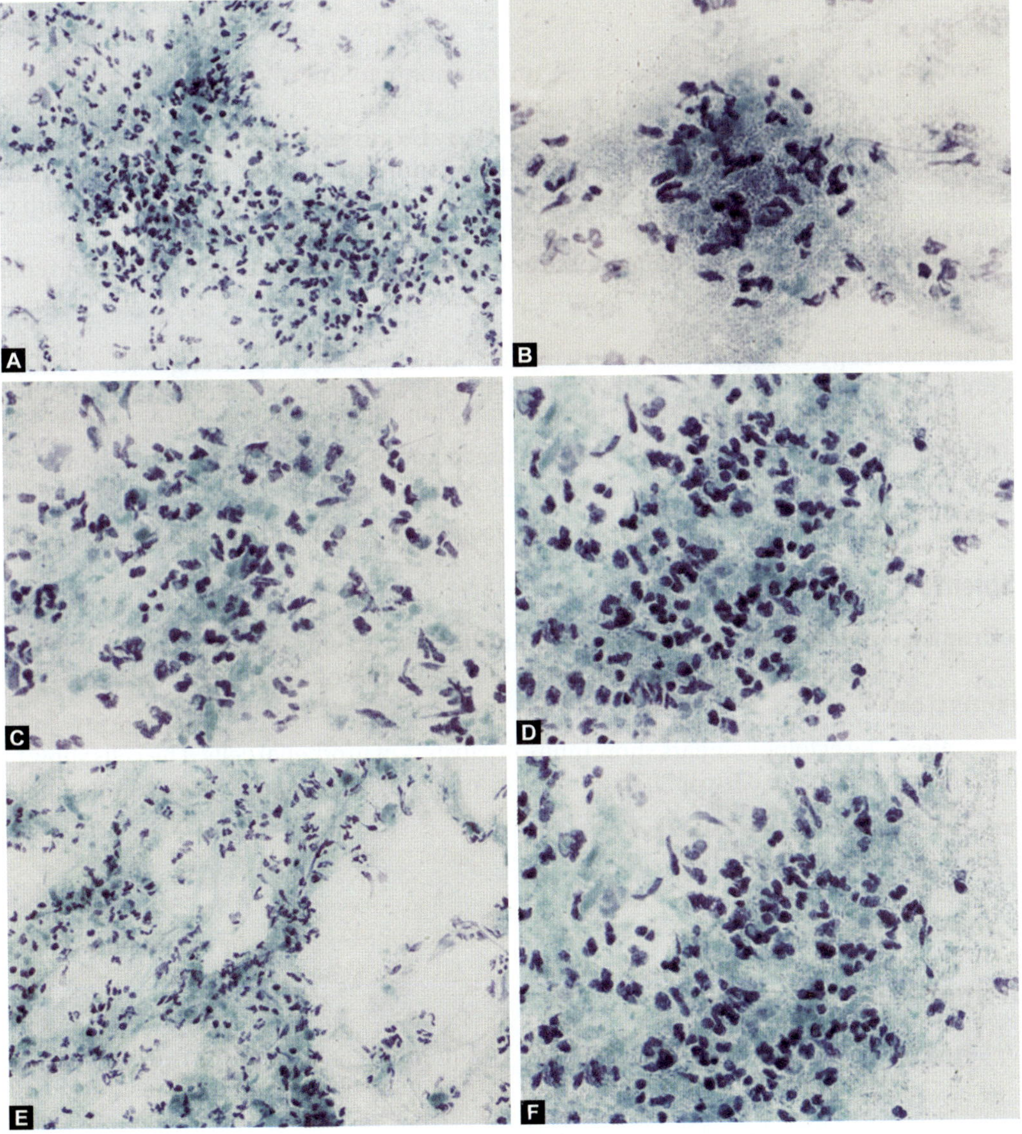

Figures 6.22A to F

ANSWERS

Problem 6.1

A.	4. Seminal vesicle cells
B.	2. Umbrella cells
C.	1. Urothelial cell
D.	3. Chemotherapy changes
E.	5. Atypical urothelial cell

Discussion

A. **Seminal vesicle cells:** These cells are uncommonly found in the urine cytology smear. The cells are large with enlarged hyperchromatic nuclei. The cells often contain lipofuscin pigment and sperm. The seminal vesicle cell may be mistaken as malignant cell.

B. **Umbrella cells:** The umbrella cells are large round to oval in shape with abundant eosinophilic finely granular cytoplasm. The cells contain single to multiple nuclei with very low N/C ratio. The chromatin is granular with inconspicuous nucleoli.

C. **Urothelial cells:** These are columnar cells with moderate to abundant cytoplasm and central to eccentric nuclei. Nuclei show normal non descriptive chromatin. N/C ratio of the cells are low.

D. **Chemotherapy changes:** The cells with chemotherapy induced changes show enlarged nuclei and cytoplasm. The N/C ratio of the cells are low. The nuclei often show degenerative changes.

E. **Atypical urothelial cell:** The cells show enlarged nuclei and N/C ratio is more than 0.5. The cells show mild hyperchromasia, however the cells cannot be classified as malignant cell on morphological ground.

Problem 6.2

B. Urothelial cell atypia: Undetermined significance

Discussion

Description: The cytology smears show multiple clusters of urothelial cells. The cells show high nucleocytoplasmic ratio (0.5 to 0.7). The nuclei are mildly hyperchromatic with irregular margin.

Diagnostic criteria in this index case:

- Nucleocytoplasmic ratio 0.5 to 0.7
- Hyperchromatic nuclei
- Irregular nuclear margin.

Further Reading

1. Epstein JI, Amin MB, Reuter VE, Mostofi FK. The World Health Organization/International Society of Urological Pathology consensus classification of urothelial (transitional cell) neoplasms of the urinary bladder. Am J Surg Pathol 1998;22:1435-48.
2. Harnden P, Parkinson MC. Transitional cell carcinoma of the bladder: Diagnosis and prognosis. Curr Diagn Pathol. 1996;3:109-21.

Problem 6.3

B. Tuberculosis

Discussion

Description: The cytology smears show loose clusters of mildly atypical urothelial cells along with epithelioid cell granulomas (Figs 6.3B and C). The rod shaped acid fast organism is seen in the Ziehl-Neelsen stain.

Diagnostic features in the index case: In the index case the following features indicate the diagnosis of tuberculosis:

- Epithelioid cell granuloma
- Acid fast bacilli in the ZN stain.

Differential diagnosis: The index case shows urothelial cells with enlarged nuclei and hyperchromasia. The reactive urothelial cells may be present in the urine cytology smear in tuberculosis. The presence of epithelioid cell granuloma usually gives warning of tuberculosis.

Further Reading

1. Adhya AK, Dey P. Cytologic detection of urinary tract tuberculosis. Acta Cytol. 2010;54(4):653-4.

Problem 6.4

C. Low grade UCC

Discussion

Description: The cytology smears show multiple loose clusters of urothelial cells. The cells show moderately enlarged, monomorphic nuclei with high N/C ratio. Nuclear margin is regular with inconspicuous nucleoli.

Diagnostic features: In the index case the features suggestive of low grade urothelial cell carcinoma (UCC) include:

- Predominantly cohesive clusters of cells
- Cells with moderately enlarged nuclei
- High N/C ratio
- Monomorphic to mildly pleomorphic nucleoli
- Small inconspicuous nucleoli
- Moderate hyperchromasia.

Differential diagnosis: The possibility of atypical urothelial cells is not considered as the cells show moderate hyperchromasia and pleomorphism. The clean background and predominant clusters of cells exclude high-grade UCC.

Problem 6.5

C. Reactive atypia only

Discussion

Description: The cytology smears show discrete and loose clusters of urothelial cells. The cells show mild nuclear enlargement and high nucleocytoplasmic ratio (0.5 to 0.7). The cells have mild hyperchromatic nuclei with regular membrane.

Diagnostic criteria:

- ❑ High nucleocytoplasmic ratio (0.5 to 0.7)
- ❑ Mild hyperchromatic nuclei with regular membrane.

Problem 6.6

B. Low-grade UCC

Discussion

Description: The cytology smears show discrete atypical urothelial cells with mildly enlarged hyperchromatic nuclei and increased N/C ratio. Background shows many polymorphs.

Diagnostic criteria: In the index case the features suggestive of low-grade urothelial cell carcinoma (UCC) include:

- ❑ Occasional discrete atypical urothelial cells
- ❑ Cells with moderately enlarged nuclei
- ❑ High N/C ratio
- ❑ Mild to moderately pleomorphic nuclei
- ❑ Absent nucleoli
- ❑ Moderate hyperchromasia.

The smear also shows many polymorphs. This may be due to secondary infection.

Problem 6.7

D. High-grade UCC

Discussion

Description: The cytology smears show abundant discrete urothelial cells. The cells show enlarged nuclei with high N/C ratio. The nuclei are hyperchromatic, moderately pleomorphic with irregular nuclear margin. The nuclear chromatin is coarse with inconspicuous nucleoli.

Diagnostic features: In this present case the following features are suggestive of high-grade UCC:

- ❑ High cellularity
- ❑ Predominantly discrete cells
- ❑ Moderately pleomorphic nuclei with marked nuclear hyperchromasia
- ❑ High N/C ratio
- ❑ Irregular nuclear membrane
- ❑ Coarse granular chromatin.

Differential diagnosis: The diagnosis of cystitis and atypical urothelial cell are not considered as these cells show significant nuclear pleomorphism and hyperchromasia. The large numbers of discrete atypical cells eliminate the possibility of low-grade urothelial cell carcinoma.

Problem 6.8

B. BCG induced changes

Discussion

Description: The cytology smears show many scattered urothelial cells with mild nuclear atypia. The background shows multiple epithelioid cell granulomas along with polymorphs and lymphocytes.

Diagnostic features in the index case: BCG is currently used as chemotherapeutic agents for bladder cancer. In this index case the smears show following changes of BCG treatment:

- Polymorphs, lymphocytes and degenerated urothelial cells
- Epithelioid cell granulomas
- Atypical transitional cells: Nuclear enlargement, hyperchromasia
- No change in N/C ratio

Differential diagnosis: The possibility of urothelial cell carcinoma is not considered in this index case as the cells show both nuclear and cytoplasmic enlargement and N/C ratio is not much altered.

Further Reading

1. Takashi M, Schenck U, Koshikawa T, Nakashima N, Ohshima S. Cytological changes induced by intravesical Bacillus Calmette-Guérin therapy for superficial bladder cancer. Urol Int. 2000;64(2):74-81.

Problem 6.9

Urinary Fusariosis

Discussion

In this index case many sickle shaped organisms consistent with morphology of *Fusarium* species are seen. Fusariosis infection is now increasingly seen in human. These are the saprophytic organisms and predominantly infect the plant. The three types of *Fusarium* species are seen: *F. solani, F. moniliforme* and *F. oxysporum*. In human, *F. solani* is the predominant organism. Fusariosis in the urine is rarely encountered. It has been reported in immunocompromised patient and in renal stone.[1,2]

References

1. Cheng-Chuan Su, Hui-Jine Hsu, Jiunn-Jong Wu, Chien-Wen Chou. Diagnosis of fusariosis in urine cytology. J Clin Pathol. 2007;60:422-4.
2. Su H, Zhang Q, Li L, Zhao Y, Zhu J, Zhu M. Urinary tract infection (UTI) caused by *Fusarium* proliferatum in an agranulocytosis Patient and a Review of Published Reports. Mycopathologia. 2015;7. [Epub ahead of print].

Problem 6.10

C. UCC, recurrence

Discussion

Description: The cytology smears show ill formed epithelioid cell granulomas, inflammatory cells and many clusters of urothelial cells. The cells show moderate nuclear enlargement, high N/C ratio, nuclear pleomorphism and hyperchromasia.

Diagnostic features: In the index case the smears show polymorphs, lymphocytes and ill formed epithelioid cell granulomas. In addition, the clusters of cells show moderate nuclear enlargement, high N/C ratio, nuclear pleomorphism and hyperchromasia. All these features indicate UCC, recurrence. The nuclear atypia in the index case is significant and therefore the possibility of reactive atypia is excluded.

Problem 6.11

A. Non-Hodgkin lymphoma

Discussion

Description: The FNAC smears show abundant discrete monomorphic immature lymphoid cells with background lymphoglandular bodies (Figs 6.11A and B). The urine cytology smears show monomorphic lymphoid cells of similar morphology (Fig. 6.11C to F). The cells have scanty cytoplasm, round nuclei with fine nuclear chromatin.

Diagnostic features in the index case: In this index case the FNAC of the retroperitoneal lymph node shows abundant monomorphic round lymphoid cells with scanty cytoplasm and fine nuclear chromatin. The features are that of NHL. Subsequently the patient had hematuria. The smear shows many monomorphic lymphoid cells in the cytology smear of urine. The individual cells are round with scanty cytoplasm and monomorphic nuclei having fine nuclear chromatin. The features are consistent with Non-Hodgkin lymphoma.

Differential diagnosis: The cytomorphology of urothelial cell carcinoma usually shows clusters and discrete cells. The cells show mild to moderate amount of cytoplasm. The nuclei of the cells are pleomorphic. The cells in the smear are larger than mature lymphocytes and chromatin is fine. Therefore the possibility of chronic inflammation is not considered. The index case is a secondary involvement of NHL and with history it is relatively easy to diagnosis such case. However, in case of primary NHL, the detailed immunochemistry is needed for the confirmation of NHL.

Problem 6.12

B. Squamous cell carcinoma

Discussion

Description: The cytology smears show many clusters and discrete cells with polyhedral in appearance. The cells show central hyperchromatic enlarged moderately pleomorphic nuclei. Occasional fiber like cells are also seen (Fig. 6.12B).

Diagnostic features: In the index case following features suggest the diagnosis of squamous cell carcinoma:

- Discrete and clusters of malignant cells
- The cells with polyhedral appearance and centrally placed nuclei
- Hyperchromatic nuclei with moderate pleomorphism
- Cytoplasmic orangeophilia indicating squamoid differentiation.

Differential diagnosis: UCC usually are round cells and cytoplasm do not show any evidence of keratinization. The cells in this case are too pleomorphic and nuclei are significantly atypical to label them only atypical cells.

Problem 6.13

B. High-grade UCC

Discussion

Description: The cytology smears show abundant discrete urothelial cells with moderately enlarged nuclei having high N/C ratio. The nuclei are hyperchromatic, pleomorphic with irregular margin and clumped chromatin.

Diagnostic features: In the index the features suggestive of high-grade urothelial cell carcinoma include:

- Abundant dissociated tumor cells
- Moderate pleomorphic nuclei
- Moderate to marked nuclear hyperchromasia
- High N/C ratio
- Irregular nuclear membrane
- Coarse clumped chromatin.

Differential diagnosis: The large number of dissociated cells with marked nuclear abnormality eliminates the possibility of low-grade urothelial cell carcinoma in this index case.

Problem 6.14

B. Urothelial cell carcinoma with villoglandular differentiation

Discussion

Description: The urine cytology smears (Figs 6.14A to F) show urothelial cells in clusters and discretely. The gland like arrangement of cells is also seen. The individual cells are moderately pleomorphic with hyperchromatic nuclei. Occasional cells show moderate amount of vacuolated cytoplasm. The histopathology of the resected tumor shows urothelial carcinoma with villoglandular differentiation (Figs 6.14G to L). The cells are arranged in diffuse sheets, cribriform gland and also in papillae. The lining cells of papillae are cuboidal with moderate nuclear atypia.

Diagnostic features: In this index case, the following features suggest the diagnosis of urothelial cell carcinoma with villoglandular differentiation:

- Clusters and discrete urothelial cells showing features of UCC
- Typical glandular arrangement of cells
- Histopathology shows features of UCC with glandular arrangement.

Disease description: Urothelial cell carcinoma with villoglandular differentiation is an uncommon variant of UCC. The lesions are usually associated with high-grade UCC. The patients usually show poor prognosis due to coexistent high-grade UCC and detrusor muscle invasion. Primary or secondary adenocarcinoma of the bladder may show similar features. Histopathology is essential to identify such variant of UCC.

Further Reading

1. Lim M, Adsay NV, Grignon D, Osunkoya AO. Urothelial carcinoma with villoglandular differentiation: a study of 14 cases. Mod Pathol. 2009;22:1280-6.
2. Mitra S, Sundaram AK, Gupta K, Devana SK. Urothelial carcinoma with villoglandular differentiation: A Rare Entity. Int J Surg Pathol. 2015.
3. Mitra S, Chatterjee D, Gupta K, Radotra BD, Dey P. Urine cytology of urothelial carcinoma with villoglandular differentiation. Diagnostic cytopathology. (in press).

Problem 6.15

D. Metastatic adenocarcinoma

Discussion

Description: The cytology smears show tight and loose clusters of tumor cells. The cells are round with scanty cytoplasm and moderately pleomorphic vesicular nuclei.

Diagnostic features:

- Clinical history of rectal carcinoma
- Individual tumor cell morphology is consistent with adenocarcinoma.

Problem 6.16

D. Other: Spindle cell tumor possibly sarcoma

Discussion

Description: The cytology smears show abundant discrete round to oval cells. The individual cells show scanty cytoplasm and elongated moderately pleomorphic spindle shaped nuclei. The background shows many inflammatory cells.

Diagnostic features:

- History of leiomyosarcoma
- Abundant spindle cells
- Spindle shaped pleomorphic nuclei.

Differential diagnosis: The differential diagnosis of abundant spindle cells in the urine cytology consists of predominantly sarcomatoid changes in urothelial carcinoma, or renal cell carcinoma and metastatic sarcoma. Clinical history is very helpful in such cases. Spindle cell variant of urothelial carcinoma and renal cell carcinoma are positive for cytokeratin, whereas sarcoma is positive for vimentin and negative for cytokeratin.

Problem 6.17

B. Small cell carcinoma

Discussion

Description: The cytology smears show many discrete round cells. The individual cells show round nuclei with scanty cytoplasm and high N/C ratio. The nuclei are monomorphic with condensed chromatin and inconspicuous nucleoli.

Diagnostic features:

- Round cells
- Monomorphic nuclei
- Condensed chromatin.

Disease description: The small cell carcinoma of bladder is very rare. The tumor simulates small cell carcinoma of lung. The tumor is aggressive and rapidly spreads in other part of the body. The tumor cells are positive for neuron-specific enolase, chromogranin A and synaptophysin.

Further Reading

1. Swanson PE, Brooks R, Pearse H, Stenzel P. Small cell carcinoma of urinary bladder. Urology. 1988;32(6):558-63.
2. Blomjous CE, Thunnissen FB, Vos W, de Voogt HJ, Meijer CJ. Small cell neuroendocrine carcinoma of the urinary bladder. An immunohistochemical and ultrastructural evaluation of 3 cases with a review of the literature. Virchows Arch A Pathol Anat Histopathol. 1988;413(6):505-12.

Problem 6.18

D. Adenocarcinoma of urinary bladder

Discussion

Description: The cytology smears show loose clusters and discrete cells. The cells show moderate to abundant cytoplasm with central to peripherally pushed nuclei. The nuclei are moderately pleomorphic and hyperchromatic with occasional prominent nucleoli.

Diagnostic features:

- Cells with abundant cytoplasm
- Eccentric nuclei
- Hyperchromatic and pleomorphic nuclei
- Occasional prominent nucleoli.

Glandular component may be present in UCC. Primary adenocarcinoma of urinary bladder is rare and represents less than 2% of primary bladder malignancies.

Problem 6.19

B. Atypia suspicious for high-grade UCC

Discussion

Description: The cytology smears show multiple loose clusters of cells. The cells have high N/C ratio (more than 0.7). The nuclei are hyperchromatic with irregular nuclear margin.

Diagnostic features:

- N/C ratio more than 0.7
- Hyperchromatic nuclei
- Irregular nuclear margin.

Problem 6.20

A. Squamous cell carcinoma

Discussion

Description: The cytology smears show predominantly discrete and occasional loose clusters of malignant cells. The cells are polyhedral in appearance with central placed enlarged hyperchromatic nuclei.

Diagnostic features:

- Previous history of squamous cell carcinoma
- Polyhedral shaped cells
- Hyperchromatic enlarged nuclei
- Elongated tadpole like cells.

Problem 6.21

D. Seminal vesicle cells

Discussion

Description: The cytology smears show many discrete large cells with abundant cytoplasm. The nuclei are central to eccentrically placed, hyperchromatic and pleomorphic. The cytoplasm of the cells contain sperm head. The background also shows many spermatozoa.

Diagnostic features:

- Large cells
- Sperm head within the cell
- Background spermatozoa.

Differential diagnosis: Seminal vesicle cells may be mistaken as carcinoma. However, the N/C ratios of these cells are not altered. Malignant cells show high N/C ratio and hyperchromatic nuclei. The tumor cells do not contain any spermatozoa.

Problem 6.22

D. Consistent with sarcoma

Discussion

Description: The cytology smears show many loose clusters and discrete tumor cells admixed with polymorphs. The individual cells show spindle shaped nuclei with moderate pleomorphism.

Diagnostic features:

- Known case of leiomyosarcoma
- Spindle cells with moderate nuclear pleomorphism.

CHAPTER 7

Cerebrospinal Fluid

Introduction

Cerebrospinal fluid (CSF) is helpful in diagnosis of various infective conditions of leptomeninges, lymphoma, leukemia, metastatic tumors and rarely primary brain tumors.

Metastatic Solid Tumor

Adenocarcinoma from lung, breast and gastrointestinal tract are the common sources of metastatic carcinomas in CSF. Proper clinical history and histological type of primary tumor are very important. Occult cases of breast or gastrointestinal tract (GIT) tumor are very uncommon. However, occasionally pulmonary tumor may be small and asymptomatic. Metastatic adenocarcinoma of the lung is more common than small cell carcinoma and squamous cell carcinoma. Carcinoma of breast uncommonly metastasizes to CSF.

Leukemia

Leukemic infiltration of leptomeninges may occur without involving the brain parenchyma. The patients with leukemic infiltration may be asymptomatic. Therefore, the routine CSF examination is very useful for diagnosis of leukemic infiltration. Both acute lymphoblastic and myeloblastic leukemia may involve the leptomeninges. The morphology depends on the type of leukemia.

Lymphomas

Lymphoma predominantly involves leptomeninges without any parenchymal involvement. About 10% of non-Hodgkin lymphoma (NHL) involves central nervous system (CNS) at the time of first presentation. The common types of NHL in CSF are diffuse large B-cell lymphoma, Burkitt's and lymphoblastic lymphomas.

Primary CNS Lymphoma

Primary CNS lymphoma represents 10 to 15% of all primary CNS tumors. Diffuse large B-cell NHL is the most common type of primary CNS lymphoma.

Primary CNS Tumor

Medulloblastoma

This is a childhood tumor and may invade the CSF through ventricles. Cytology smears show small clusters and dissociated cells. The individual cells have scanty cytoplasm and round to oval hyperchromatic nuclei having prominent nucleoli. Nuclear molding may be seen. The cells are positive for synaptophysin and negative for glial fibrillary acidic protein (GFAP).

QUESTIONS

Problem 7.1

A 19-year-old boy presented with gum bleeding, pallor and high fever. Microphotographs of the routine CSF smear are shown in Figure 7.1. What is the most likely diagnosis in this case?

A. Viral infection
B. Acute leukemia
C. Lymphocyte only
D. Metastatic carcinoma

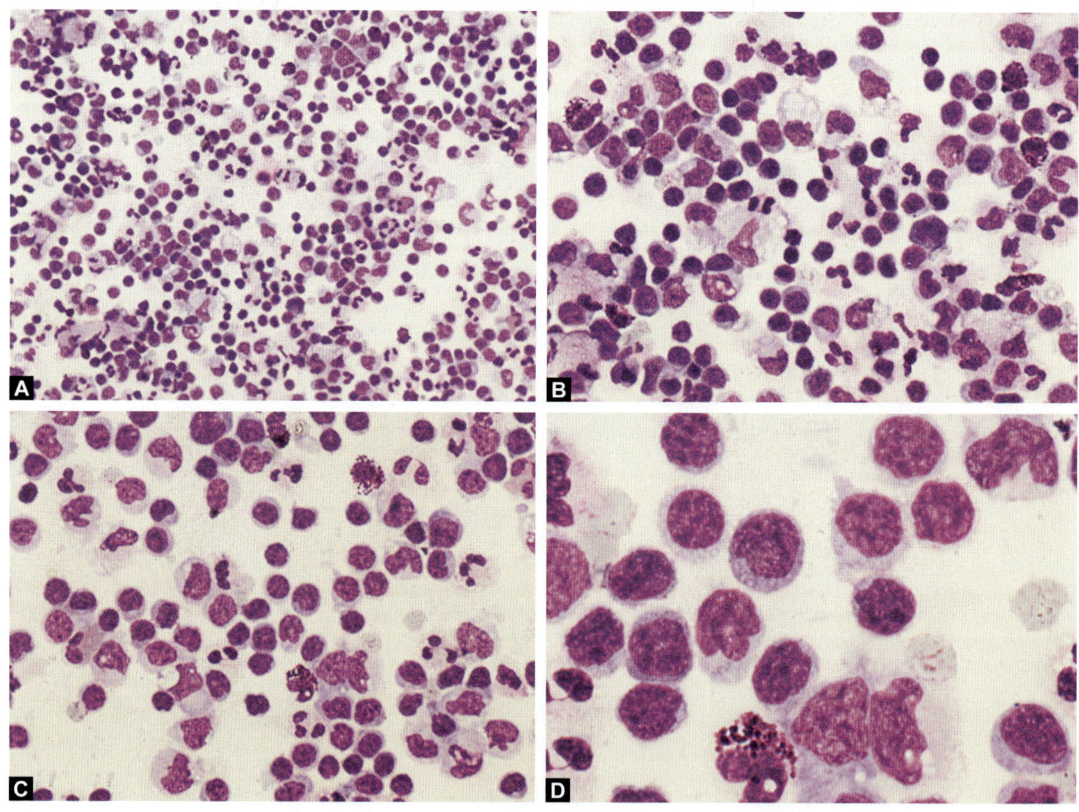

Figures 7.1A to D

Problem 7.2

A 28-year-old male presented with severe headache and stiff neck. The most likely diagnosis from CSF microphotographs (Fig. 7.2) is:

A. Cryptococci
B. Lymphocytosis
C. Lymphoma
D. Tuberculosis

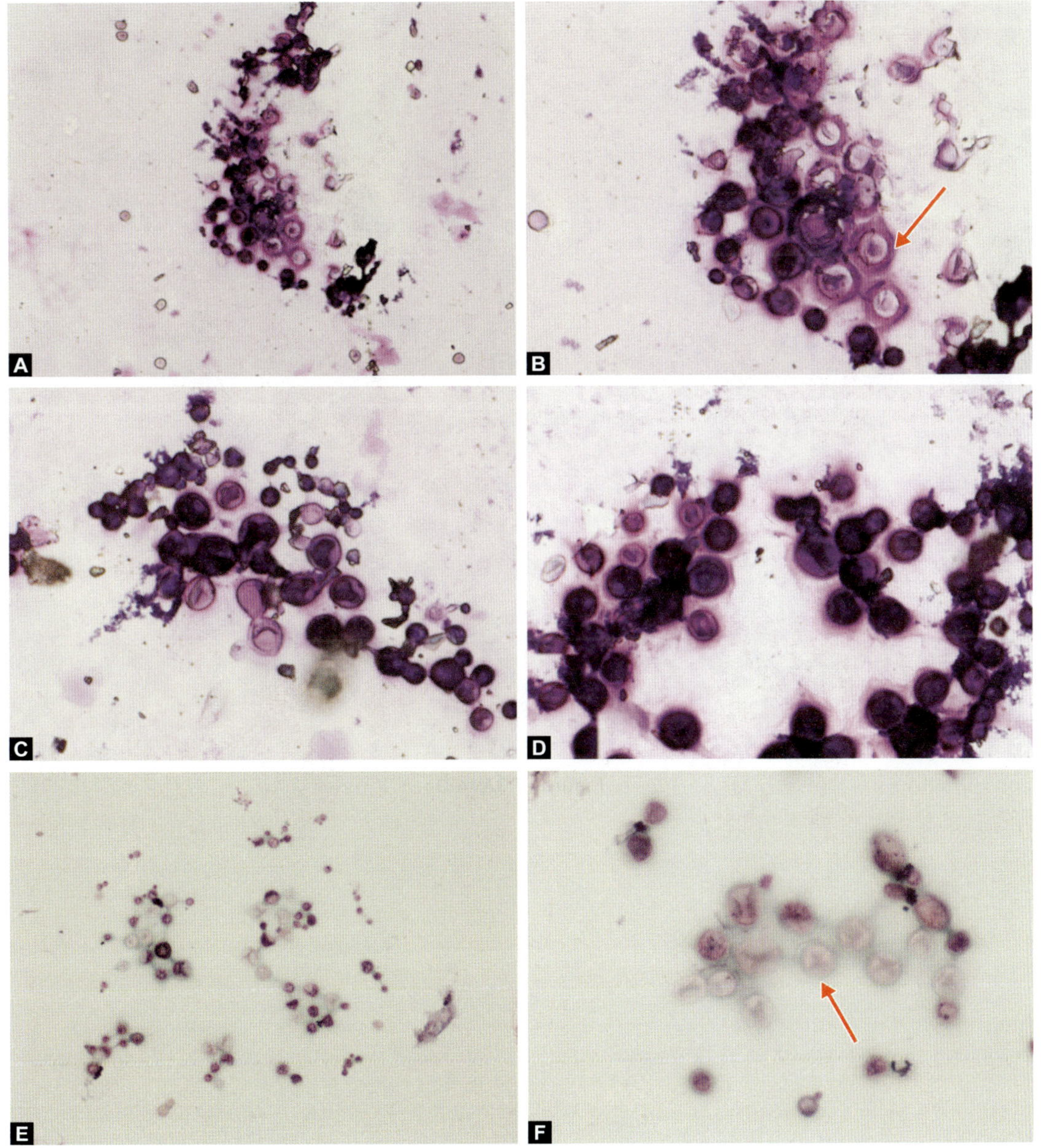

Figures 7.2A to F

Problem 7.3

A 28-year-old male presented with headache and low-grade fever. The most likely diagnosis from the microphotograph (Fig. 7.3) of his CSF is:

A. Acute leukemia
B. Lymphocytosis: No malignancy
C. Non-Hodgkin lymphoma
D. Metastatic carcinoma

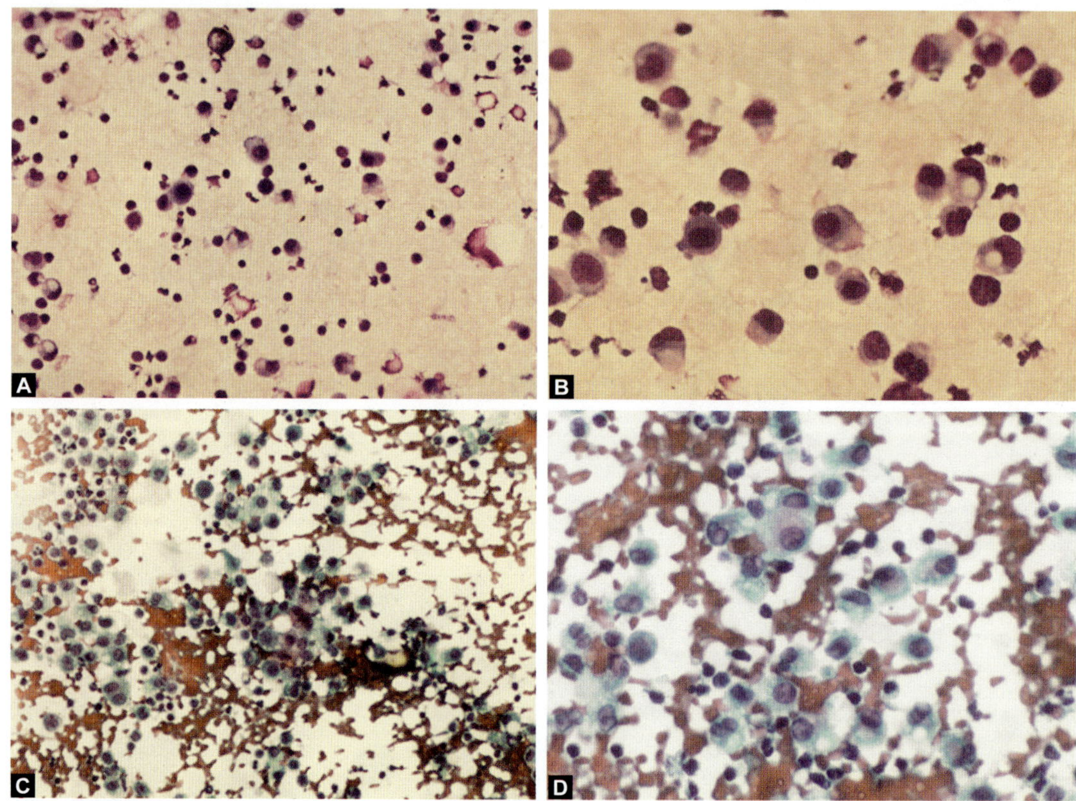

Figures 7.3A to D

Problem 7.4

A 38-year-old female was operated for carcinoma of left breast 2 year back. Now the patient complaints of severe headache for 2 weeks. The CSF cytology picture is shown in Figure 7.4. The most likely diagnosis is:

A. Metastatic carcinoma of breast
B. Lymphocytes only
C. Lymphoma
D. Primary brain tumor

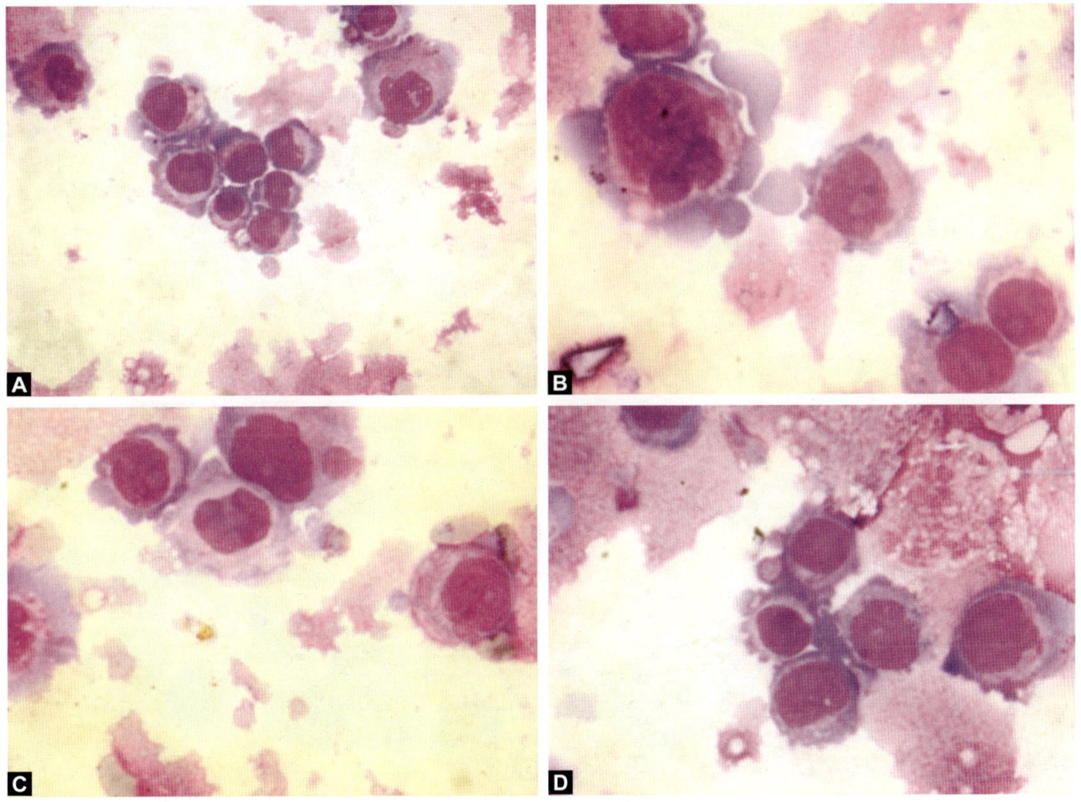

Figures 7.4A to D

Problem 7.5

A 58-year-old male had neurological symptoms. He had history of back pain and serum electrophoresis showed M band. CSF examination was done. The most likely diagnosis from the microphotographs (Fig. 7.5) of CSF smear is:

A. Lymphocytosis
B. Plasma cell tumor (multiple myeloma)
C. Metastatic carcinoma
D. Acute inflammation

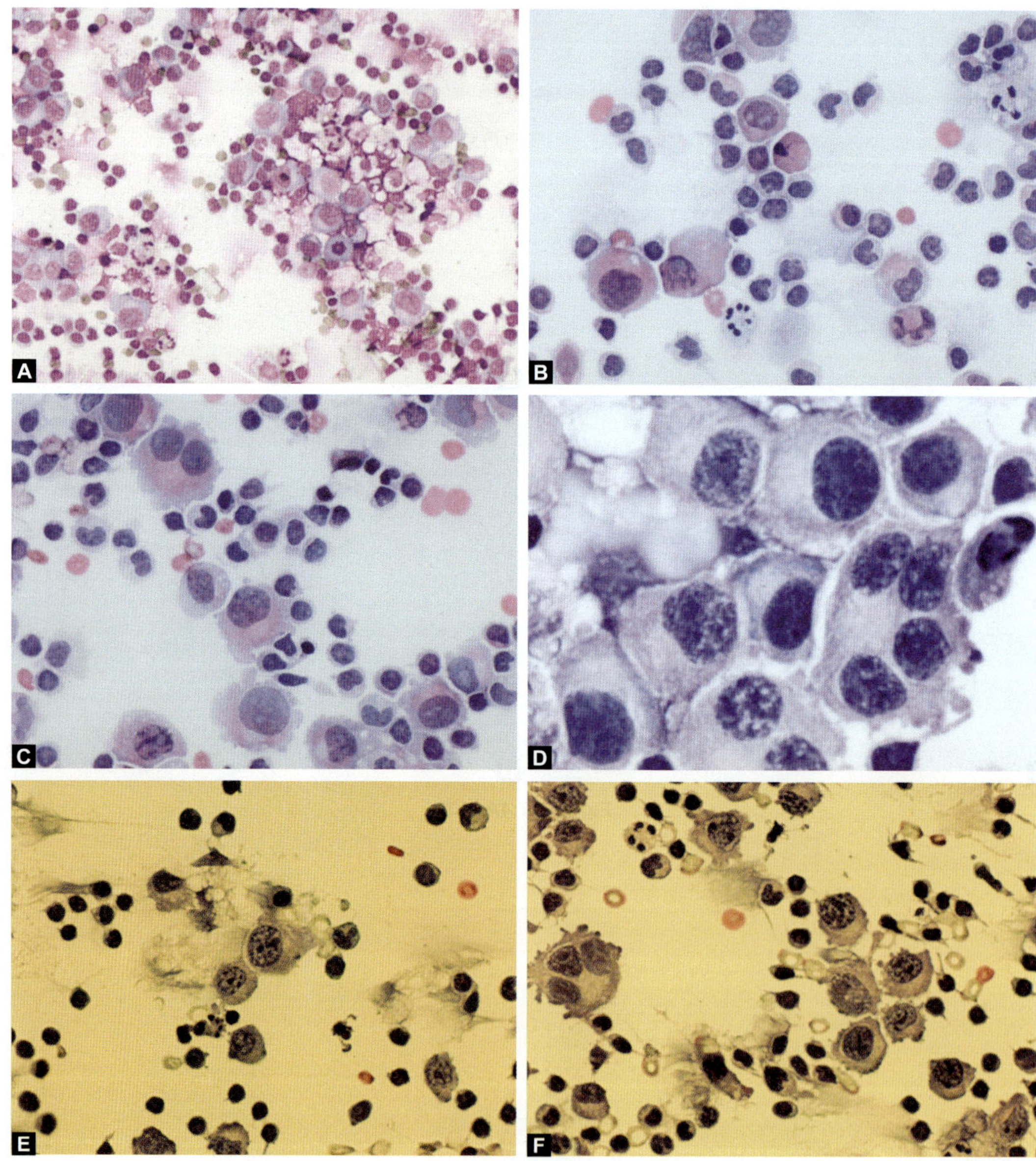

Figures 7.5A to F

Problem 7.6

A 3-year-old male child was operated for eye tumor 2 months back for retinoblastoma. Now the patients develop meningeal irritation in the form of stiff neck. CSF cytology was done. The most likely diagnosis from Figure 7.6 is:

A. Lymphocytosis
B. Retinoblastoma
C. Lymphoma
D. Cryptococcal infection

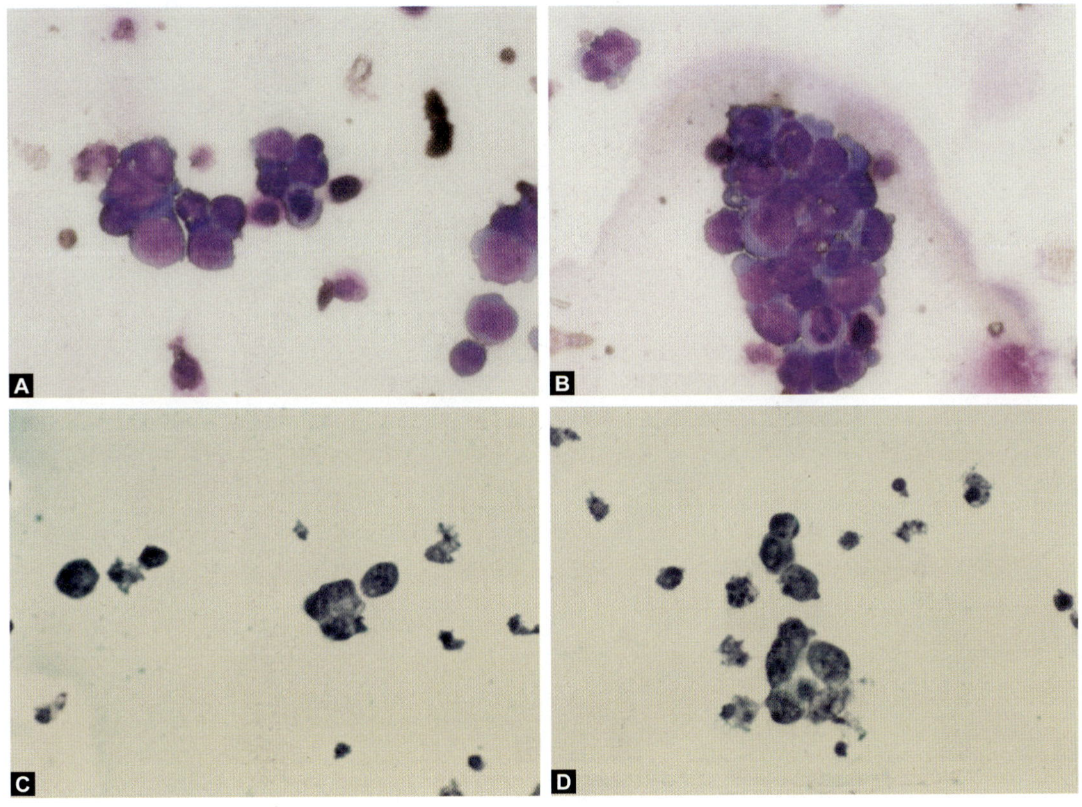

Figures 7.6A to D

Problem 7.7

A 10-year-old female presented with anemia, bleeding gum, weakness and headache. Her CSF examination was done. The most likely diagnosis from Figure 7.7 is:

A. Acute myeloid leukemia
B. Lymphocytosis
C. Metastatic round blue cell tumor
D. Primary CNS tumor

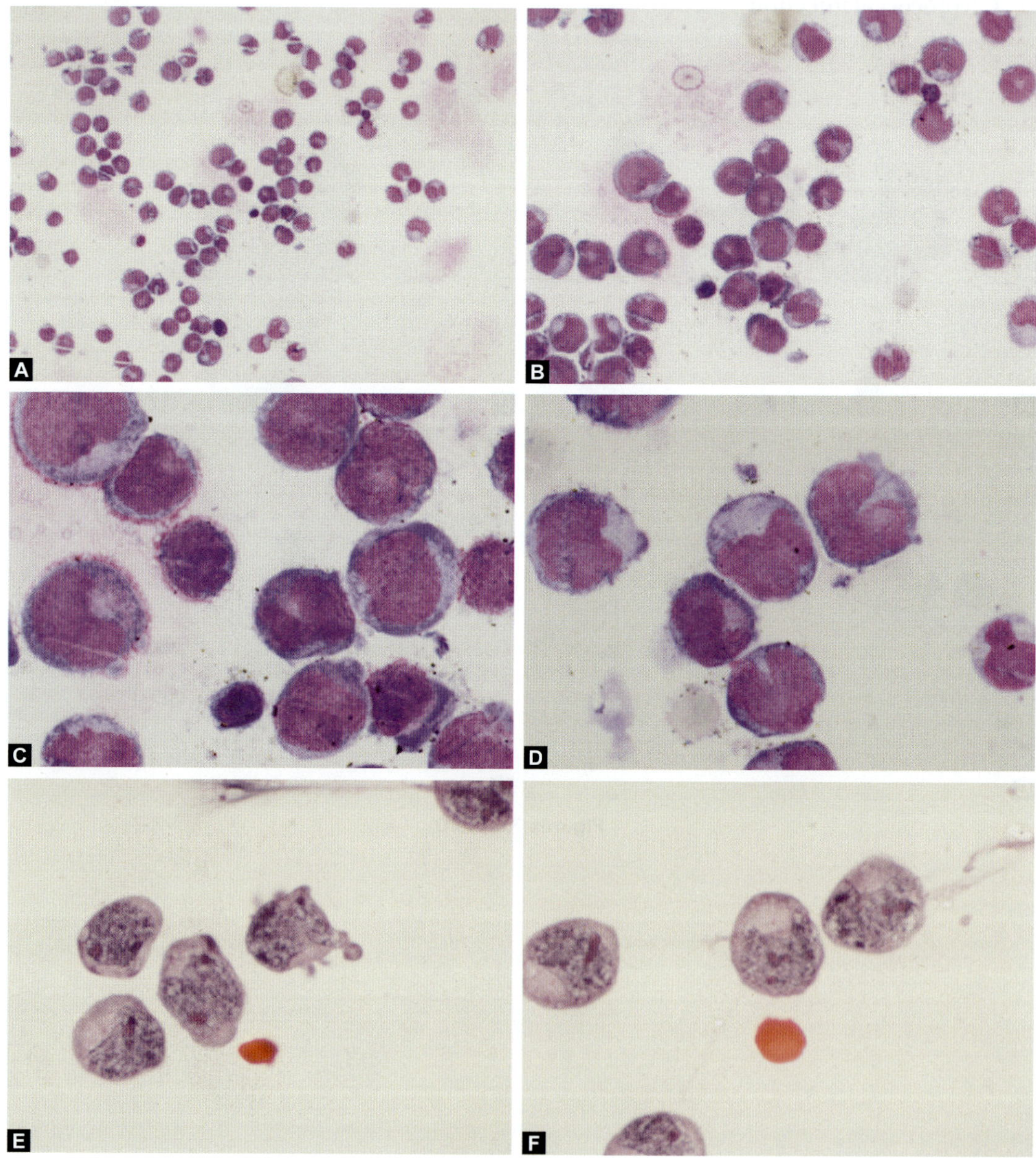

Figures 7.7A to F

Problem 7.8

A 12-year-old male presented with headache for 2 months. Routine CSF examination was done. The most likely diagnosis from Figure 7.8 is:

A. Acute leukemia
B. Medulloblastoma
C. Lymphocytosis
D. Viral infection

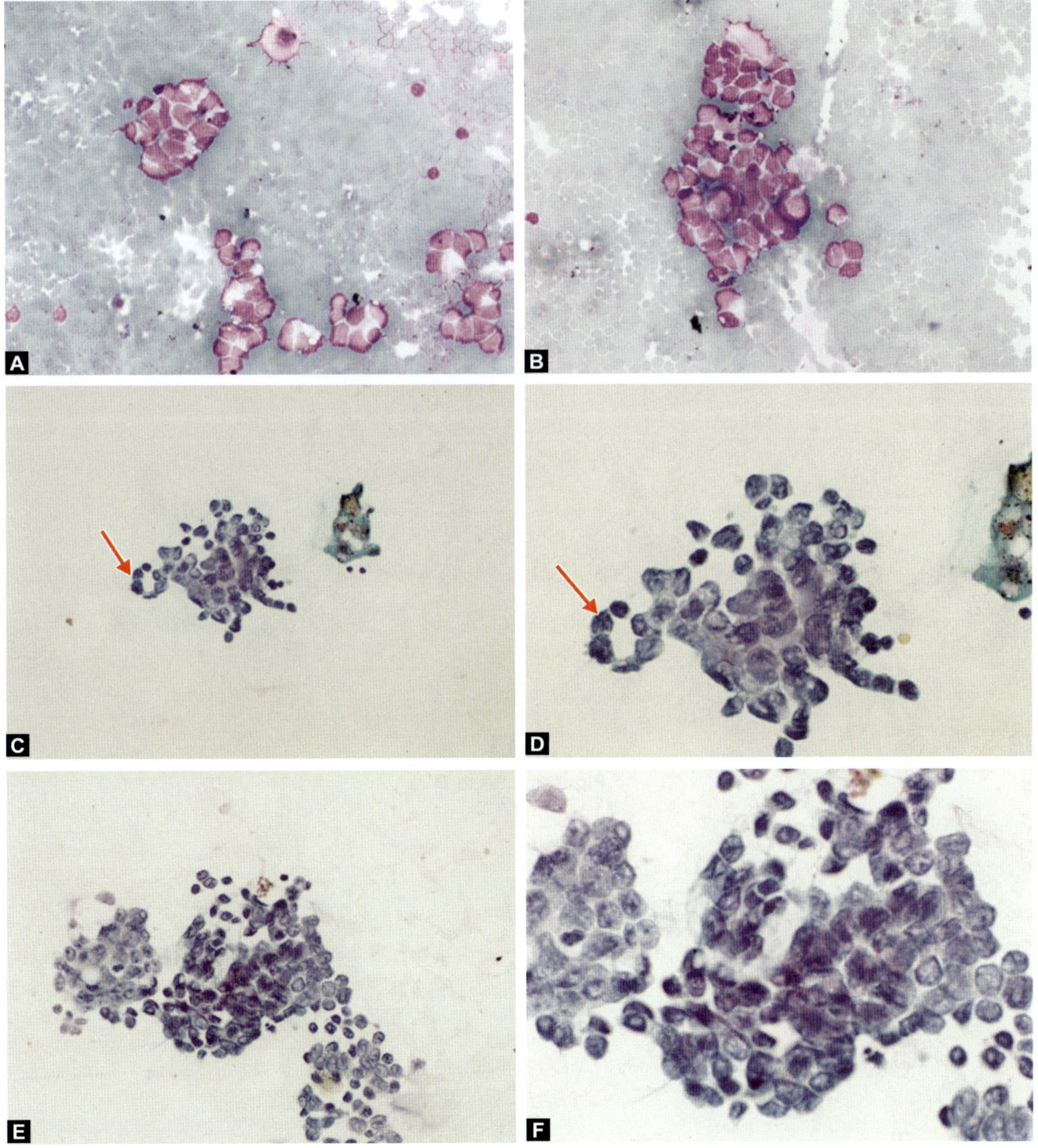

Figures 7.8A to F

Problem 7.9

A 13-year-old male complaints of severe headache and stiff neck. CSF cytology picture is shown. The most likely diagnosis from Figure 7.9 is:

A. Viral infection
B. Tuberculosis
C. Ependymoma
D. Lymphoma

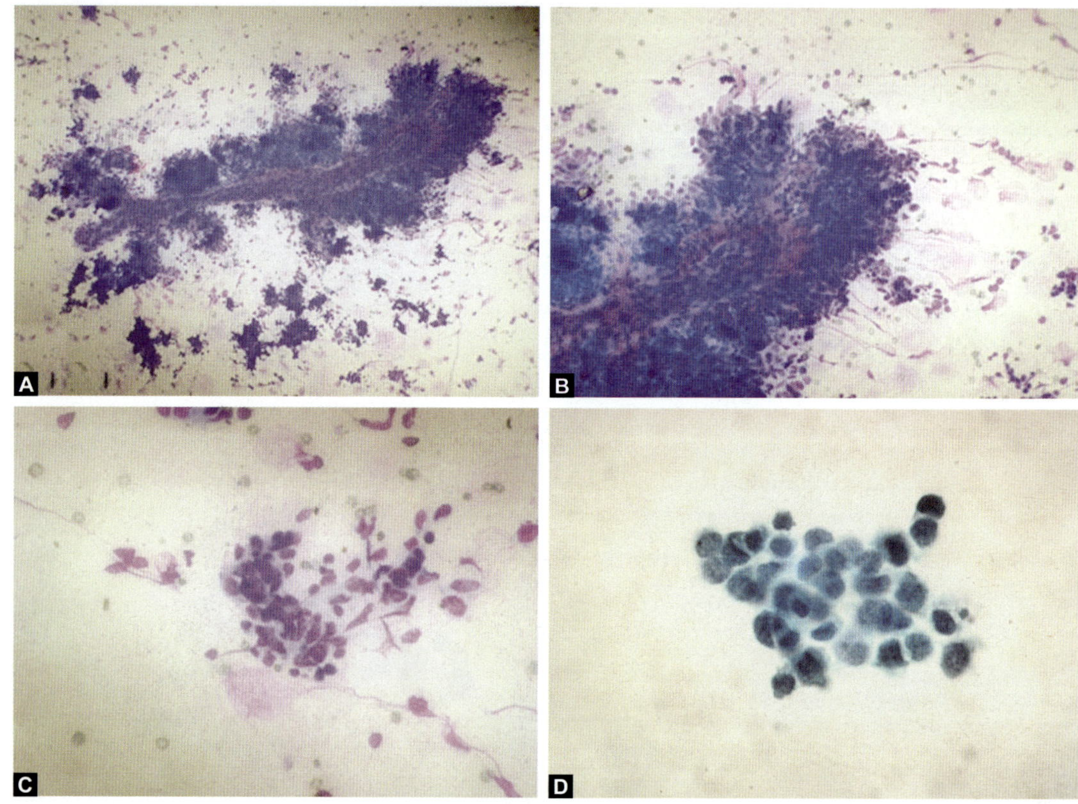

Figures 7.9A to D

Problem 7.10

A 65-year-old male had cough, hemoptysis and chest pain for 6 months. He also had consolidation in left upper lobe of lung. Presently the patient has severe headache. The microphotographs of the CSF cytology smear are shown in Figure 7.10. The most likely diagnosis is:

A. Metastatic carcinoma from lung
B. Primary tumor of brain
C. Tuberculosis
D. No malignant cells

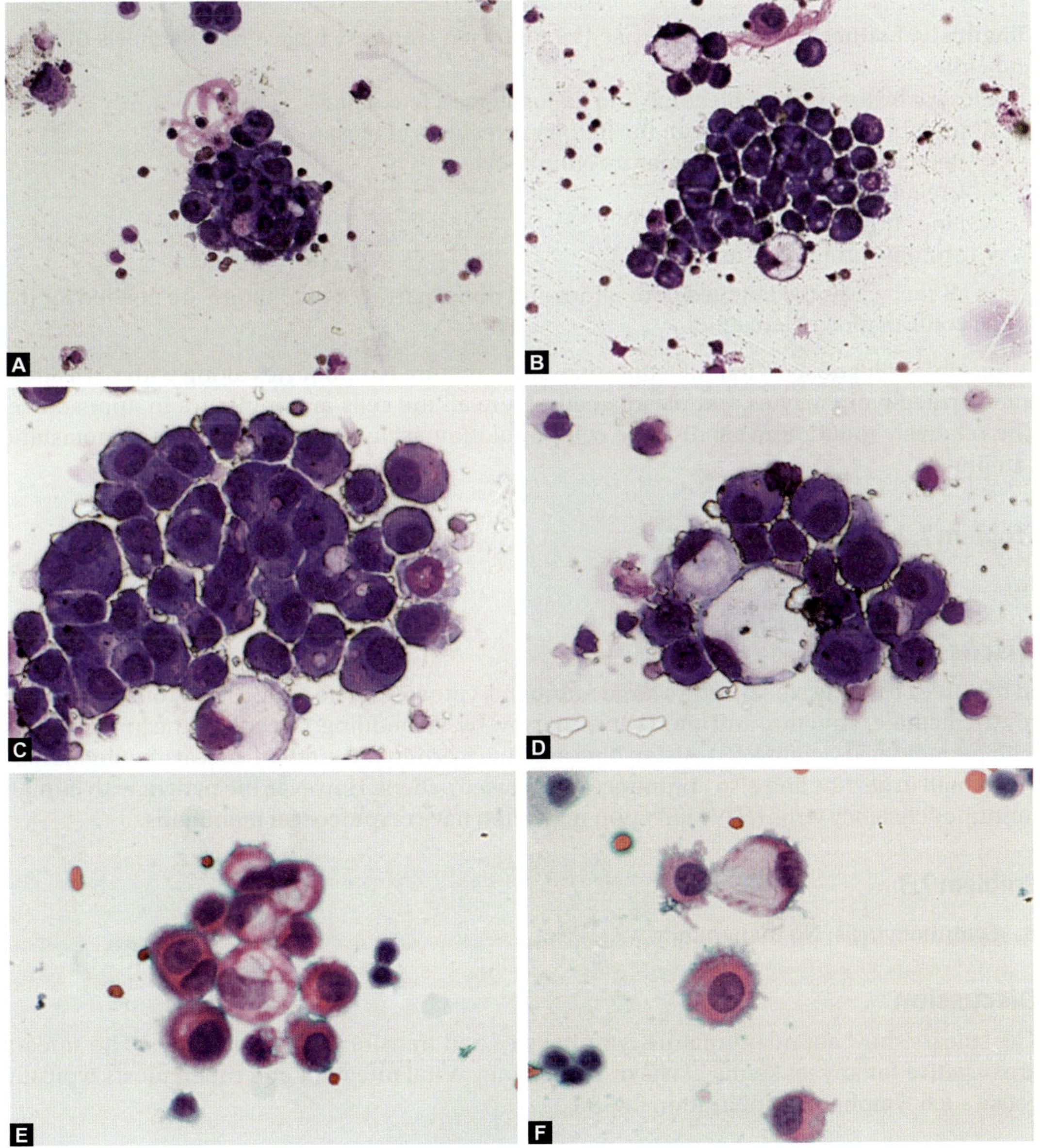

Figures 7.10A to F

ANSWERS

Problem 7.1

B. Acute leukemia

Discussion

Description: The cytology smears show abundant immature cells. The cells show scanty cytoplasm having enlarged nuclei with fine nuclear chromatin and irregular nuclear margin.

Diagnostic features: In the index case the following features suggest the diagnosis of acute leukemia:

- Clinical history of gum bleeding and pallor suggest leukemia
- Abundant immature cells with the following features of blasts
  - Relatively large cells than mature lymphocytes
  - Irregular nuclear margin
  - Fine chromatin
  - Multiple prominent nucleoli.

In all such cases the detailed hemogram and bone marrow examination are needed for the initial confirmation of leukemia.

Differential diagnosis: Viral infection may show mature lymphocytes along with occasional transformed lymphocytes and plasma cells. Here all the cells are immature in appearance. The relatively monomorphic discrete cell population excludes the possibility of metastatic carcinoma.

Problem 7.2

A. Cryptococci

Discussion

In the index case the smear shows large number cryptococci. The organism is round shaped 5 to 10 μ diameter structure and often shows narrow based budding. Capsule of the fungus stains with periodic acid-Schiff (PAS), Alcian blue and mucicarmine. The cryptococcal meningitis may occur both in healthy and also immunocompromised patient. However, the patient with human immunodeficiency virus (HIV) infection may often have cryptococcal meningitis.

Problem 7.3

B. Lymphocytosis: No malignancy

Discussion

The smears show abundant mature lymphocytes and transformed lymphocytes. The smears are negative for any leukemic blasts or malignancy. Viral infection and tuberculosis typically cause such lymphocytic infiltration in CSF.

Problem 7.4

A. Metastatic carcinoma of breast

Discussion

Description: The cytology smears show many discrete and loose clusters of tumor cells with round to oval moderately pleomorphic nuclei having prominent nucleoli and irregular nuclear margin.

Diagnostic features: In this index case the features suggestive of metastatic carcinoma of breast include:

- Loose clusters and discrete cells
- Round to oval cells with moderate amount of cytoplasm
- Large moderately pleomorphic nuclei
- Clinical history of breast carcinoma.

Differential diagnosis: Loose clusters of cells and moderate amount of cytoplasm exclude the possibility of lymphoma. Cytomorphology is not consistent with any other primary neoplasm of brain.

Problem 7.5

B. Plasma cell tumor (multiple myeloma)

Discussion

Description: The cytology smears show abundant plasma cells. The cells are mono, bi and multinucleated. Individual cells have moderate amount of cytoplasm and nuclei show typical cartwheel chromatin pattern. Occasional cells show enlarged nuclei with prominent nucleoli.

Diagnostic features: In the index case the following features favor the diagnosis of plasma cell tumor in CSF:

- Clinical history of back pain and serum electrophoresis displaying M band
- Abundant cells with central to eccentric nuclei and typical cartwheel chromatin indicating the morphology of plasma cell.

The nuclear morphology of these cells excludes the possibility of lymphocytosis or carcinoma.

Disease description: Multiple myeloma is a neoplasm of plasma cell. It is usually localized in the bone marrow. Involvement of meninges and demonstration of plasma cells in CSF is rare. Approximately only in 1% of multiple myeloma case may involve central nervous system[1]. It has been noted that the patient with CSF involvement may also show circulating atypical plasma cells. Flow cytometry of CSF may be helpful in diagnosis of multiple myeloma in CSF.[2] The overall prognosis of multiple myeloma with CNS involvement is poor.

References

1. Cavanna L, Invernizzi R, Berte' R, Vallisa D, Buscarini L. Meningeal involvement in multiple myeloma: report of a case with cytologic and immunocytochemical diagnosis. Acta Cytol. 1996;40(3):571-5.
2. Marini A, Carulli G, Lari T, Buda G, Lambelet P, Ciancia EM, et al. Myelomatous meningitis evaluated by multiparameter flow cytometry: report of a case and review of the literature. J Clin Exp Hematop. 2014;54(2):129-36.

Problem 7.6

B. Retinoblastoma

Discussion

Description: The cytology smears show cohesive clusters and discrete round cells with scanty cytoplasm. The cells show round nuclei with moderate pleomorphism and fine chromatin.

Diagnostic features:
- History of retinoblastoma
- Malignant round cells with scanty cytoplasm
- Round nuclei with fine chromatin.

Problem 7.7

A. Acute myeloid leukemia

Discussion

Description: The cytology smears show many discrete immature cells. The cells have scanty cytoplasm with round pleomorphic nuclei, irregular nuclear membrane and prominent nucleoli.

Diagnostic criteria:
- The history of anemia, bleeding gum, weakness and headache suggests a leukemic process
- Round nuclei with irregular nuclear membrane and prominent nucleoli.

Problem 7.8

B. Medulloblastoma

Discussion

Description: The cytology smears show tumor cells in many tight clusters and also in rosette like arrangement. The individual cells show scanty cytoplasm, round nuclei and fine chromatin.

Diagnostic features:
- Cohesive clusters
- Rosette like arrangement
- Cell with scanty cytoplasm, round nuclei and fine chromatin.

Problem 7.9

C. Ependymoma

Discussion

Description: The cytology smears show clusters and discretely arranged cells. Perivascular arrangements of cells are also seen. The tumor cells are small, round with scanty cytoplasm having relatively monomorphic nuclei.

Diagnostic features: In the index case the following features favor the diagnosis of ependymoma:

- Cell are arranged in small clusters, perivascular and discretely
- Individual cells are small, round with scanty cytoplasm and relatively monomorphic nuclei.
- Occasional cells with elongated nuclei.

Ependymoma is one of the most common primary brain tumors in children. However, the tumor cells are rarely exfoliated in the CSF. High grade ependymoma and myxopapillary ependymoma are commonly detected in CSF.

Problem 7.10

A. Metastatic carcinoma from lung

Discussion

Description: The cytology smears show multiple cohesive clusters of tumor cells. The cells are large and round with abundant vacuolated cytoplasm. The nuclei are central to eccentric having moderate pleomorphism.

Diagnostic features: In the index case the features suggestive of adenocarcinoma of lung include:

- Multiple cohesive clusters of cells
- Cells with moderate amount of vacuolated cytoplasm
- Moderate nuclear enlargement and pleomorphism
- Cells with eccentric nuclei and cytoplasmic vacuolation
- Clinical history of cough, hemoptysis, chest pain and consolidation of the left lung.

On cytomorphology it is difficult to diagnose the primary site of the tumor. Immunocytochemistry such as TTF-1, napsin and CK7 are helpful for further confirmation.

Differential diagnosis: The cytomorphology of the cells of the CSF is not consistent with any primary neoplasm of brain. Tuberculosis of the central nervous system usually provokes lymphocytic reaction.

CHAPTER 8

Miscellaneous Areas

QUESTIONS

Problem 8.1

A 45-year-old female presented with right sided pleural effusion along with dysphagia and anorexia for 2 months. Upper gastrointestinal (GI) endoscopy showed elevated ulcerated lesion in gastroesophageal junction. Effusion cytology fluid examination was done. The most likely diagnosis from Figure 8.1 is:

A. Reactive mesothelial cells
B. Tuberculous effusion
C. Mesothelioma
D. Signet ring cell carcinoma

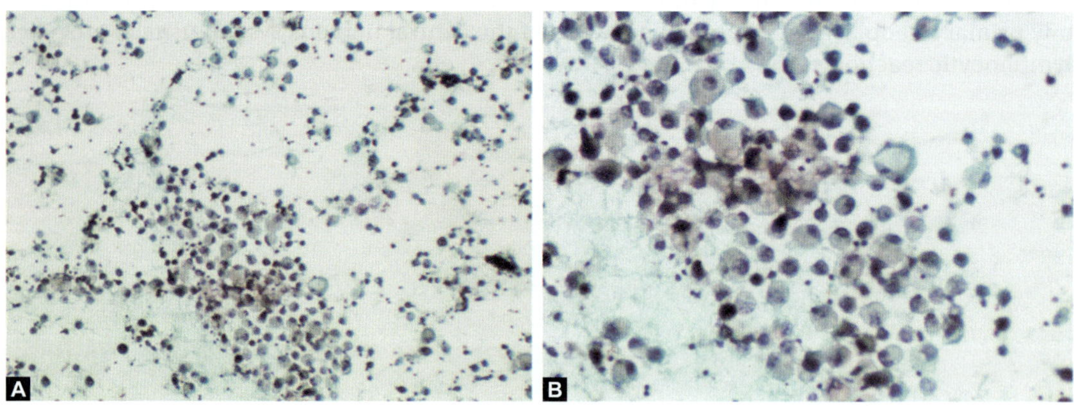

Figures 8.1A and B

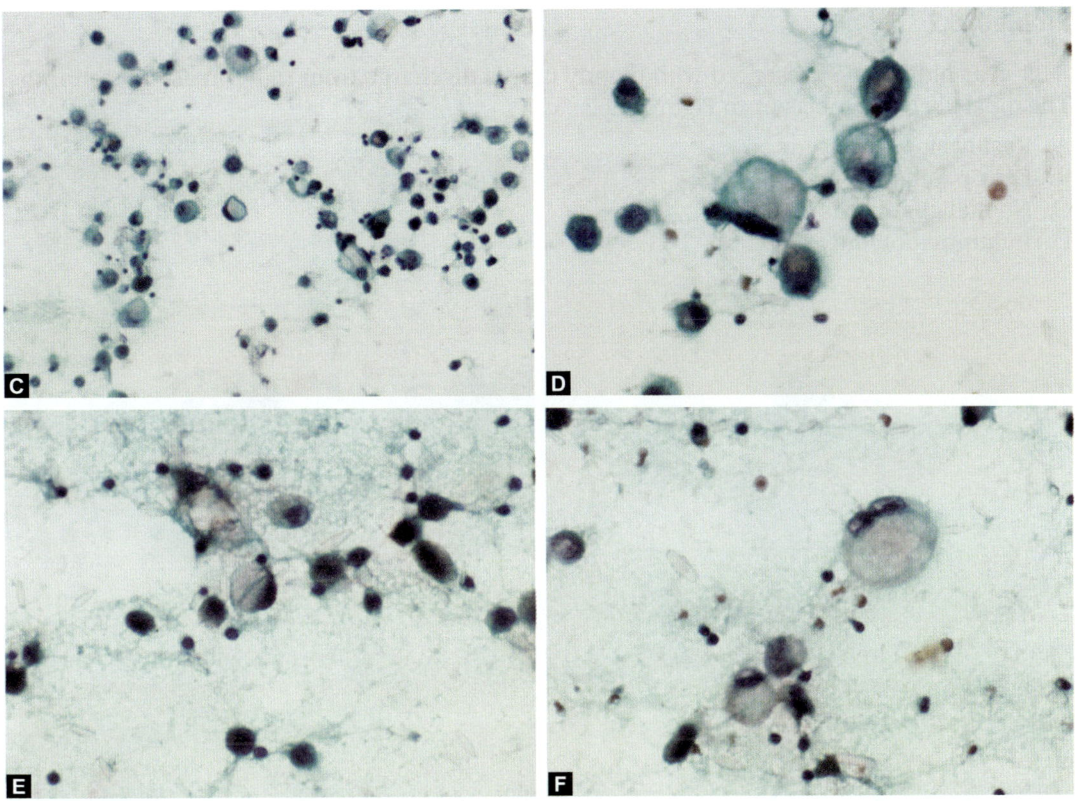

Figures 8.1C to F

Problem 8.2

A 39-year-old female presented with bloody nipple discharge from right breast for 3 months. The most likely diagnosis from Figure 8.2 is:

A. Positive for carcinoma
B. Duct papilloma
C. No malignant cells
D. Inflammation only

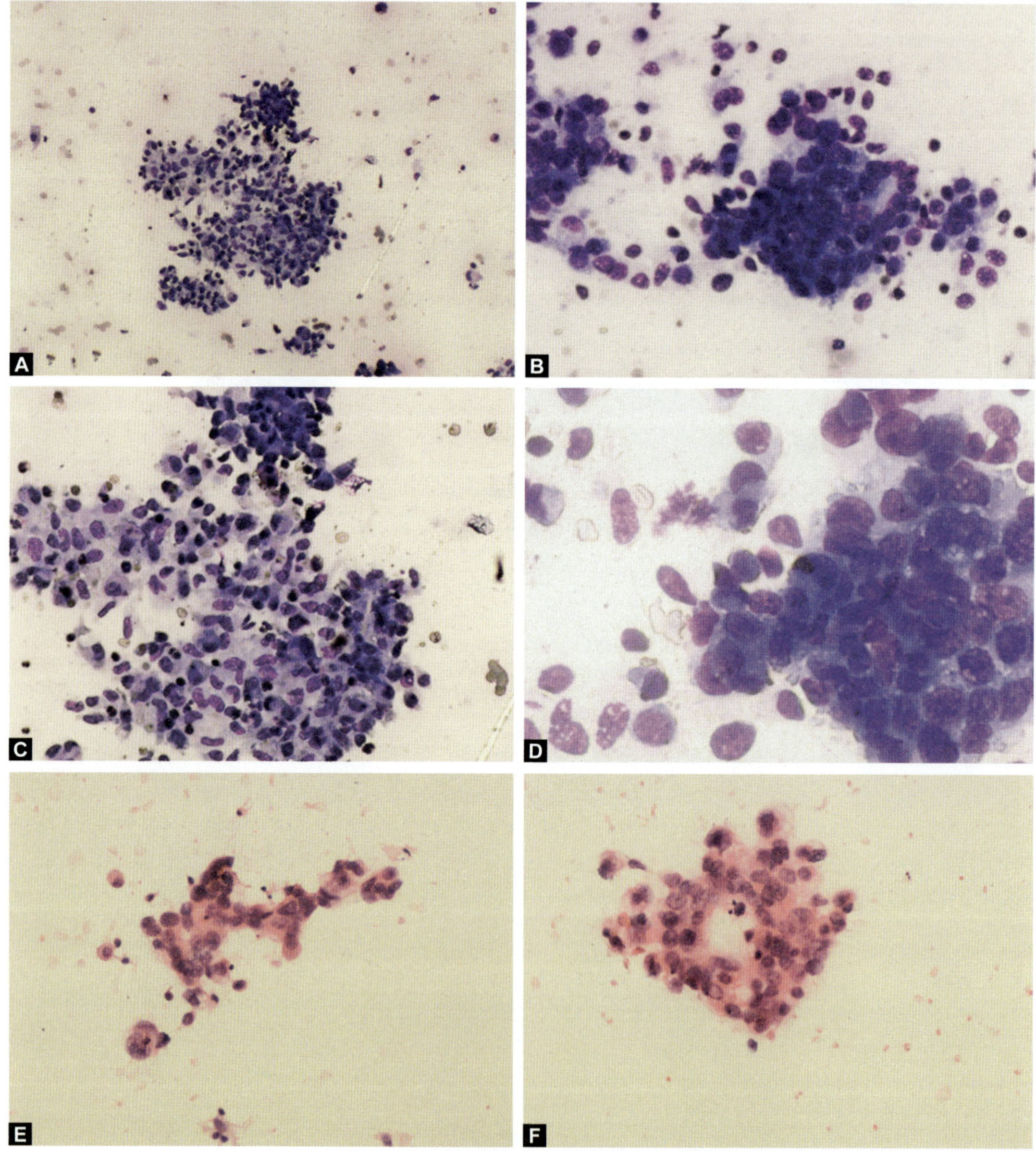

Figures 8.2A to F

Problem 8.3

A 12-year-old boy presented with progressive blindness of right eye and white eye reflex. The vitreous fluid of the right eye was aspirated and sent for cytological examination. The most likely diagnosis from Figure 8.3 is:

A. Retinoblastoma
B. Chronic inflammation
C. Acute leukemia
D. Viral infection

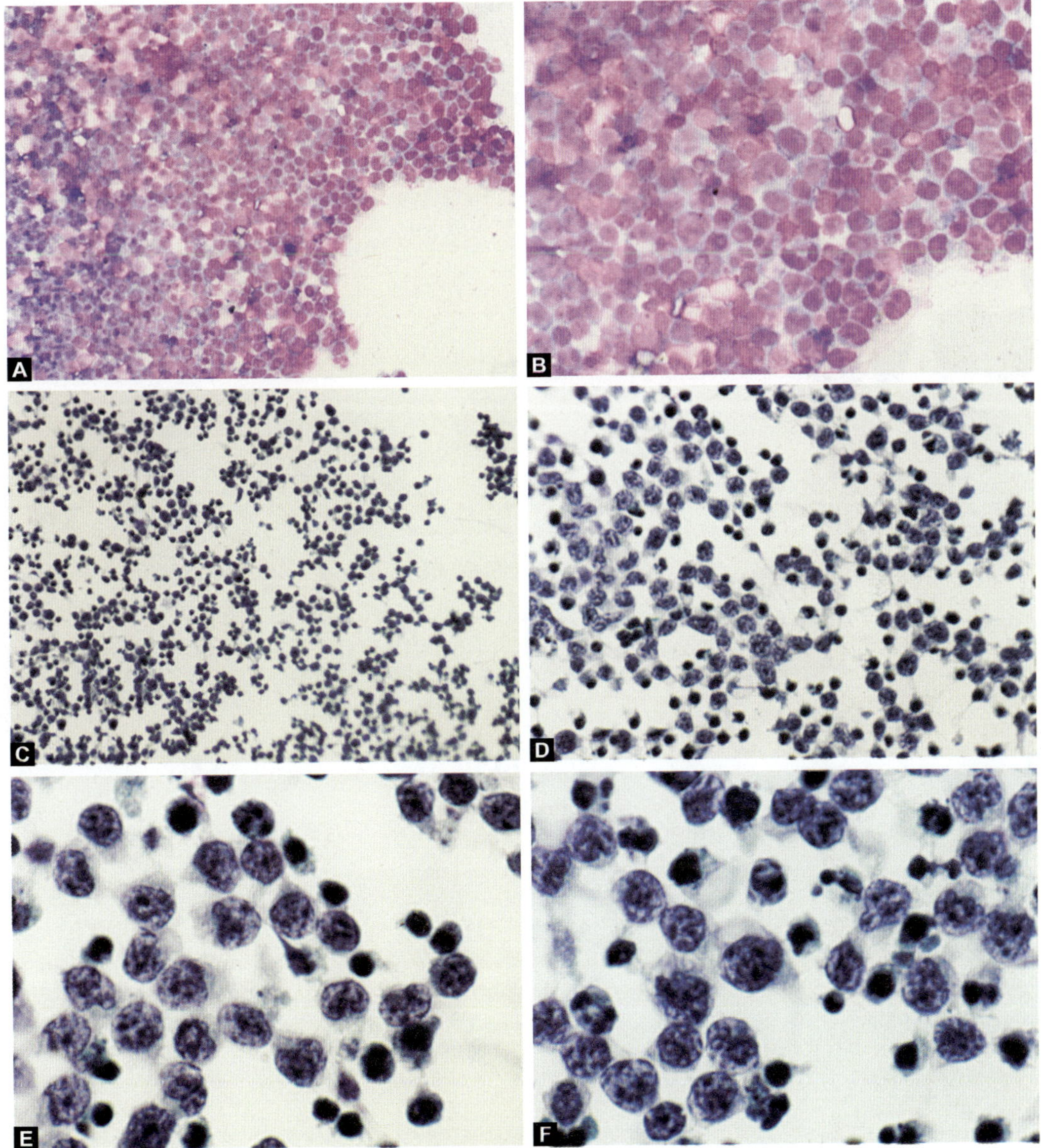

Figures 8.3A to F

Problem 8.4

A 39-year-old female presented with blood stained nipple discharge of right breast for 2 months. No palpable lump was noted. Cytology examination of the nipple discharge was done. The most likely diagnosis from Figure 8.4 is:

A. Normal cytology smear
B. Paget's disease
C. Carcinoma
D. Papillary neoplasm

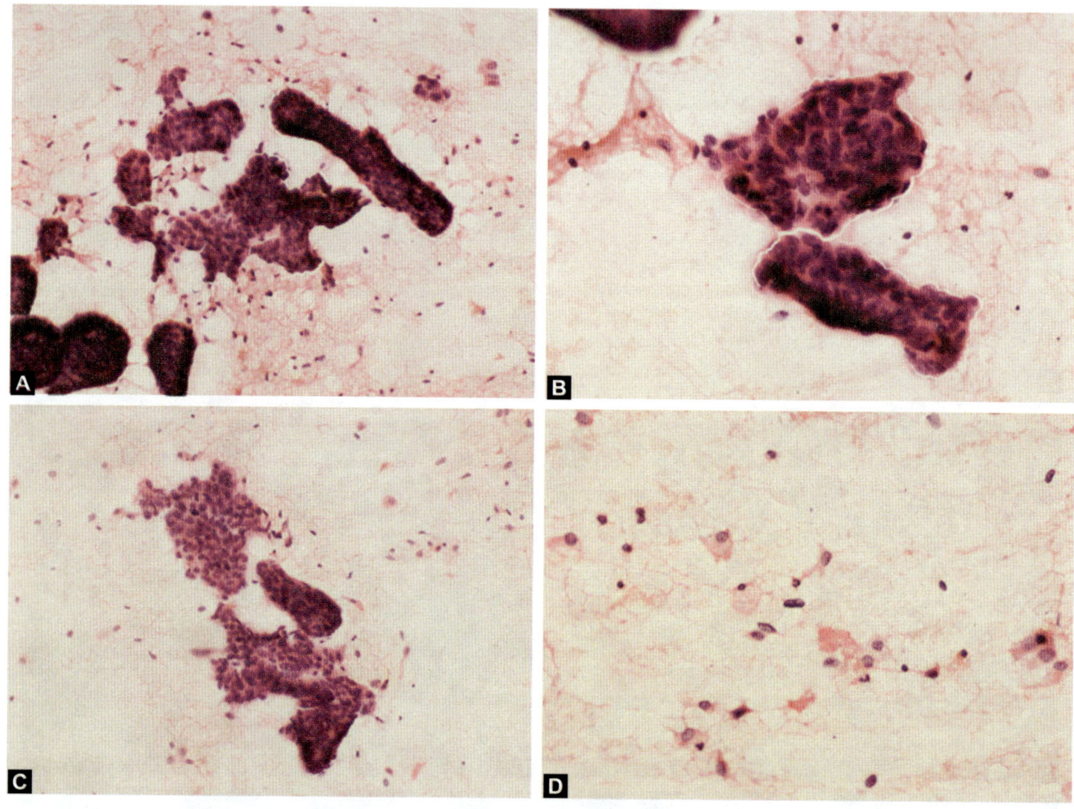

Figures 8.4A to D

Problem 8.5

A 52-year-old male presented with obstructive jaundice for 15 days. There was obstruction in free flow of the bile in the common bile duct in the cholangiogram. The brush cytology of the bile duct was done. The most likely diagnosis from Figure 8.5 is:

A. Adenocarcinoma in common bile duct
B. Dysplastic cells only
C. Stone induced atypia
D. Normal bile duct cells

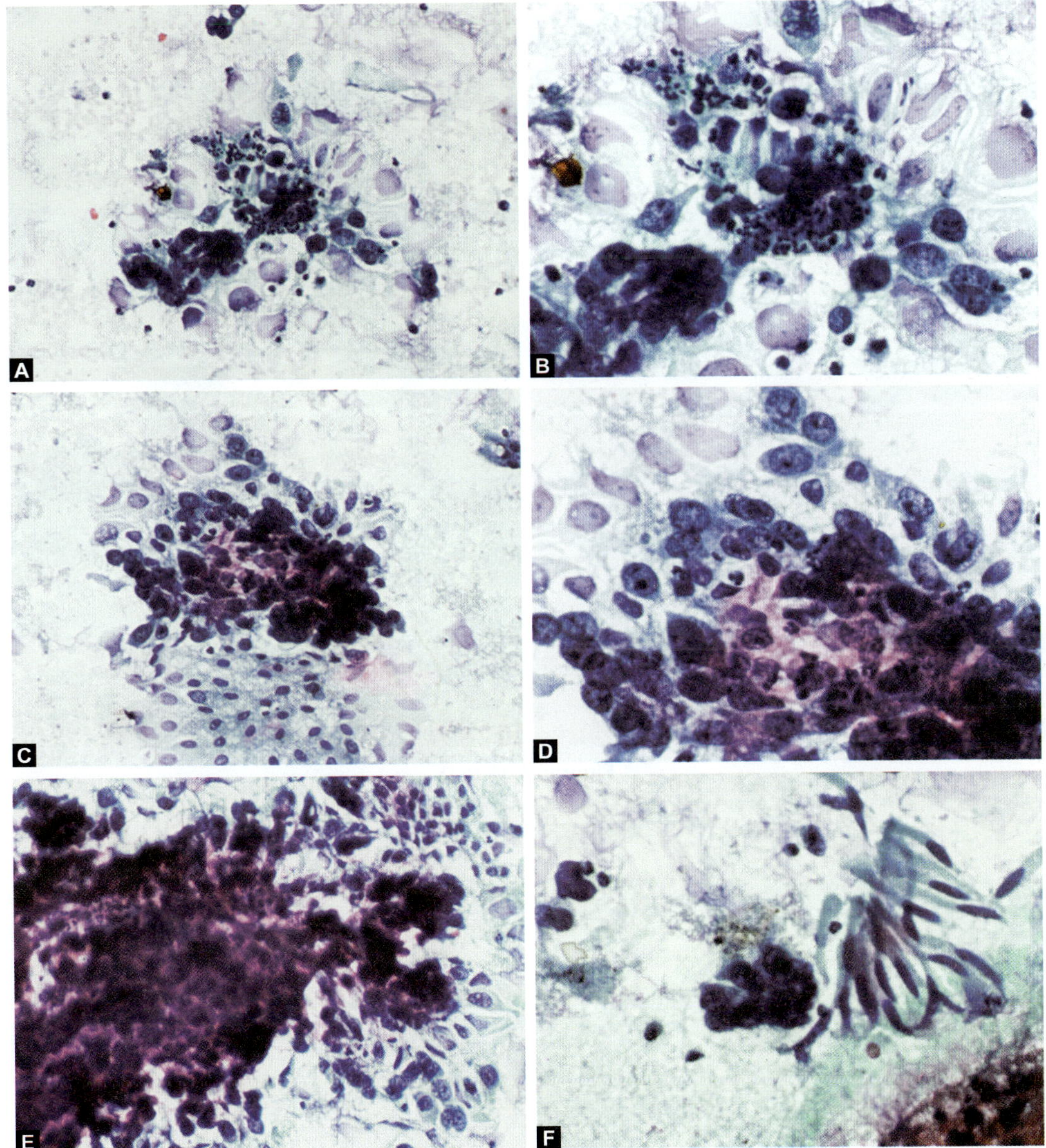

Figures 8.5A to F

Problem 8.6

A 36-year-old female with routine cervical smear. SurePath preparation done. The most likely diagnosis from Figure 8.6 is:

A. *Candida* present
B. Sperm present
C. Both *Candida* and sperm head present
D. *Trichomonas vaginalis*

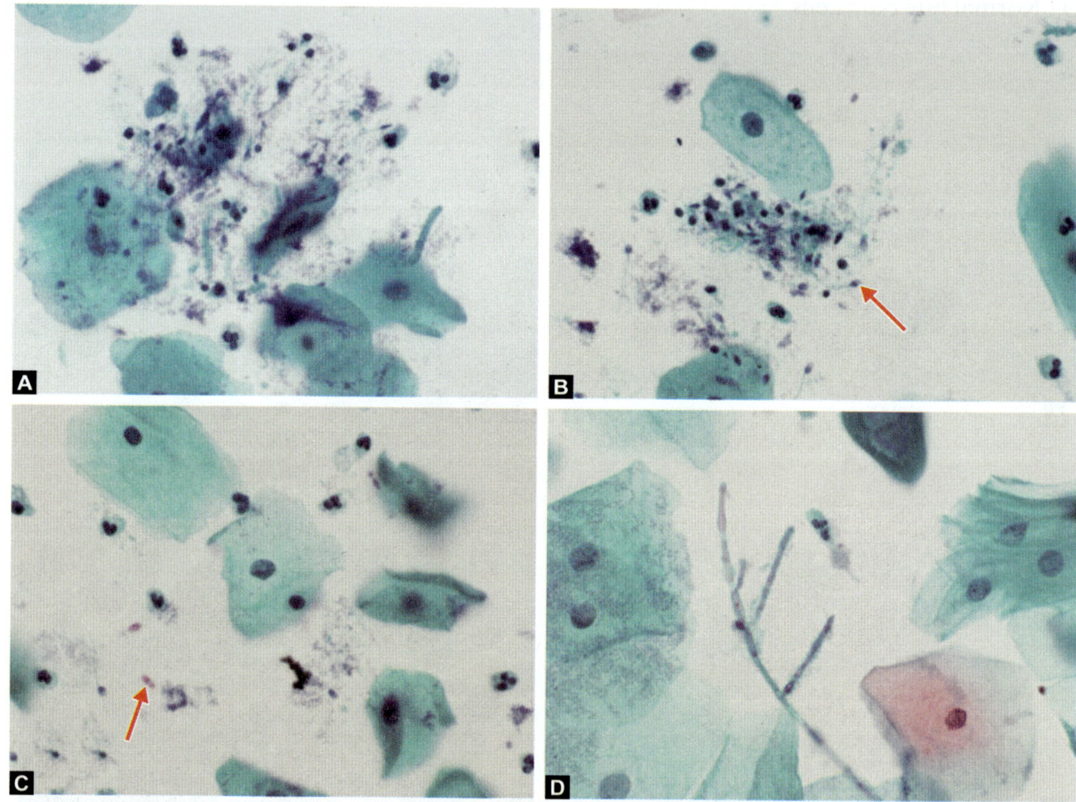

Figures 8.6A to D

Problem 8.7

A 65-year-old male presented with cough and hemoptysis for 6 months. He also had constipation. The patient had anemia and enlarged mediastinal lymph nodes. He also had left sided pleural effusion. The cytological examination (Figs 8.7A to F) along with immunocytochemistry in the cell block was done (Figs 8.7G to J). The most likely diagnosis is:

A. Metastatic adenocarcinoma from lung
B. Metastatic adenocarcinoma from colon
C. Metastatic carcinoma from the breast
D. Reactive mesothelial cells

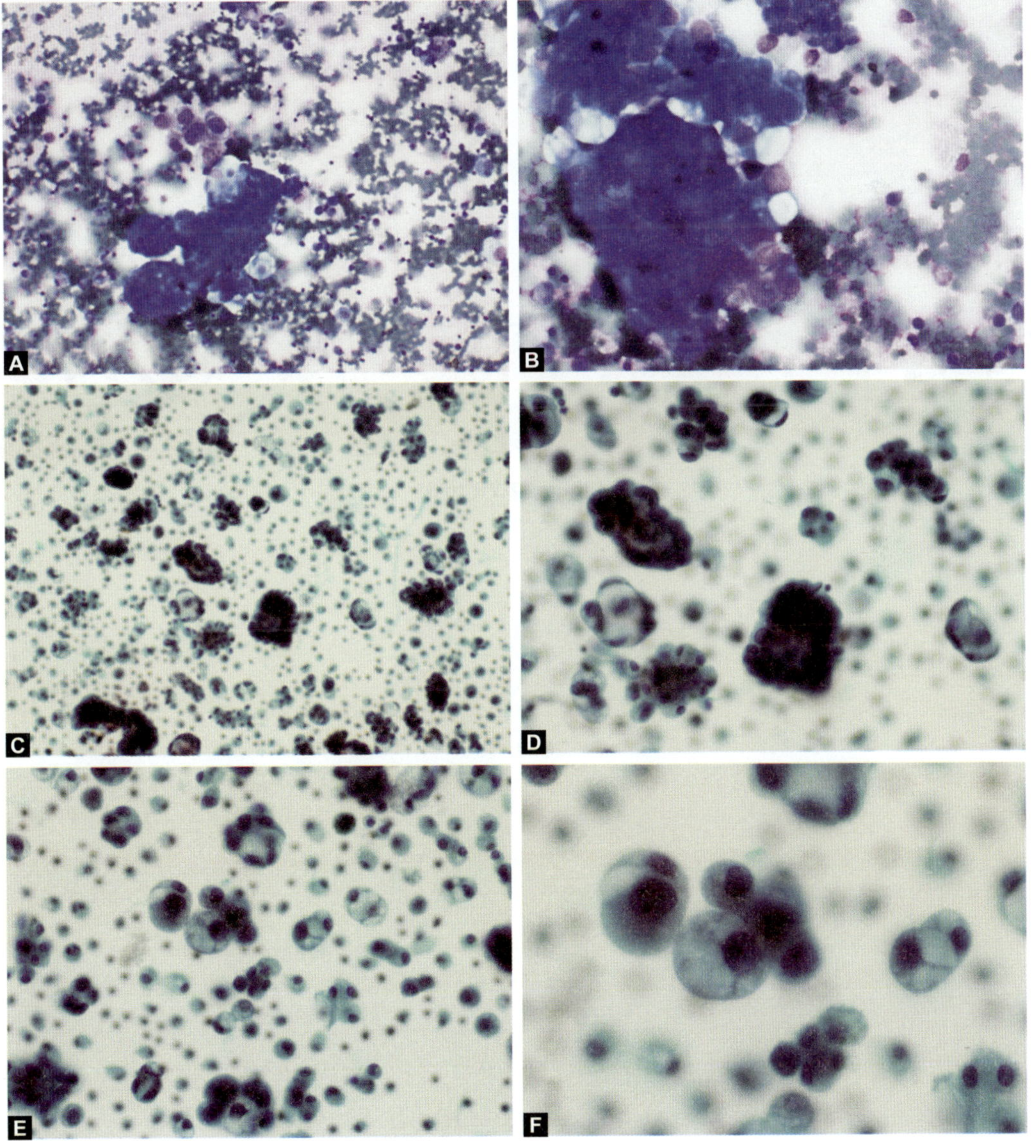

Figures 8.7A to F

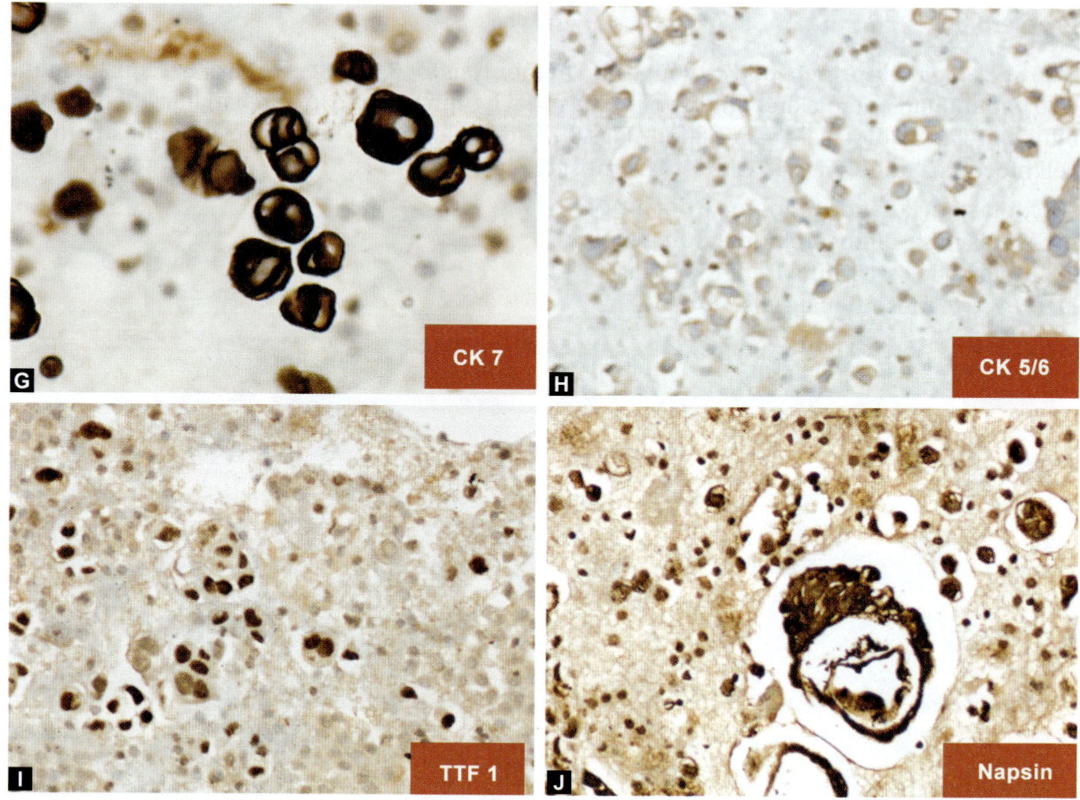

Figures 8.7G to J

Problem 8.8

A 49-year-old male presented with dysuria and increased frequency of urination. He was a diagnosed and treated case of low grade UCC. The most likely diagnosis from Figure 8.8 is:

A. BCG granuloma
B. Positive for UCC
C. Granulomatous inflammation: tuberculosis
D. Atypical cells present

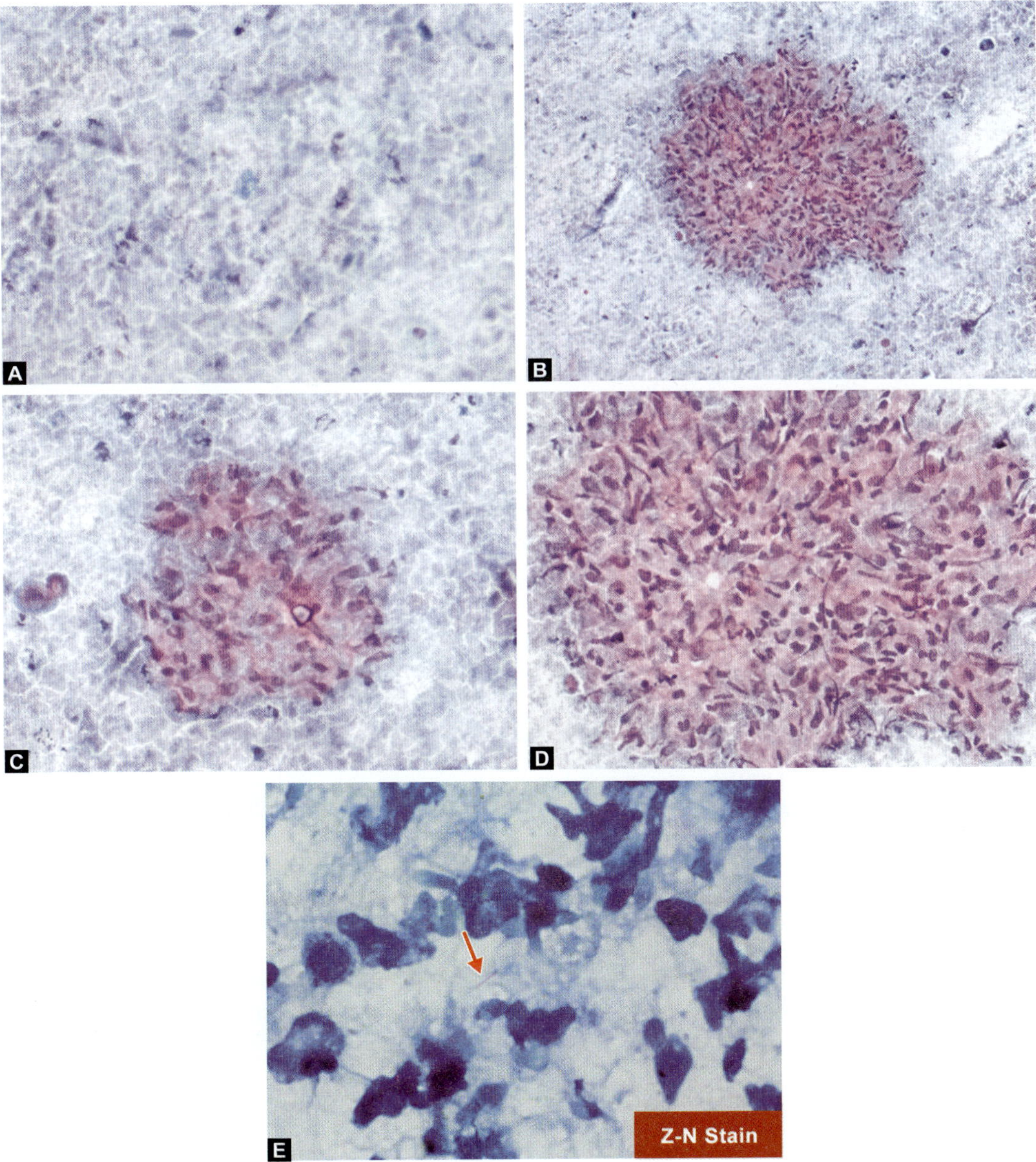

Figures 8.8A to E

Problem 8.9

A 39-year-old female presented with irregular menstrual bleeding. Her cervical smear was done. The most likely diagnosis from Figure 8.9 is:

A. AGCUS
B. Adenocarcinoma
C. HSIL
D. ASCUS

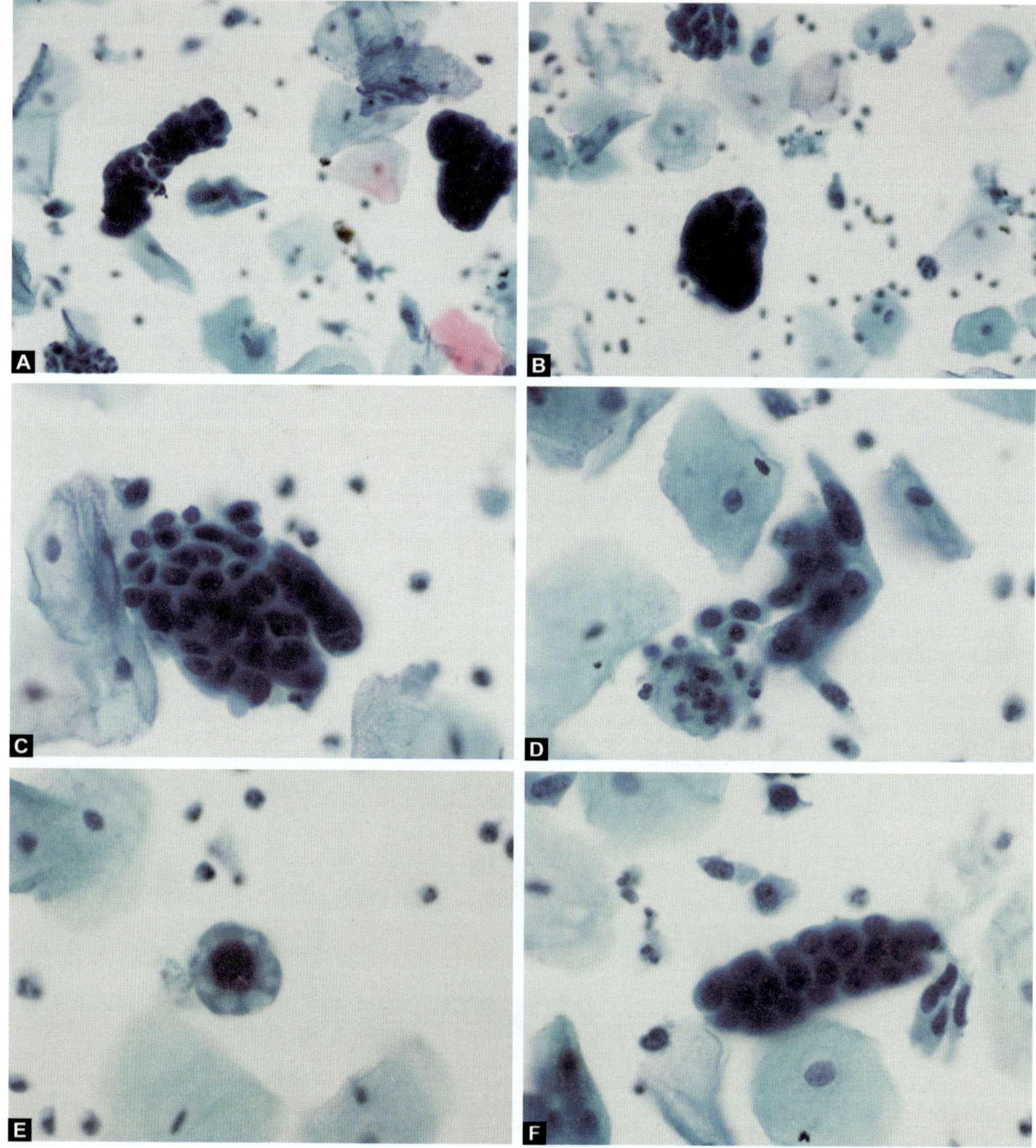

Figures 8.9A to F

Problem 8.10

A 35-year-old HIV positive male presented with fever and chest pain. His bronchoalveolar lavage was done. The most likely diagnosis from Figure 8.10 is:

A. Acute non-specific inflammation
B. Aspergillosis
C. Mucormycosis
D. Candidal infection

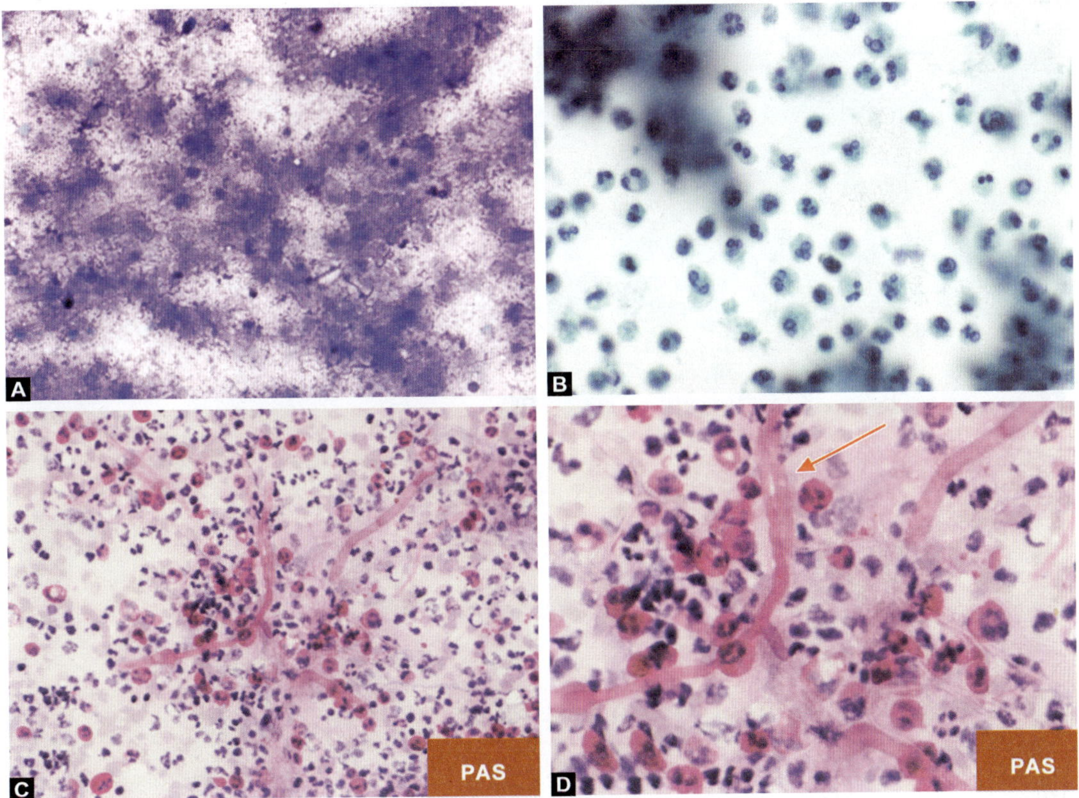

Figures 8.10A to D

Problem 8.11

A 41-year-old female was diagnosed as cervical carcinoma one year back. The patient received adequate therapy. Now routine cervical smear was done from the vault. The most likely diagnosis from Figure 8.11 is:

A. Squamous cell carcinoma
B. HSIL
C. AGCUS
D. Adenocarcinoma

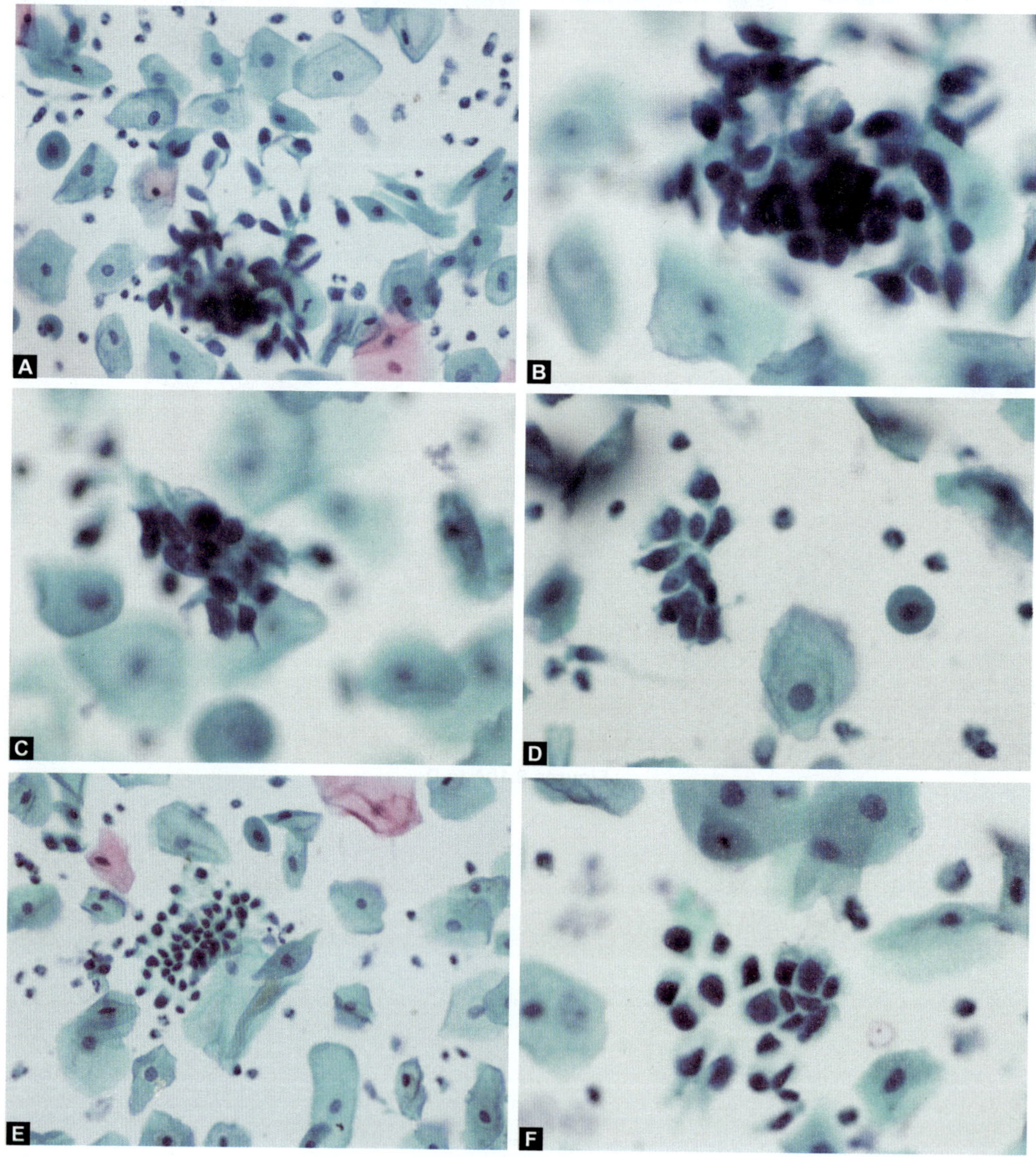

Figures 8.11A to F

Problem 8.12

Routine cervical smear in a 48-year-old female. The most likely diagnosis from Figure 8.12 is:

A. ASCUS
B. ASC-H
C. LSIL
D. HSIL

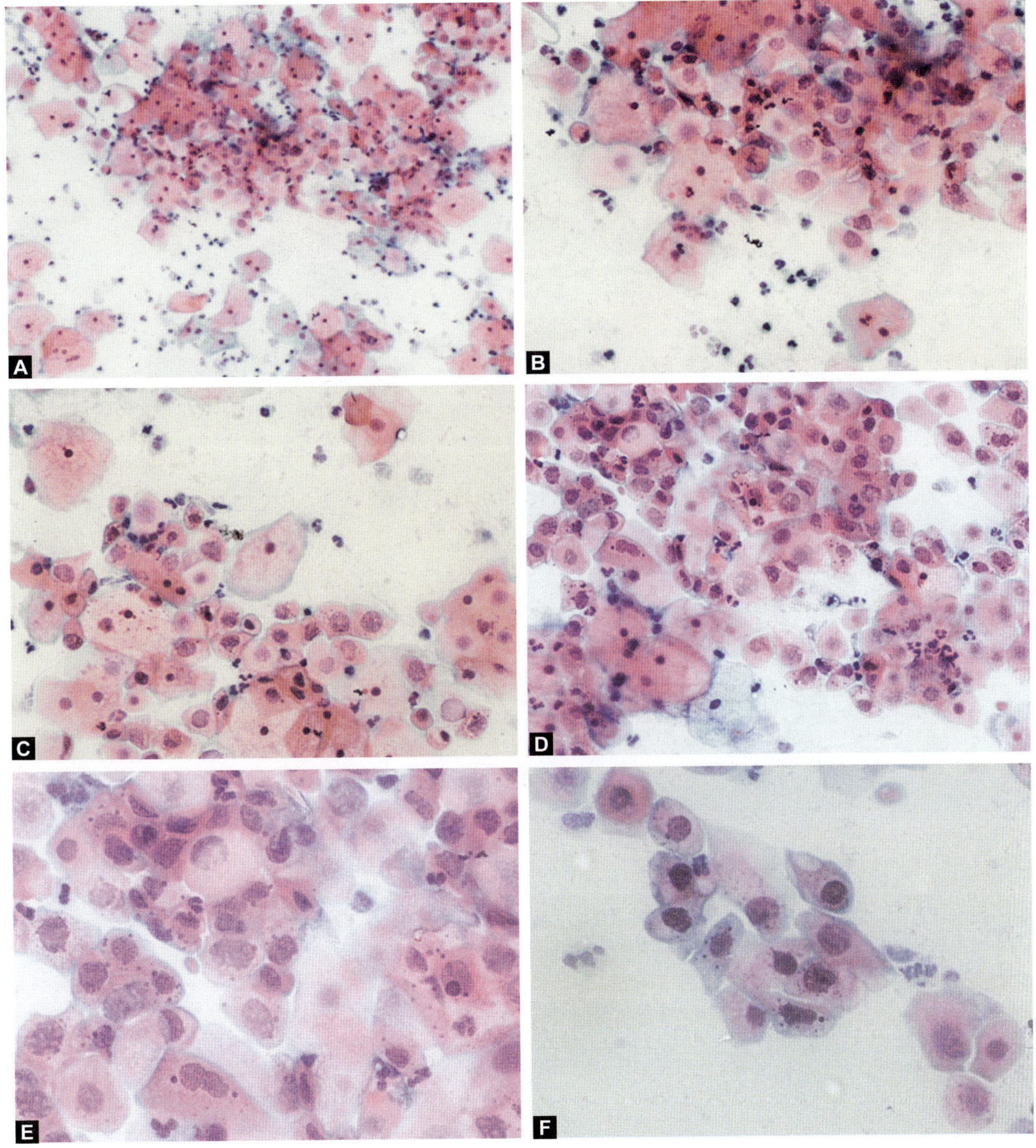

Figures 8.12A to F

Problem 8.13

Routine cervical smear in a 38-year-old female. The most likely diagnosis from Figure 8.13 is:

A. ASCUS
B. LSIL
C. HSIL
D. ASC-H

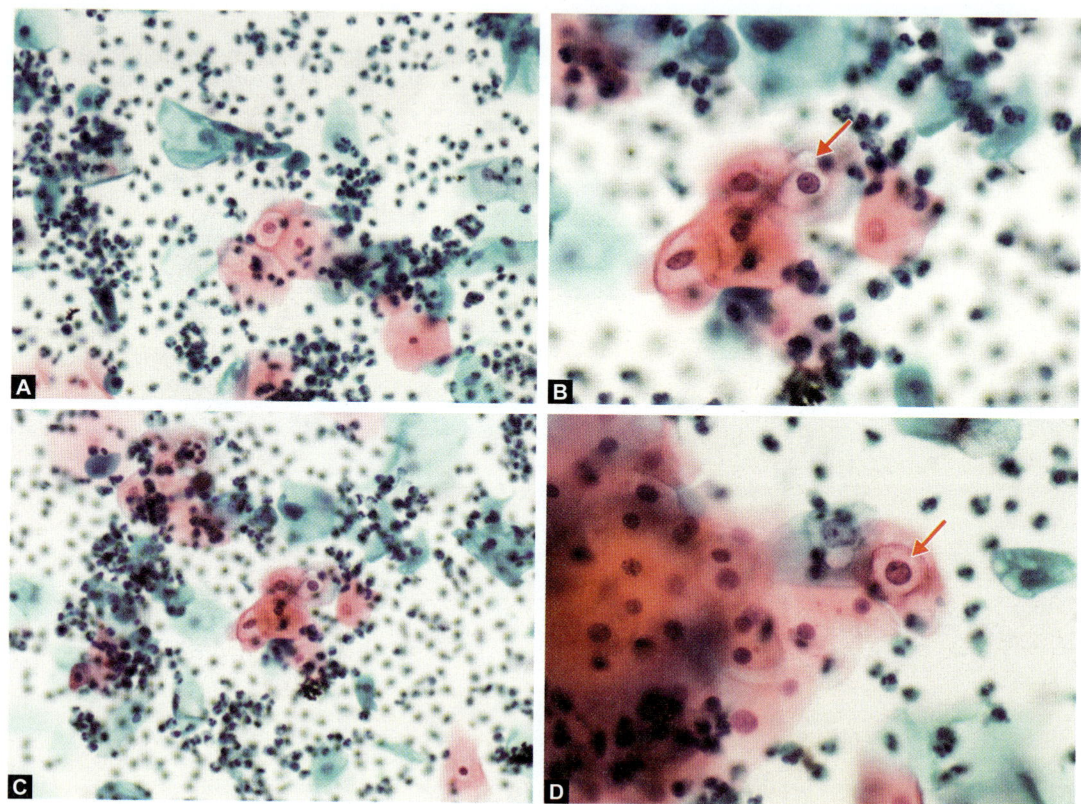

Figures 8.13A to D

Problem 8.14

A 42-year-old female had bilateral adnexal mass along with ascites. The peritoneal aspirate yielded thick gelatinous material. The most likely diagnosis from Figure 8.14 is:

A. Reactive mesothelial cells
B. Pseudomyxoma peritonei
C. Adenocarcinoma
D. Mesothelioma

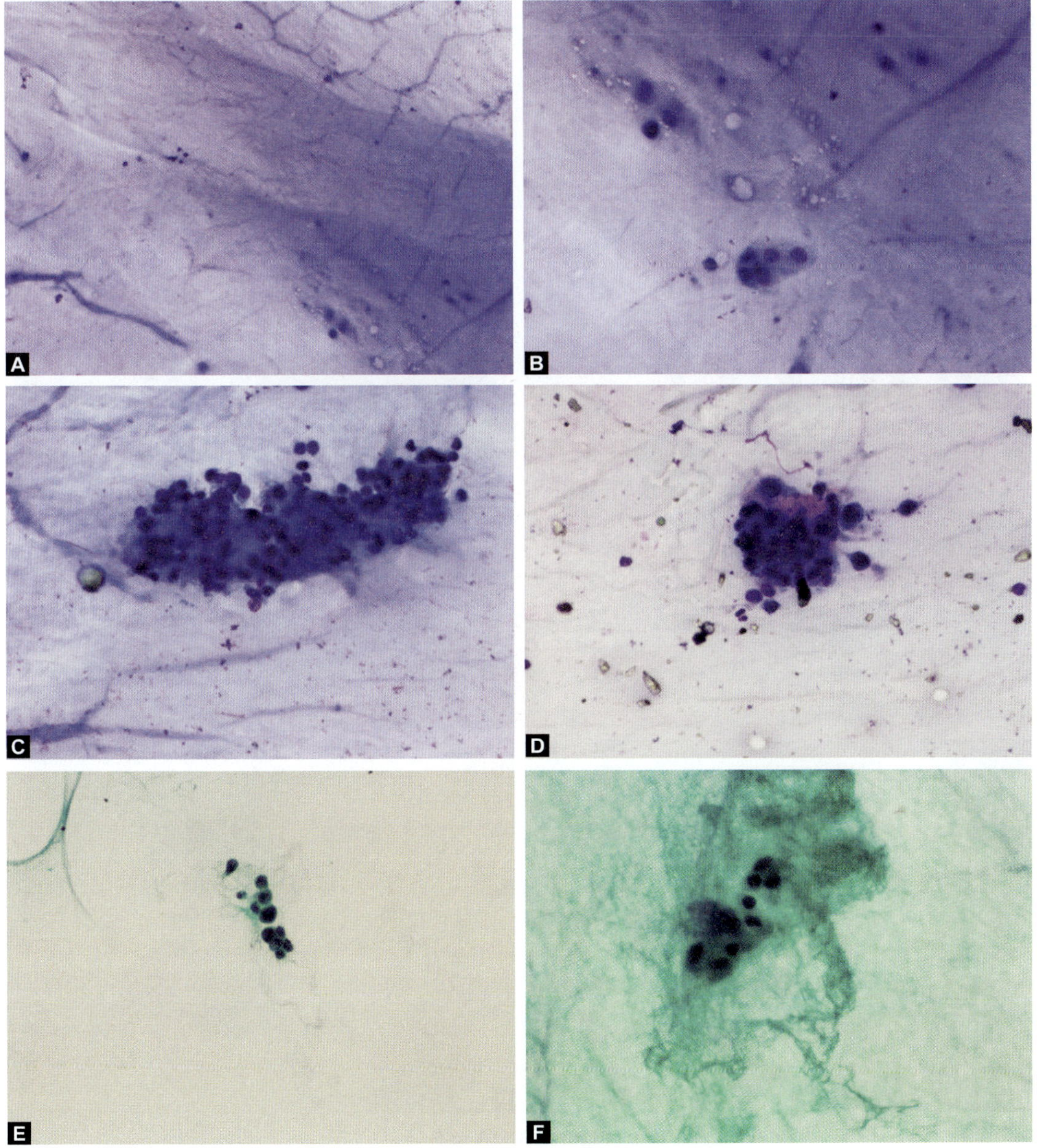

Figures 8.14A to F

Problem 8.15

Routine cervical smear in a 41-year-old female. The most likely diagnosis from Figure 8.15 is:

A. HSIL
B. ASC H
C. AGC favors neoplasia
D. Adenocarcinoma

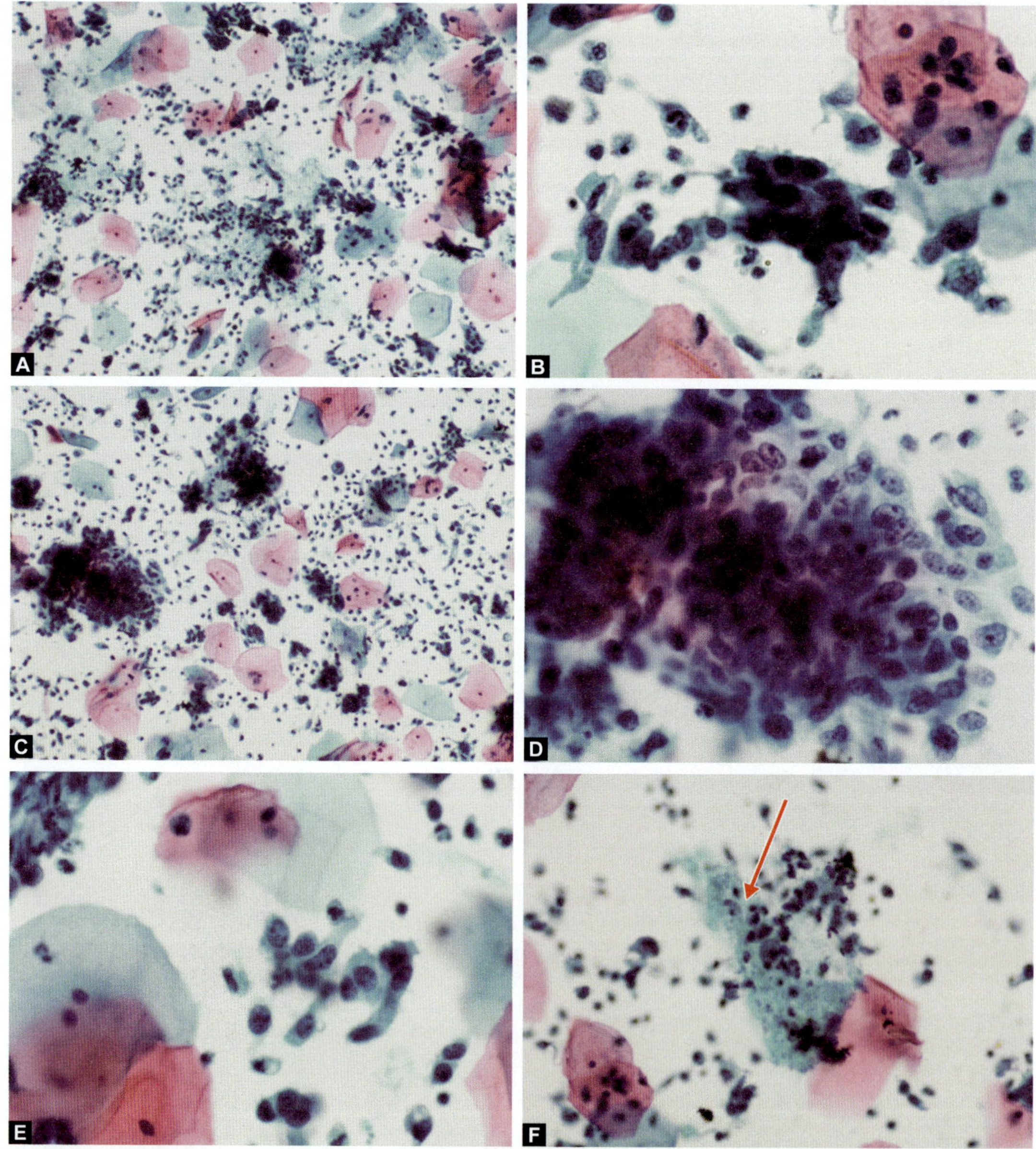

Figures 8.15A to F

Problem 8.16

Routine cervical smear in a 39-year-old female. The most likely diagnosis from Figure 8.16 is:

A. ASCUS
B. LSIL
C. ASC-H
D. HSIL

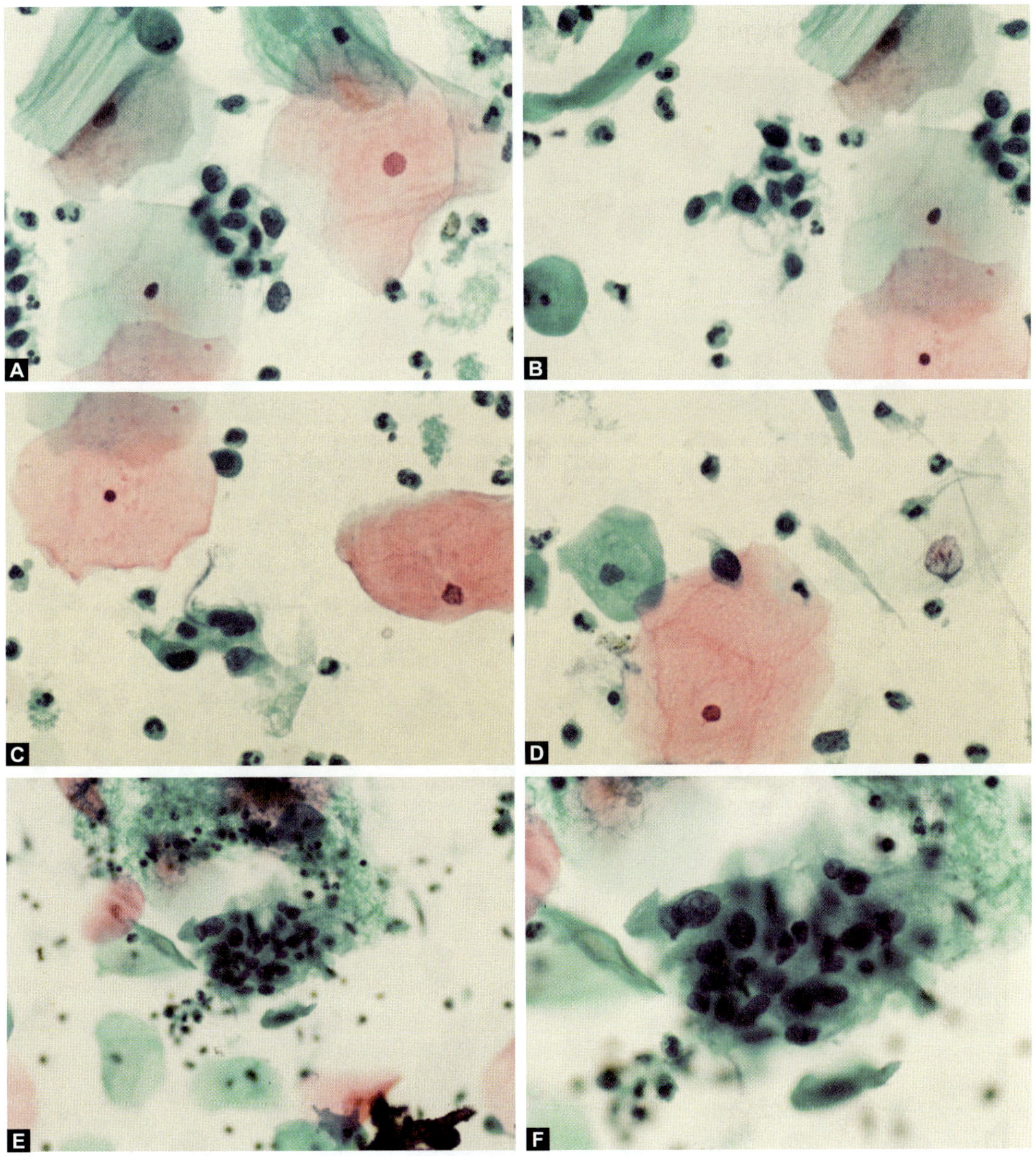

Figures 8.16A to F

Problem 8.17

A 41-year-old female was treated by radiation therapy for squamous cell carcinoma of cervix. Cervical smear is taken after one year of radiation therapy. The most likely diagnosis from Figure 8.17 is:

A. Squamous cell carcinoma
B. Radiation induced changes
C. No abnormality seen
D. Radiation induced atypia

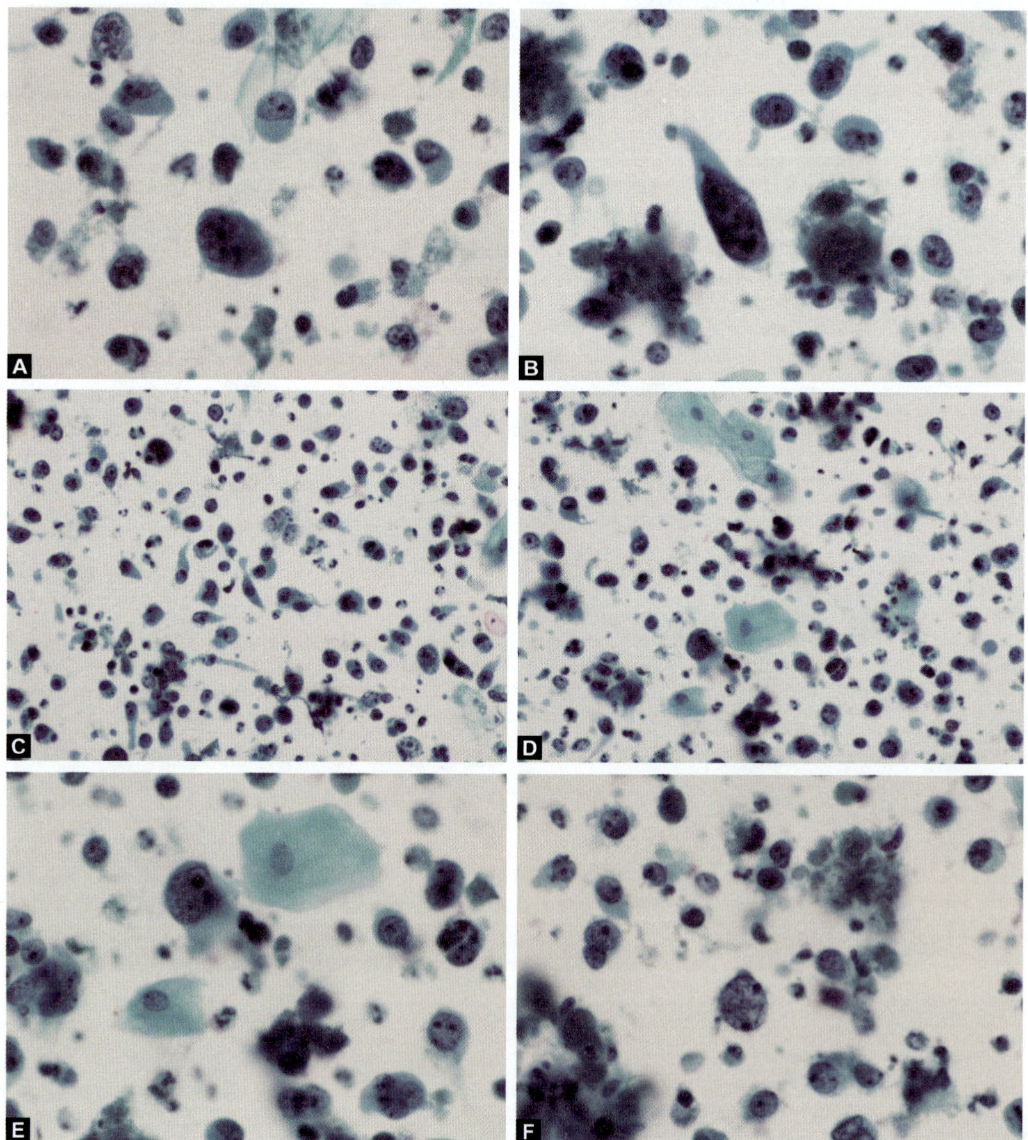

Figures 8.17A to F

Problem 8.18

A 39-year-old female had cervical biopsy which was diagnosed as CIN1 on cone biopsy. Routine cervical smear after one year was done. The most likely diagnosis from Figure 8.18 is:

A. LSIL
B. ASCUS
C. Tubal metaplasia
D. AGCUS

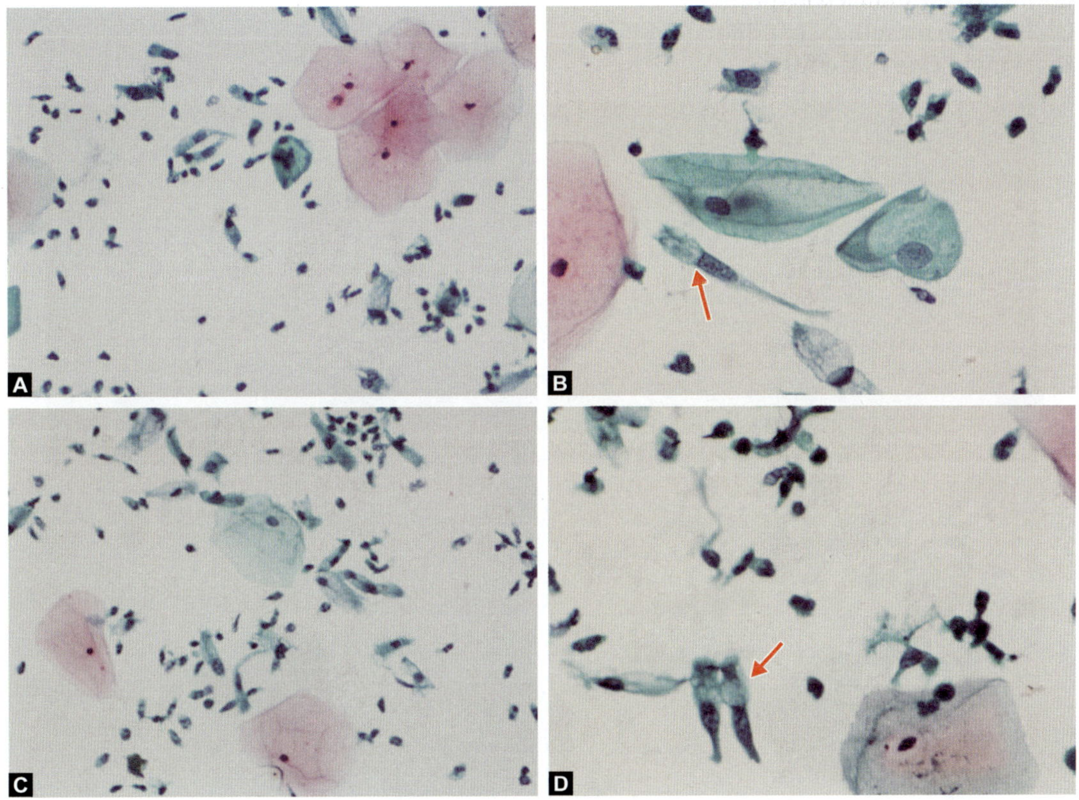

Figures 8.18A to D

Problem 8.19

A 37-year-old female had 3 cm diameter firm mass in her right upper part of breast. Microphotographs of the histopathology of the swelling are shown in Figures 8.19A and B. The patient was treated accordingly. After 6 months the patient presented with right sided pleural effusion. The microphotographs of the effusion cytology are shown in Figures 8.19C to F. The most likely diagnosis is:

A. No malignant cell seen
B. Metastatic ductal carcinoma
C. Reactive mesothelial proliferation
D. Lymphocyte rich effusion

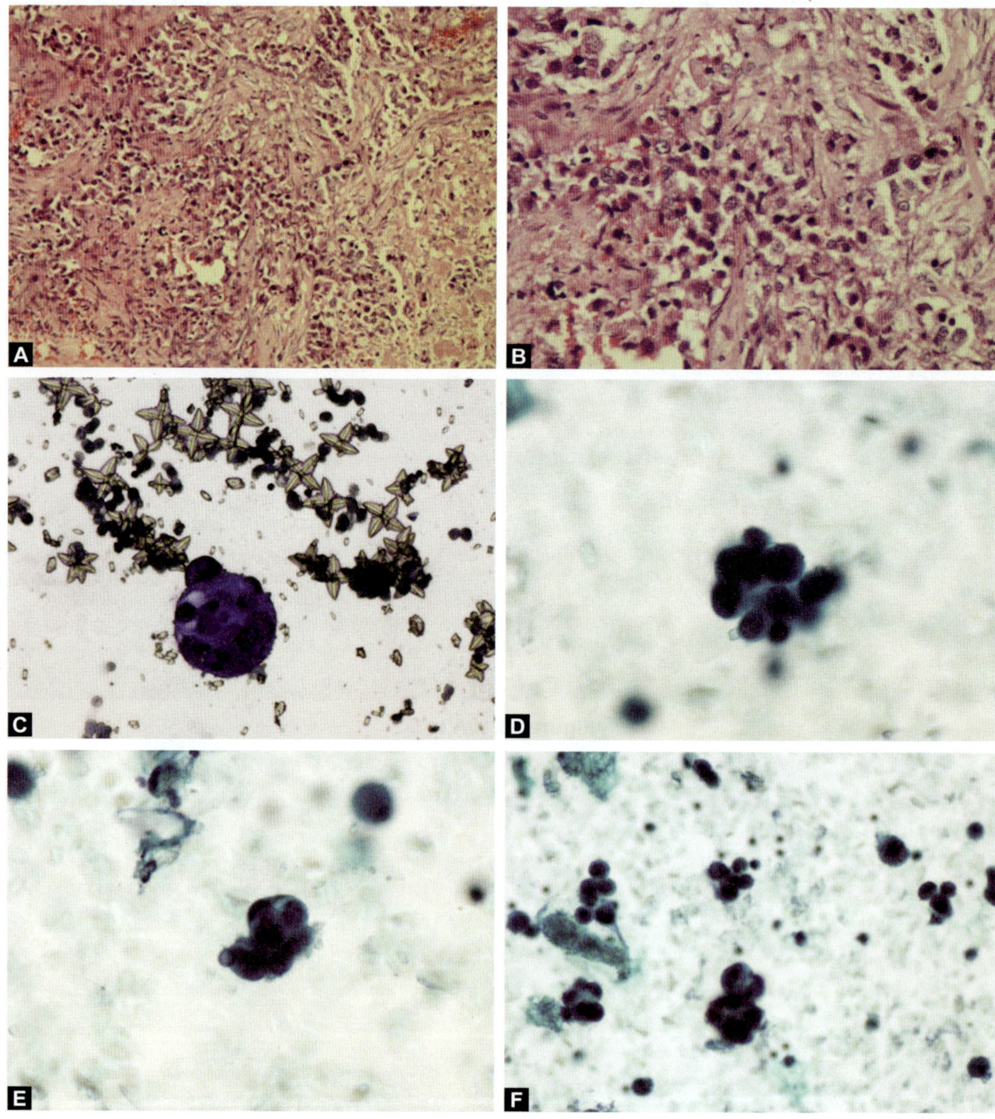

Figures 8.19A to F

Problem 8.20

Question1. Routine cervical smear in a 36-year-old female. The most likely diagnosis from Figure 8.20A to F is:

A. LSIL
B. ASCUS
C. HSIL
D. AGCUS

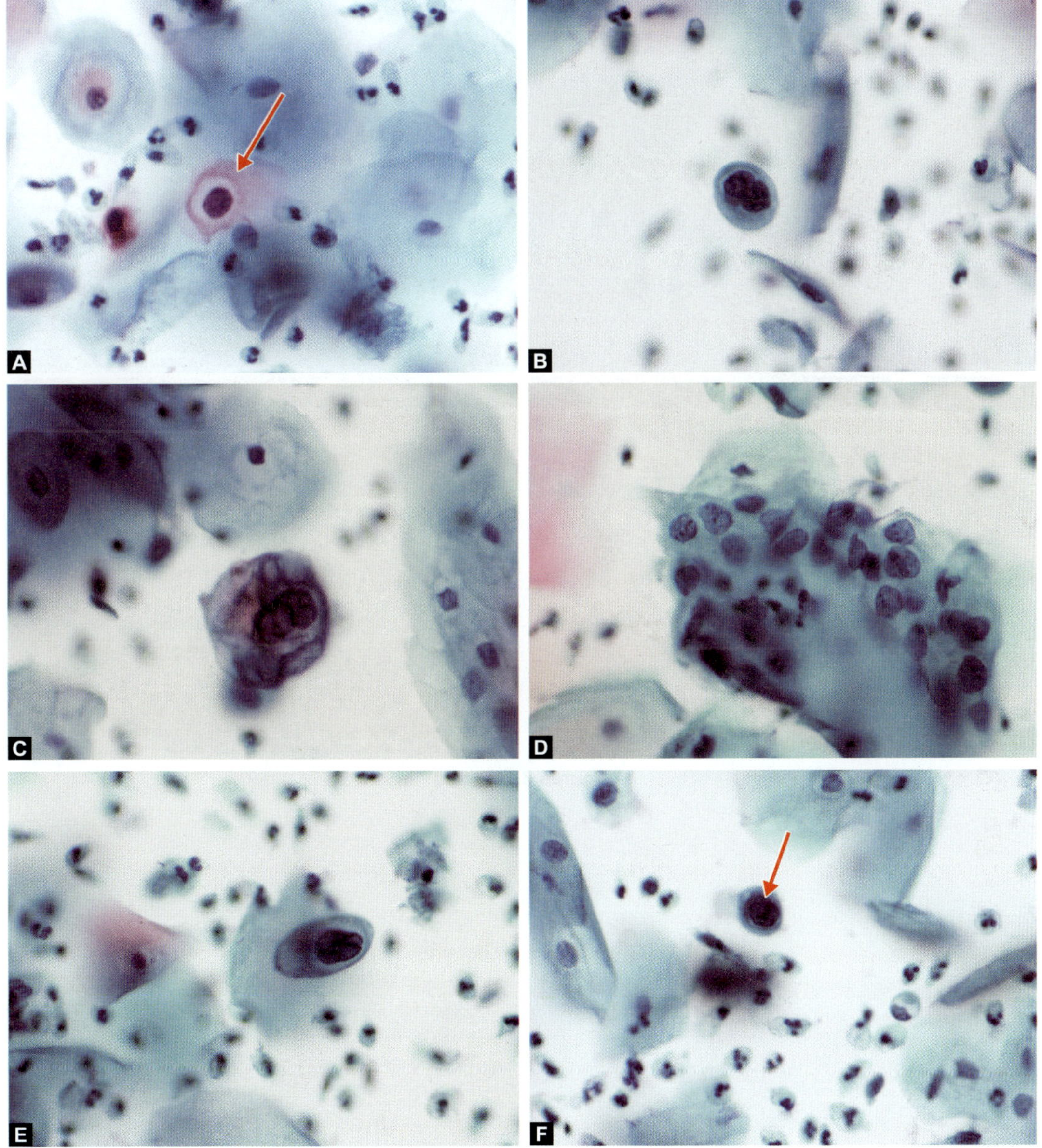

Figures 8.20A to F

Question 2: What should be the recommended management?

A. Colposcopy followed by colposcopy guided biopsy
B. Simple follow up after 6 month
C. Radiation therapy
D. Visual inspection under acetic acid
E. Hysterectomy

Question 3: Match with possible outcome from Figures 8.20G to J.

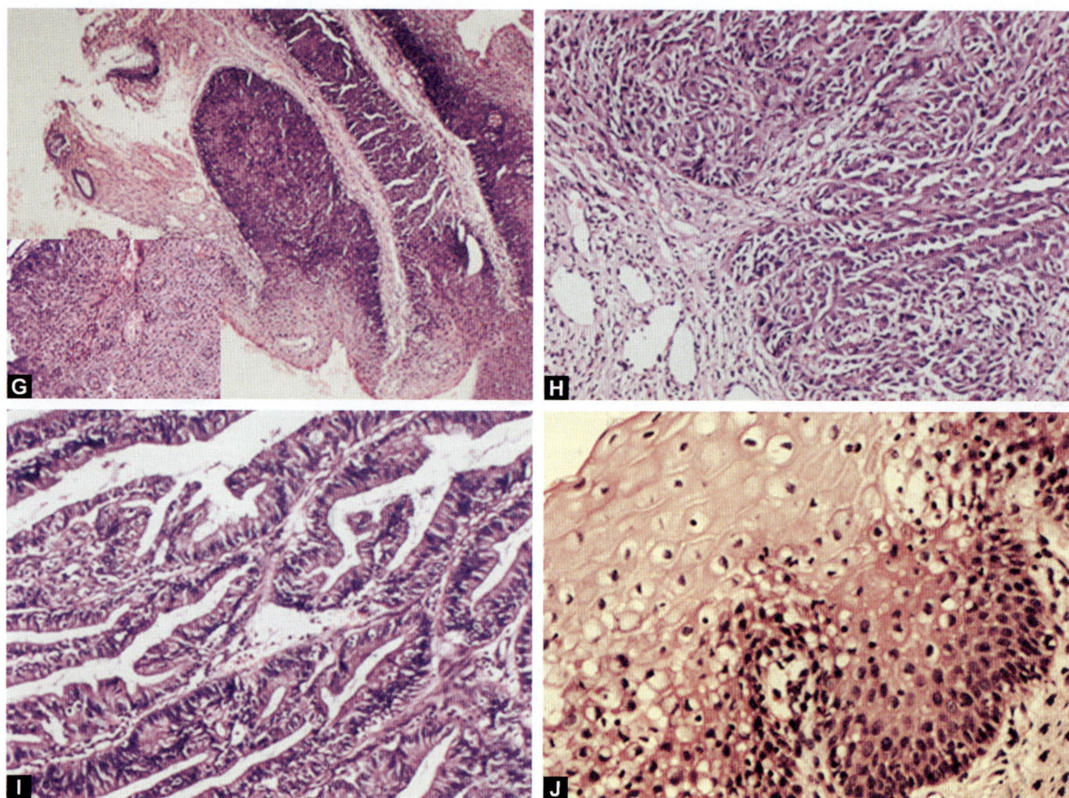

Figures 8.20G to J

Problem 8.21

Routine cervical smear in a 56-year-old female. The most likely diagnosis from Figure 8.21 is:

A. Atrophic smear
B. ASCUS
C. ASC-H
D. Squamous cell carcinoma

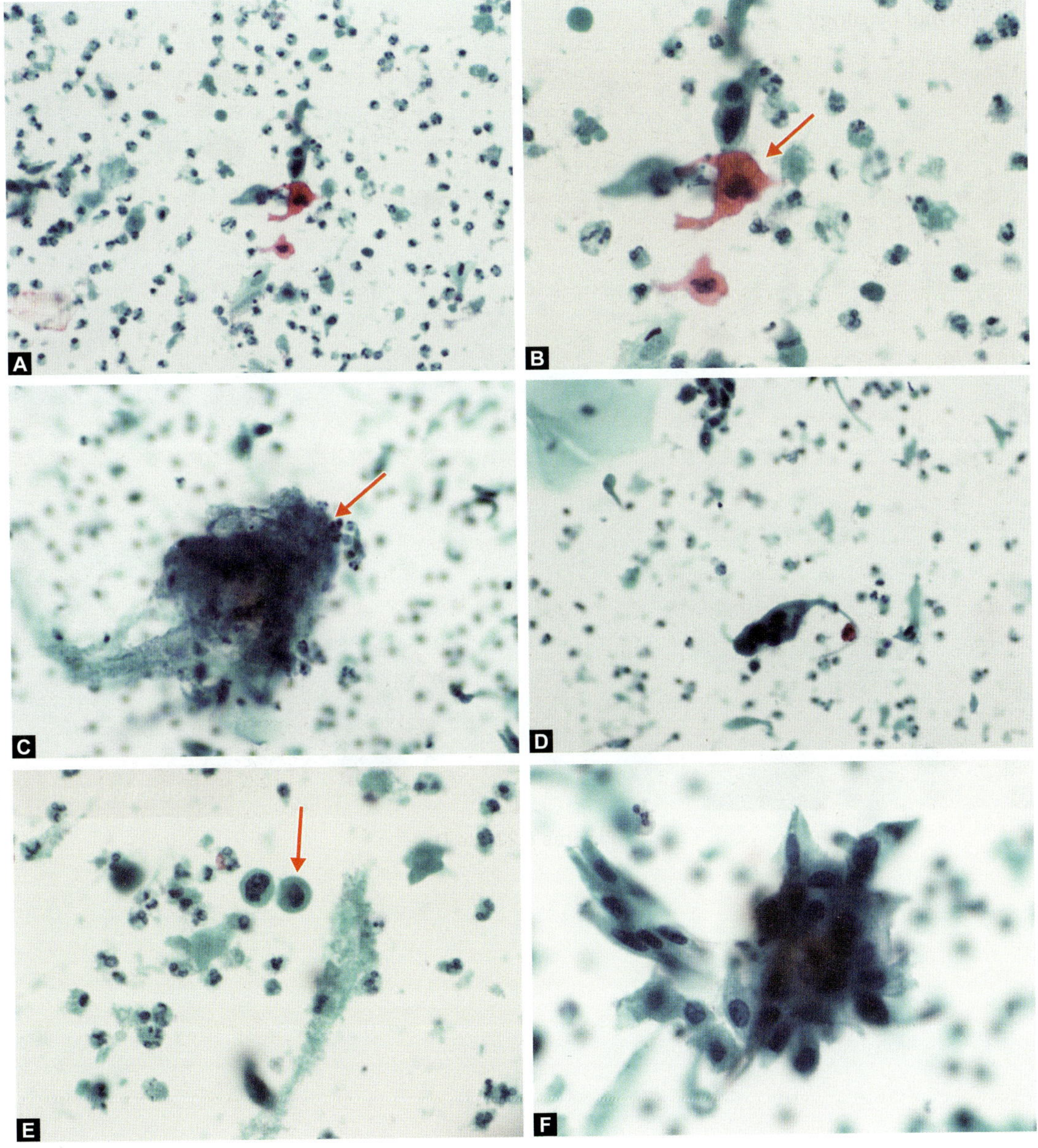

Figures 8.21A to F

Problem 8.22

A 45-year-old male had renal transplantation for chronic renal failure. The patient had fever, cough and mild chest pain. Chest X-ray showed diffuse hazy opacity. His BAL was done. The most likely diagnosis from Figure 8.22 is:

A. Pneumocystis carinii
B. Histoplasma
C. Cryptococci
d. Normal cytology

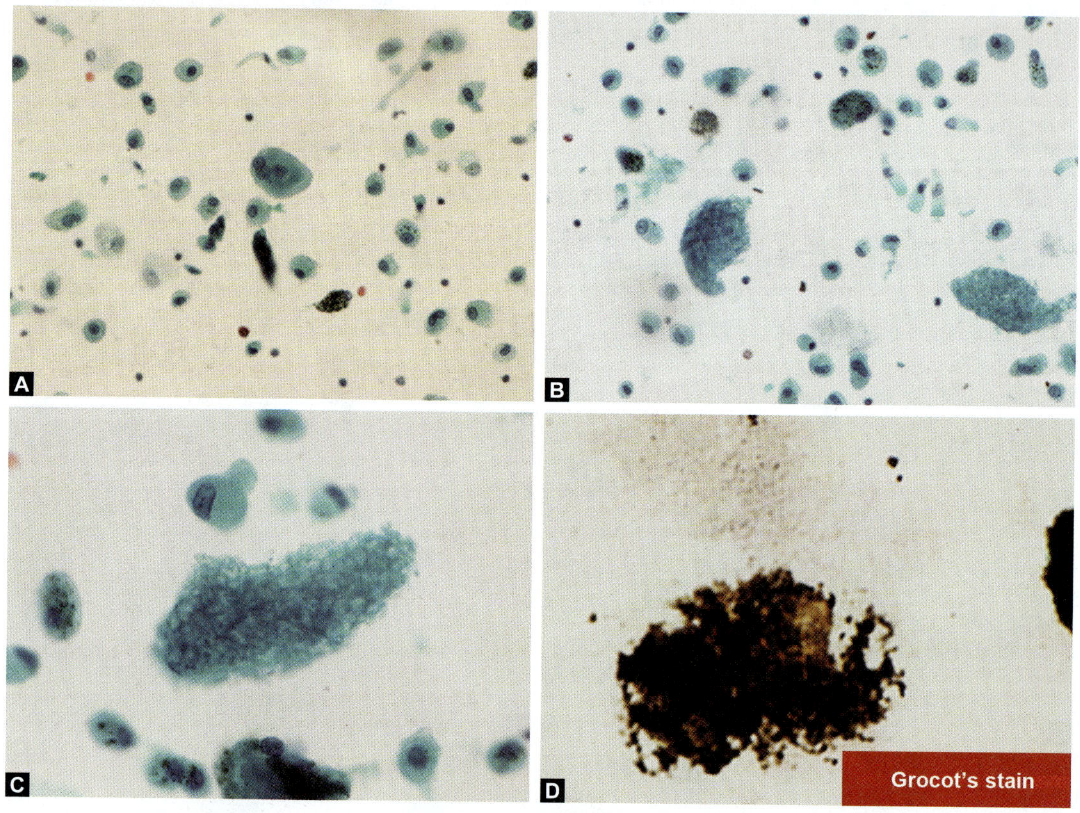

Figures 8.22A to D

Problem 8.23

Urethral wash cytology in a 57-year-old female who was operated for bladder tumor one year back. The most likely diagnosis from Figure 8.23 is:

A. Urothelial cell carcinoma (recurrence)
B. No evidence of any malignancy
C. Non-Hodgkin lymphoma
D. Metastatic carcinoma

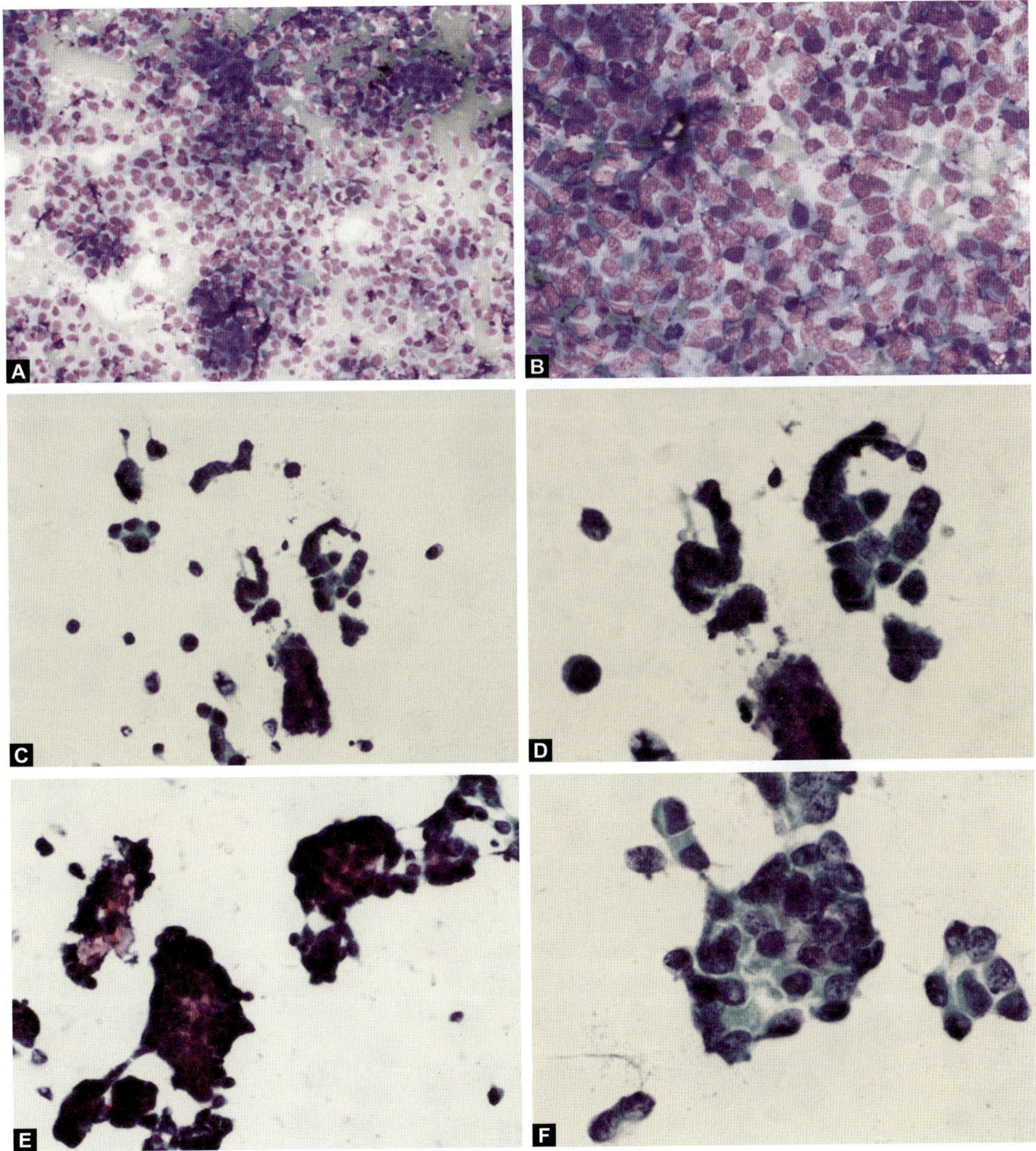

Figures 8.23A to F

Problem 8.24

A 41-year-old female presented with ovarian mass and tense ascites. SurePath preparation of the ascitic fluid was done. The most likely diagnosis from Figure 8.24 is:

A. Metastatic adenocarcinoma
B. Mesothelioma
C. Lymphoma
D. Reactive mesothelial cells

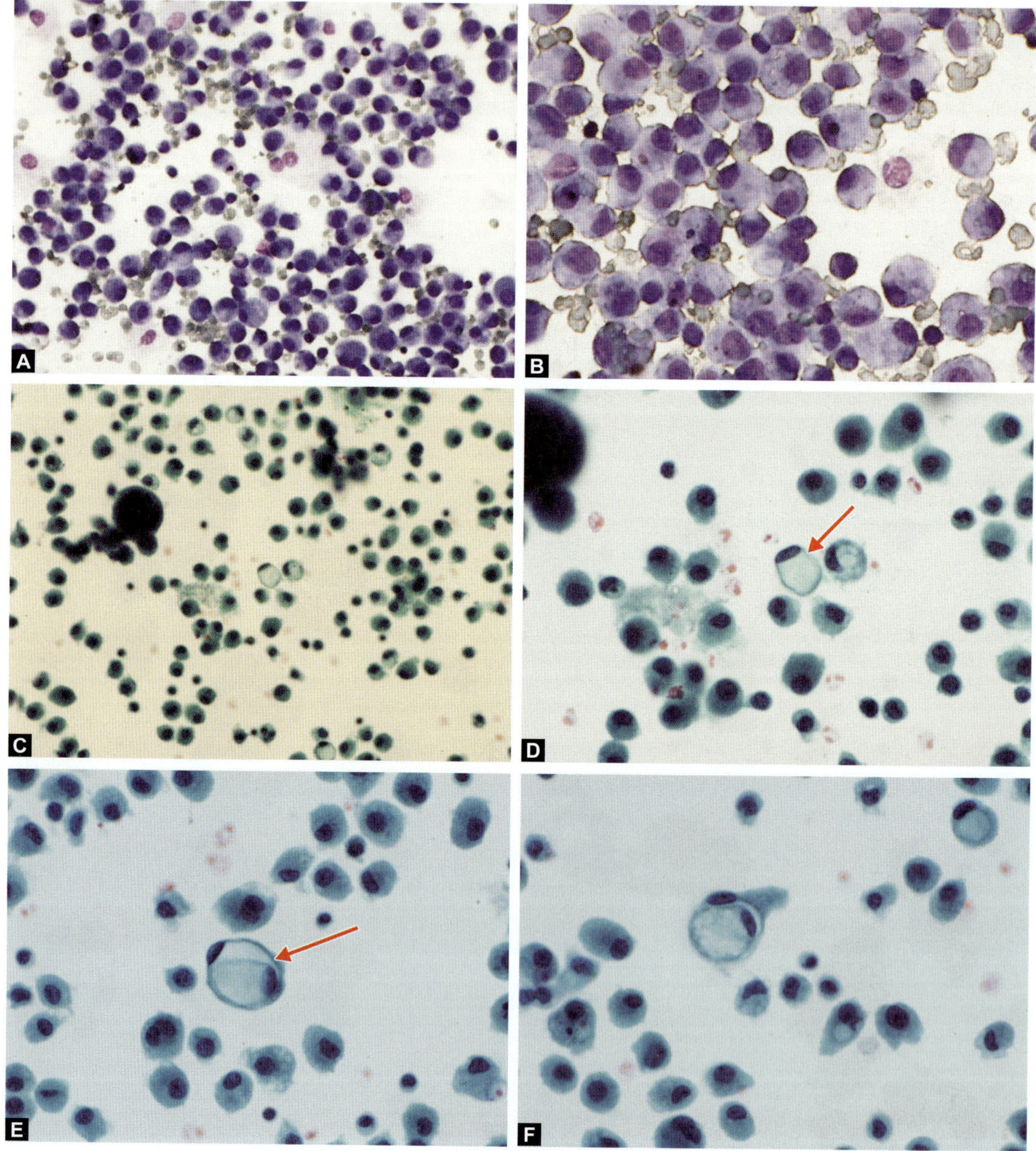

Figures 8.24A to F

Problem 8.25

A 39-year-old female with routine cervical smear. SurePath preparation was done. The most likely diagnosis from Figure 8.25 is:

A. ASCUS
B. AGCUS
C. AGCUS favors neoplasia
D. HSIL

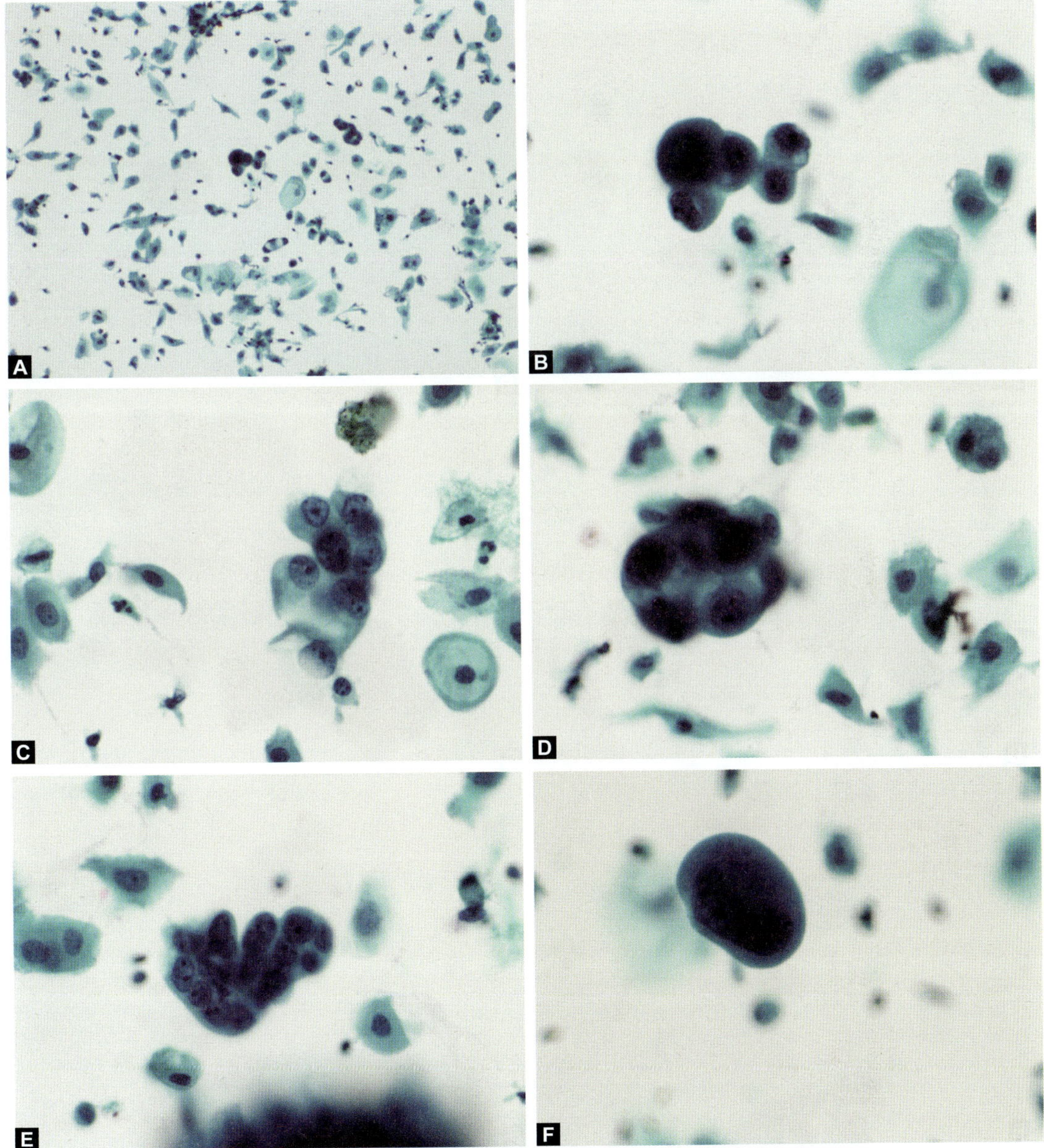

Figures 8.25A to F

Problem 8.26

A 48-year-old male was operated for cholecystectomy 6 months back. Histopathology report was not available. Now the patient presented with perihepatic fluid collection around the gallbladder operated site. The fluid was aspirated for cytology. The most likely diagnosis from Figure 8.26 is:

A. Adenocarcinoma
B. Reactive mesothelial cells
C. Inflammation only
D. Benign epithelial cells with reactive changes

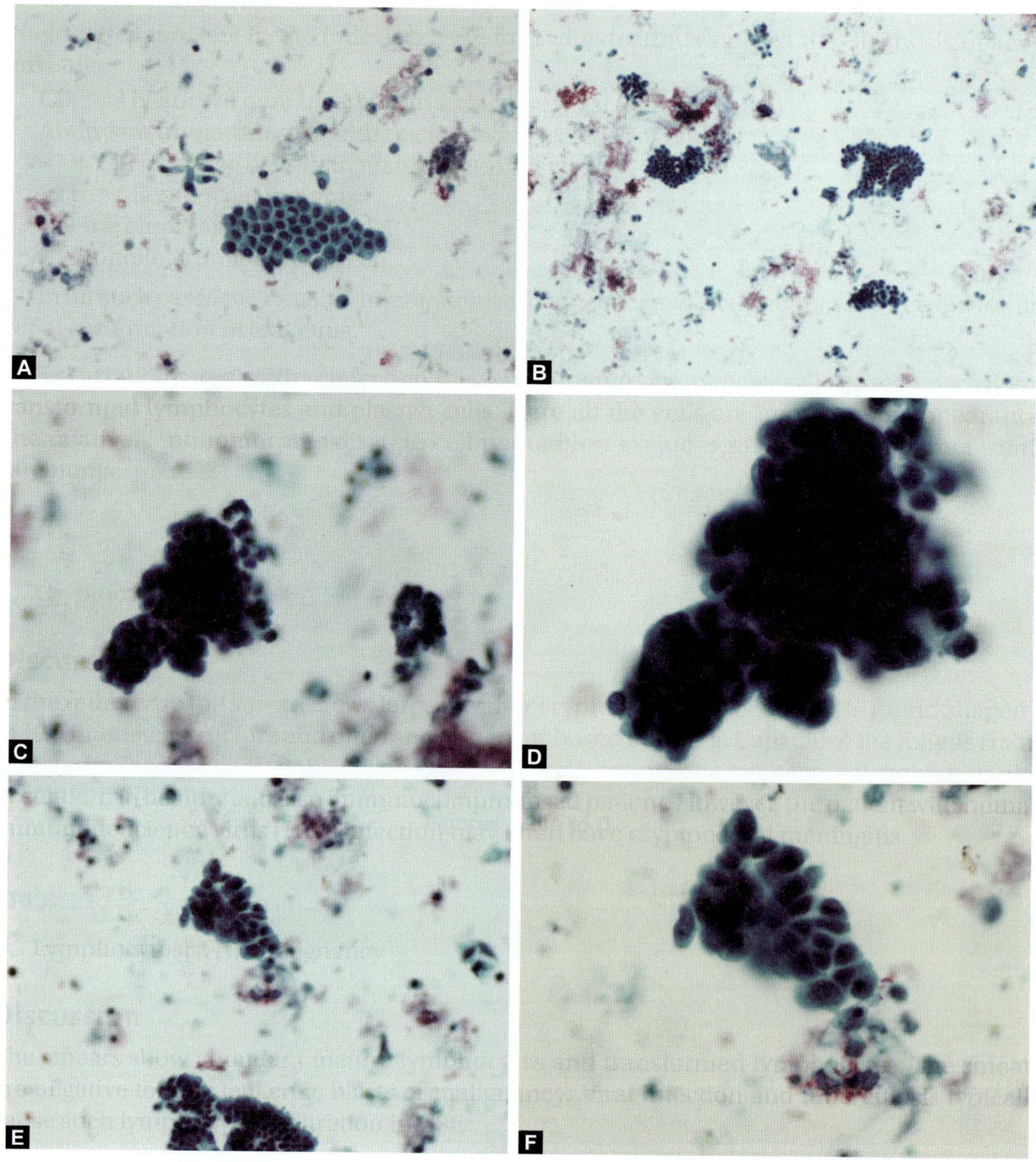

Figures 8.26A to F

Problem 8.27

A 41-year-old male was diagnosed as follicular lymphoma 5 year back. He was treated by chemotherapy and was in complete remission. Now the patient presents with severe headache. His CSF examination was done. The most likely diagnosis from Figure 8.27 is:

A. Cryptococcal infection
B. Reactive lymphocytosis
C. Metastatic carcinoma
D. Non-Hodgkin lymphoma (recurrence)

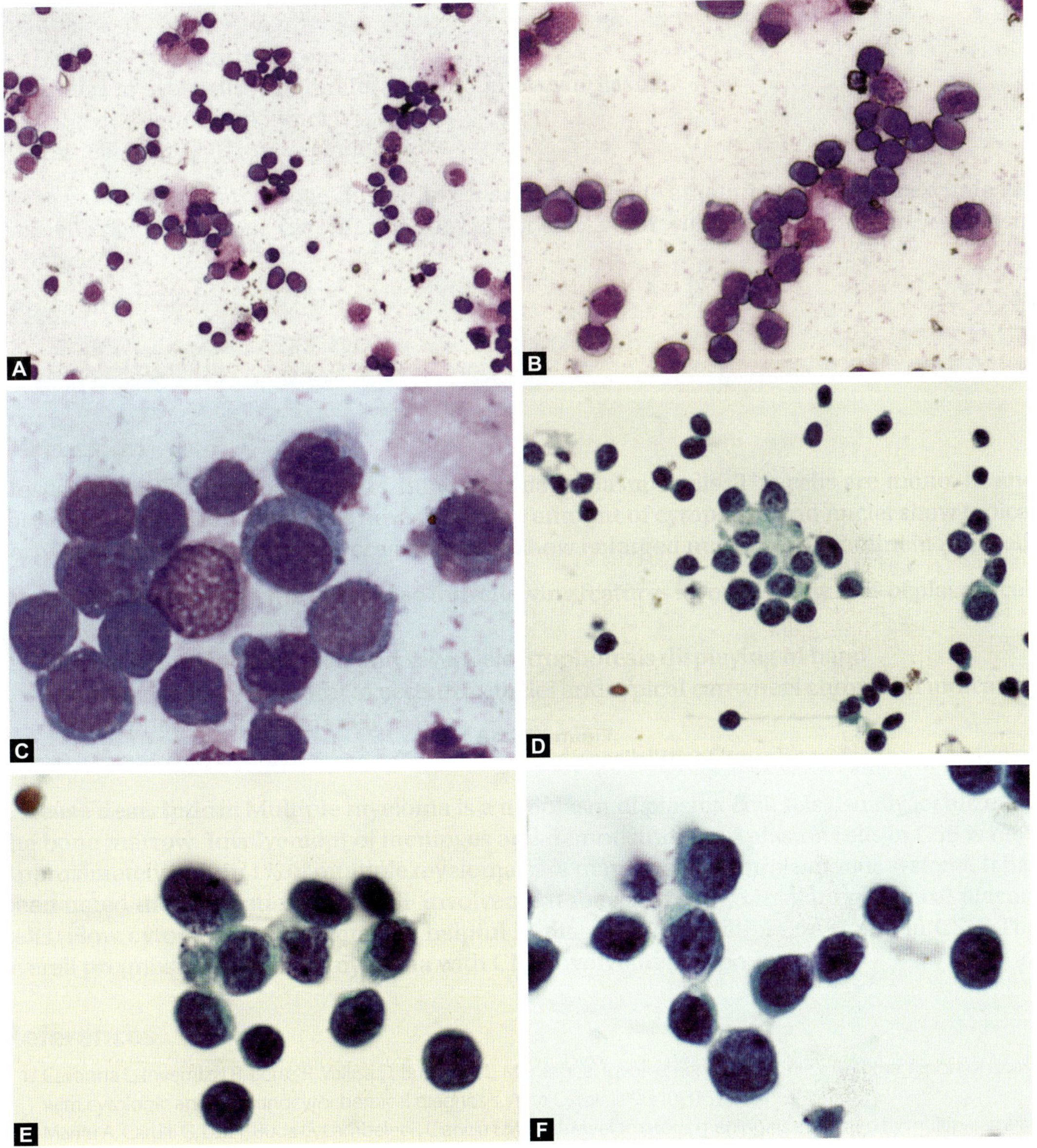

Figures 8.27A to F

Problem 8.28

Routine cervical smear in a 31-year-old female. The most likely diagnosis from Figure 8.28 is:

A. ASCUS
B. No epithelial cell abnormality
C. LSIL
D. HSIL

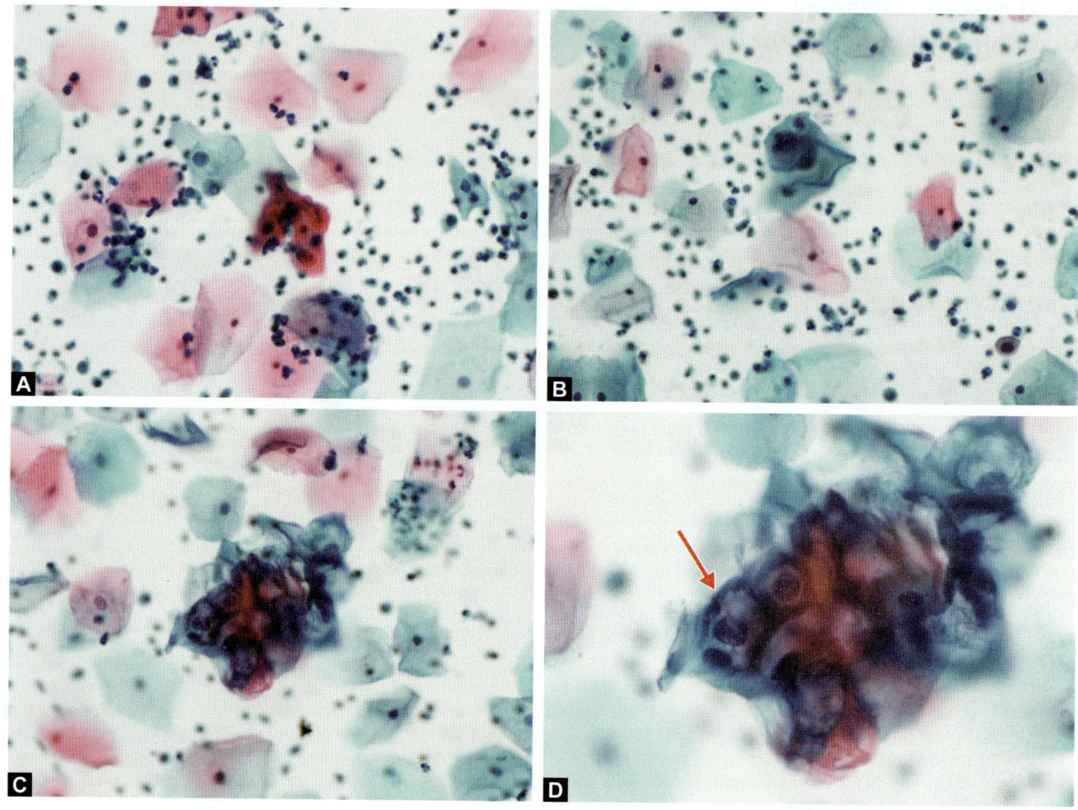

Figures 8.28A to D

Problem 8.29

Routine cervical smear in a 38-year-old female. SurePath preparation was done. The most likely diagnosis from Figure 8.29 is:

A. ASCUS
B. LSIL
C. HSIL
D. AGCUS
E. ASC-H

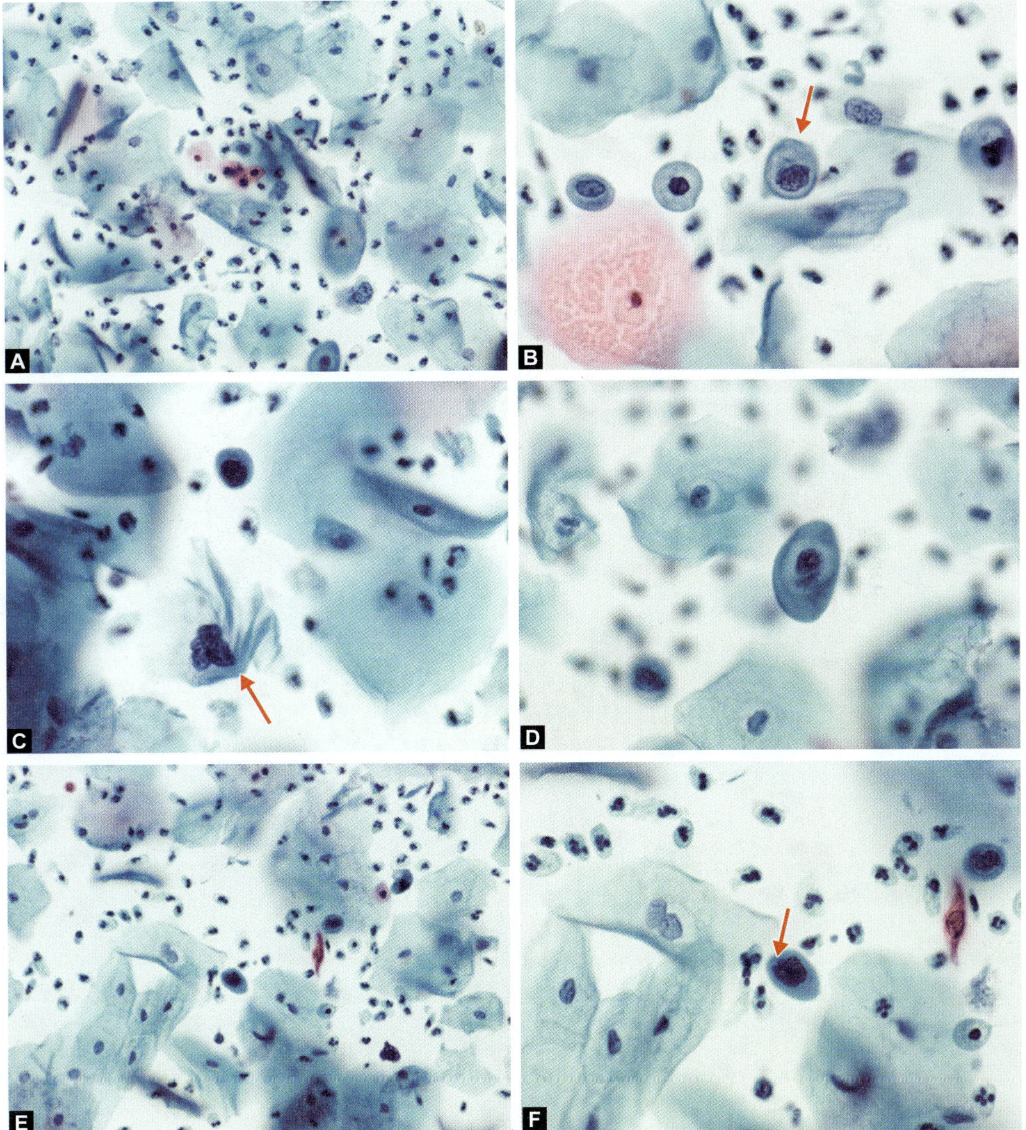

Figures 8.29A to F

Problem 8.30

Routine cervical smear in a 53-year-old female. The most likely diagnosis from Figure 8.30 is:

A. Atrophic smear
B. ASC-H
C. LSIL
D. HSIL
E. Squamous cell carcinoma

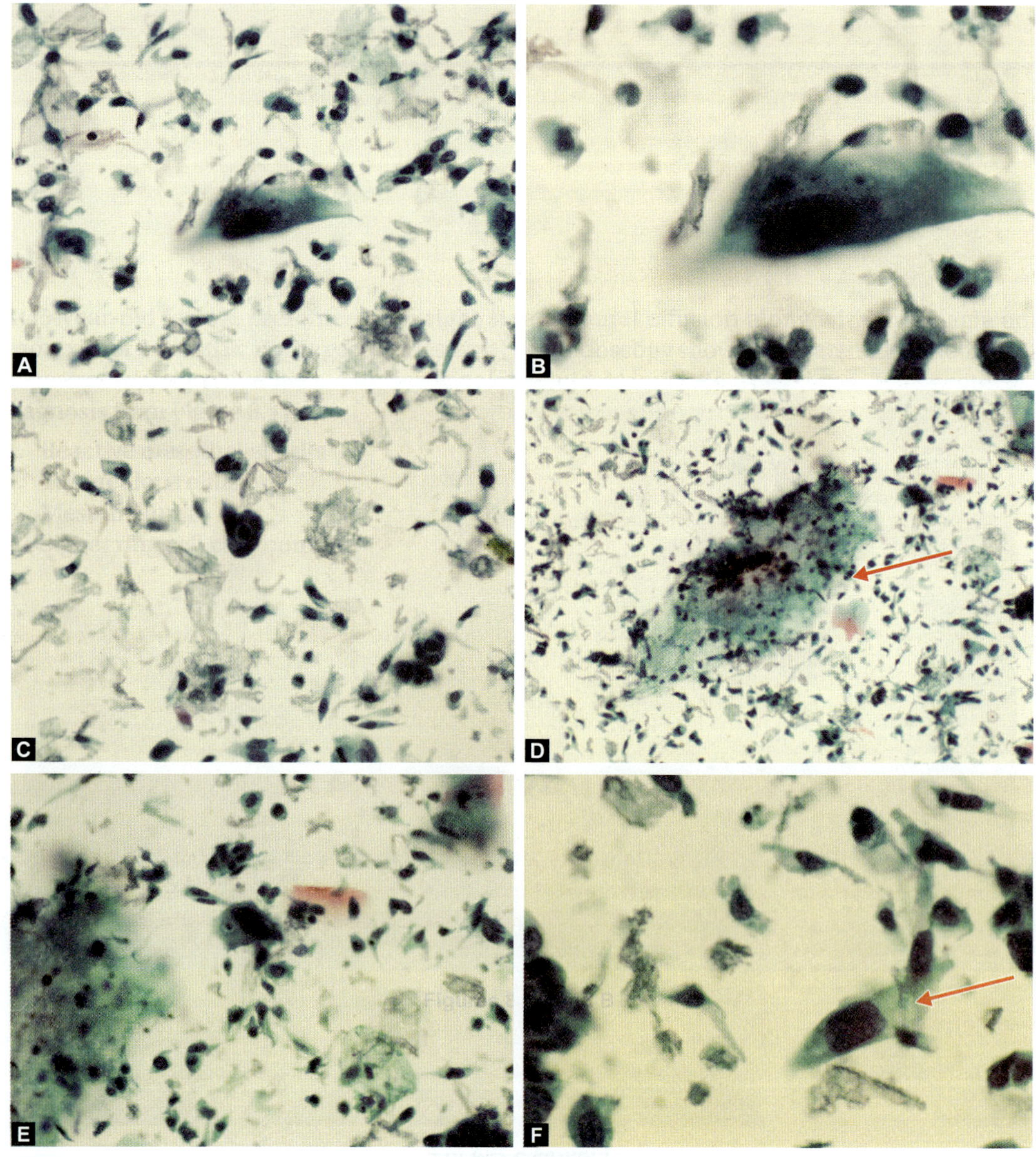

Figures 8.30A to F

ANSWERS

Problem 8.1

D. Signet ring cell carcinoma

Discussion

Description: The cytology smears show many discrete mesothelial cells admixed with signet ring cells. The individual cells show abundant vacuolated cytoplasm and eccentric nuclei. The nuclei are enlarged and mildly pleomorphic.

Diagnostic features: In this index case following features favor the diagnosis of signet ring cell carcinoma:

- Abundant discrete signet ring type of cells
- The cells show abundant cytoplasm
- Compressed crescentic nuclei
- Moderate nuclear pleomorphism.

This is a case of gastric adenocarcinoma. The diffuse type of gastric adenocarcinoma shows this type of cytomorphology.

Differential diagnosis: Occasionally reactive mesothelial cells may show signet ring type of cells. The nuclei of such signet ring type of mesothelial cells do not show any atypia. The malignant signet ring cells are mucicarmine positive, whereas benign cells are negative for mucicarmine.

Problem 8.2

A. Positive for carcinoma

Discussion

Description: The cytology smears show abundant loose clusters of cells. The cells show scanty to moderate amount of cytoplasm and enlarged moderately pleomorphic nuclei.

Diagnostic features: In the present case the features suggestive of carcinoma include:

- Multiple clusters of cells
- Round to oval cells with moderate nuclear pleomorphism
- Enlarged nuclei.

Comments: Nipple discharge cytology is not suitable for the detection of breast carcinoma as only 3% of breast carcinoma patients present with positive nipple discharge. The positive nipple discharge patients usually show bloody unilateral discharge. Milky or brownish discharge is usually negative for malignancy.

Problem 8.3

A. Retinoblastoma

Discussion

Description: The cytology smears show abundant round cells with scanty cytoplasm. The nuclei are round with mild irregular clumped chromatin and prominent nucleoli.

Diagnostic features: In this index case the following features suggest the diagnosis of retinoblastoma:

- Abundant malignant round cells
- Cells with scanty cytoplasm and high N/C ratio
- Nuclear enlargement and pleomorphism
- Irregular clumped chromatin
- Typical clinical feature of retinoblastoma such as progressive blindness and white reflex.

Differential diagnosis: The clinical history with the cytology smear excludes the possibility of acute leukemia. The cells are immature and the possibility of chronic inflammation is remote.

Problem 8.4

D. Papillary neoplasm

Discussion

Description: The cytology smears show many tight papillary clusters of cells admixed with foamy macrophages. The cells are round with scanty cytoplasm and monomorphic nuclei.

Diagnostic features: In the present case features suggestive of papillary neoplasm include:

- Multiple papillary clusters of cells
- The background foamy macrophages.

Differential diagnosis: It is not possible to label such cases as benign or malignant. The histopathological features of malignancy are usually absent in cytology smear. Paget's disease show cells with abundant cytoplasm and centrally placed pleomorphic nuclei.

Problem 8.5

A. Adenocarcinoma in common bile duct

Discussion

Description: The cytology smears show abundant loose clusters of cells. Individual cells are tall columnar with scanty cytoplasm having moderately pleomorphic nuclei. The nuclei show fine chromatin with prominent nucleoli.

Diagnostic features:

- Columnar appearance
- Cells with scanty cytoplasm having moderately pleomorphic nuclei
- Prominent nucleoli.

Differential diagnosis: The cells are moderately pleomorphic and enlarged. The nuclear abnormality is more than that of dysplastic cells. Benign columnar cells are also admixed with these malignant cells. Stone induced atypia is usually less severe.

Problem 8.6

C. Both Candida and sperm head present

Discussion

In this index case the smear show many candidal spore, hyphae along with sperm. These sperm heads may be mistaken as spores of candida. The sperm head is triangular in shape and often is attached with tail. Whereas, the candidal spore is elongated with blunted ends.

Problem 8.7

A. Metastatic adenocarcinoma from lung

Discussion

Description: The cytology smears (Figs 8.7A to F) show multiple ball like tight clusters of cells admixed with background mesothelial cells. The tumor cells have moderate amount of vacuolated cytoplasm with centrally to eccentric nuclei. The nuclei are enlarged with moderate pleomorphism. The immunocytochemistry (Figs 8.7G to J) from the cell block preparation is positive for CK 7, TTF-1 and napsin and negative for CK 5/6.

Diagnostic features:

- Ball like tight clusters of cells
- Tumor cells moderate amount of vacuolated cytoplasm and moderately pleomorphic nuclei
- CK 7, TTF-1 and napsin positivity and CK 5/6 negativity indicate adenocarcinoma of lung.

Problem 8.8

C. Granulomatous inflammation: tuberculosis

Discussion

Description: The cytology smears show multiple epithelioid cell granulomas in a necrotic background. The Ziehl–Neelsen stain for acid fast bacilli (Fig. 8.8E) is positive.

Diagnostic features:

- Necrosis
- Epithelioid cell granuloma
- Acid fast bacilli positive.

Problem 8.9

B. Adenocarcinoma

Discussion

Description: The cytology smears show multiple tight clusters of cells. The cells are round with scanty cytoplasm and moderately pleomorphic nuclei having prominent nucleoli.

Diagnostic features:
- Cohesive three-dimensional clusters
- Round cells, scanty cytoplasm, hyperchromatic nuclei with prominent nucleoli.

Problem 8.10

C. Mucormycosis

Discussion

Diagnostic feature: The cytology smears show necrosis, polymorphs and many fungal hyphae which are evident in PAS stain (Figs 8.10C and D). The fungi are broad aseptate and show wide angle branching. The morphology of the fungus is consistent with Mucormycosis.

Problem 8.11

D. Adenocarcinoma

Discussion

Description: The cytology smear shows multiple tight and loose clusters of tumor cells. The individual cells are round with moderate amount of cytoplasm. The nuclei are moderately pleomorphic hyperchromatic with occasional prominent nucleoli.

Diagnostic features:
- History of carcinoma
- Cohesive three-dimensional clusters
- Round cells, hyperchromatic nuclei with prominent nucleoli.

Problem 8.12

B. ASC-H

Discussion

Description: The cytology smears show abundant discrete oval to polyhedral cells. The cells have moderate amount of cytoplasm with enlarged nuclei occupying more than 2/3rd of the cytoplasmic area. The nuclei are hyperchromatic with regular nuclear margin.

Diagnostic features in the index case:
- Polyhedral cells
- Enlarged hyperchromatic nuclei with regular margin.

The cytology smear did not fulfil the criteria of HSIL and therefore a diagnosis of ASC-H was the most suitable. The colposcopic guided biopsy was done in this case which revealed HSIL (Fig. 8.31).

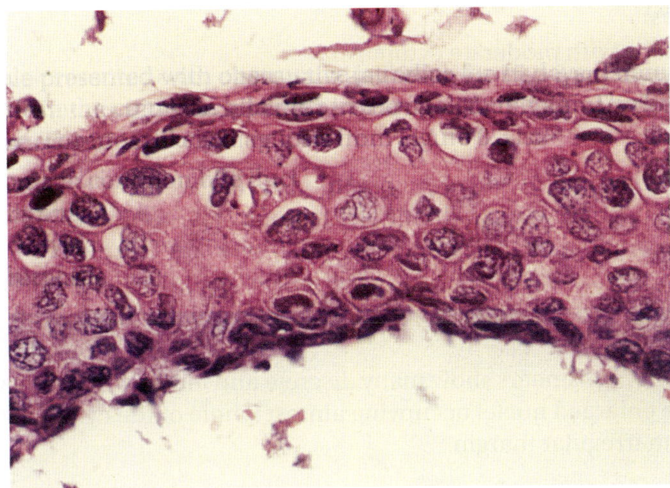

Figure 8.31: Histopathology of the cervical lesion of the index case

Problem 8.13

B. LSIL

Discussion

Description: The cytology smears show many discrete squamous cells with enlarged nuclei and perinuclear halo (koilocytotic cells). The cytoplasmic margin of the halo is thickened.

Diagnostic feature: The presence of koilocytes in the smear indicates LSIL. No other abnormal cells are seen in the smear.

Problem 8.14

B. Pseudomyxoma peritonei

Discussion

Description: The cytology smear shows abundant mucinous material and occasional clusters of epithelial cells. The cells are round with round monomorphic nuclei and scanty cytoplasm.

Diagnostic features:

- Abundant thick mucinous background
- Occasional clusters of epithelial cells.

Problem 8.15

D. Adenocarcinoma

Discussion

Description: The cytology smears show discrete and multiple loose clusters of cells. The cells are columnar in shape with round moderately pleomorphic nuclei and prominent nucleoli. In addition background shows tumor diathesis (Fig. 8.14F).

Diagnostic features:

- Columnar in shape with moderate cytoplasm
- Moderately pleomorphic nuclei
- Large prominent nucleoli
- Tumor diathesis.

Problem 8.16

D. HSIL

Discussion

Description: The cytology smears show many discrete and loose clusters of oval to polyhedral cells. The cells have enlarged nuclei occupying almost whole of the cytoplasm. The nuclei are hyperchromatic with irregular margin.

Diagnostic features:

- Oval to polyhedral cells
- Enlarged nuclei with high N/C ratio (nuclei occupying more than 2/3rd of cytoplasm)
- Hyperchromatic nuclei with irregular margin.

Problem 8.17

A. Squamous cell carcinoma

Discussion

Description: The cytology smears show many discrete oval to polyhedral cells with enlarged nuclei and high N/C ratio. The cells show moderate to marked nuclear pleomorphism and coarsely clumped chromatin. In addition, many bi and multinucleated giant cells are also seen.

Diagnostic features:

- Oval to polyhedral cells with high N/C ratio
- Oval to polyhedral cells with enlarged nuclei and high N/C ratio
- Marked nuclear pleomorphism
- Coarsely clumped chromatin
- Multinucleated giant cells (radiation induced changes).

Problem 8.18

C. Tubal metaplasia

Discussion

Diagnostic features:

- Abundant discrete columnar cells with cilia in one end.

  Tubal metaplasia is a common condition that is frequently seen after cone biopsy.

Problem 8.19

B. Metastatic ductal carcinoma

Discussion

Description: The histopathology section (Figs 8.19A and B) shows features of infiltrating ductal carcinoma of breast. The cytology smears show multiple tight ball like clusters of cells. The cells are round with moderately enlarged pleomorphic hyperchromatic nuclei.

Diagnostic features:

- Previous histopathology section shows infiltrating ductal carcinoma of breast
- Cytomorphology of the cells in effusion fluid is consistent with histopathology section of breast carcinoma.

Problem 8.20

Question 1

C. HSIL

Discussion

Description: The cytology smears show many discrete cells and occasional loose clusters of large atypical cells. The cells show moderate nuclear enlargement with high N/C ratio as the nuclei occupy more than 2/3rd of the cytoplasmic area. The nuclei are hyperchromatic with irregular margin. Occasional koilocytotic cell (Fig. 8.20A) and binucleated (Fig. 8.20C) cells are also noted.

Diagnostic features:

- High N/C ratio
- Nuclear hyperchromasia and pleomorphism
- Irregular nuclear margin
- Koilocytosis.

Question 2

A. Colposcopy followed by colposcopy guided biopsy

Discussion: In high-grade squamous intraepithelial lesion (HSIL) case colposcopy along with biopsy is recommended. In histopathology proved HSIL cases the entire transformation zone should be removed along with the lesion. The management of HSIL includes: (1) Cryotherapy, (2) Loop electrosurgical excision procedure/large loop excision of the transformation zone and (3) Cold knife conization.

Question 3

G. CIN 3

Discussion: In this index case the colposcopic guided biopsy revealed CIN 3 changes. The other three Figures are of:

Figure 8.20H: squamous cell carcinoma

Figure 8.20I: Adenocarcinoma

Figure 8.20J: CIN 1

Problem 8.21

D. Squamous cell carcinoma

Discussion

Description: The cytology smears show many large polyhedral orangeophilic cells with atypical nuclei (Fig. 8.21B). There are cells with enlarged nuclei having hyperchromatic nuclei. Background tumor diathesis is also seen (Fig. 8.21C).

Diagnostic features:
- Dyskeratotic squamous cells
- Squamous cells with enlarged hyperchromatic nuclei
- Tumor diathesis.

Problem 8.22

A. Pneumocystis carinii

Discussion

Description: The cytology smears show foamy to bubbly alveolar cast material with round shaped structures along with abundant respiratory lining epithelial cells, lymphocytes and histiocytes.

Diagnostic features:
- Bubbly alveolar cast material with ring shaped tiny organism
- Positive Grocott's methenamine silver stain confirms the diagnosis.

See problem 4.6 for more information.

Problem 8.23

A. Urothelial cell carcinoma (recurrence)

Discussion

Description: The cytology smears show abundant discrete and also multiple clusters of cells. The cells show scanty to moderate amount of cytoplasm and enlarged hyperchromatic moderately pleomorphic nuclei.

Diagnostic features:
- Abundant clusters and discrete tumor cells
- Round to oval cells with moderately pleomorphic hyperchromatic nuclei.

Differential diagnosis: In this case multiple clusters of cells and absence of lymphoglandular bodies eliminate the possibility of non-Hodgkin lymphoma. History of bladder tumor suggests a primary bladder tumor with recurrence rather than a metastatic tumor.

Problem 8.24

D. Reactive mesothelial cells

Discussion

Description: The cytology smears show abundant reactive mesothelial cells along with occasional cells with eccentric nuclei and abundant cytoplasm. No nuclear abnormality is seen in these cells.

Diagnostic features:
- Reactive mesothelial cells only
- No nuclear atypia in signet ring type of mesothelial cells.

The follow-up of this case showed no omental involvement by ovarian tumor. The primary tumor was serous cyst adenocarcinoma of ovary. The ascites in malignant ovarian tumor may be due to pressure effect on venous drainage by primary ovarian mass or due to hypoproteinemia.

Problem 8.25

C. AGCUS, favors neoplasia

Discussion

Description: The cytology smears show many loose clusters and discrete cells. The cells are round with scanty vacuolated cytoplasm. The nuclei of the cells are enlarged pleomorphic with single to multiple prominent nucleoli.

Diagnostic features:
- Small clusters of cells
- Round cells, vacuolated cytoplasm
- Enlarged nuclei with prominent nucleoli.

Problem 8.26

A. Adenocarcinoma

Discussion

Description: The cytology smears show abundant benign columnar cells along with many tight clusters of atypical cells. The cells are columnar in appearance with enlarged moderately pleomorphic hyperchromatic nuclei.

Diagnostic features:
- Columnar cells with moderate nuclear enlargement
- Hyperchromatic and pleomorphic nuclei.

Differential diagnosis: The cells do not look like mesothelial cells and the possibility of reactive mesothelial cells is ruled out. The atypia in the nucleus is too severe to call it reactive change.

Problem 8.27

D. Non-Hodgkin lymphoma (recurrence)

Discussion

Description: The cytology smears show many discrete round to oval cells with scanty deep blue cytoplasm. The nuclei are round, mildly pleomorphic having fine nuclear chromatin. Nucleoli are inconspicuous.

Diagnostic features:
- ❑ Old history of lymphoma
- ❑ Round cells with deep blue cytoplasm
- ❑ Round nuclei with fine nuclear chromatin.

Differential diagnosis: The cryptococci are bigger in size than lymphocytes and show clear round structure in Papanicolaou's stain. Reactive lymphocytosis shows mature lymphocytes only and this possibility is eliminated as these cells are looking immature. Metastatic poorly differentiated carcinoma may simulate NHL. However, clinical history and discrete monomorphic cell population exclude this possibility.

Problem 8.28

C. LSIL

Discussion

Description: The cytology smears show clusters of cells with koilocytotic changes.
Diagnostic feature: The presence of koilocytosis always indicates at least LSIL. The further grade depends on nuclear atypia of the cells.

Problem 8.29

C. HSIL

Discussion

Description: The cytology smear show many discrete round to polyhedral cells. The cells have enlarged nuclei occupying more than 2/3rd of cytoplasmic area. The nuclei are hyperchromatic, moderately pleomorphic with irregular margin.

Diagnostic features:
- ❑ Oval to polyhedral cells
- ❑ High N/C ratio, nuclei occupying more than 2/3rd of cytoplasmic area
- ❑ Hyperchromatic, moderately pleomorphic nuclei
- ❑ Irregular nuclear margin.

Problem 8.30

E. Squamous cell carcinoma

Discussion

Description: The cytology smears show abundant discrete polyhedral cells with scanty cytoplasm and enlarged hyperchromatic pleomorphic nuclei. Many elongated fiber-like cells are seen. In addition, Figure 2.30E shows tumor diathesis.

Diagnostic features:

- Polyhedral cells
- Markedly pleomorphic hyperchromatic nuclei
- Irregular nuclear margin
- Fiber cells
- Tumor diathesis.

Differential diagnosis: The presence of tumor diathesis eliminates the possibility of HSIL. The nuclear abnormalities of the cells are severe and thereby possibility of ASC-H and LSIL is remote.

Index

Page numbers followed by *f* refer to figure and *t* refer to table

D

E

F

G

H

I

K

L

M

N

O

P

R

S

T

U

V

W

Y